云计算与虚拟化技术丛书

An in-Depth Analysis of Spring Cloud

Spring Cloud微服务架构进阶

朱荣鑫 张天 黄迪璇 编著

图书在版编目（CIP）数据

Spring Cloud 微服务架构进阶 / 朱荣鑫，张天，黄迪璇编著. —北京：机械工业出版社，2018.9（2025.1 重印）

（云计算与虚拟化技术丛书）

ISBN 978-7-111-60868-4

I. S… II. ①朱… ②张… ③黄… III. 互联网络 – 网络服务器 IV. TP368.5

中国版本图书馆 CIP 数据核字（2018）第 209906 号

Spring Cloud 微服务架构进阶

出版发行：机械工业出版社（北京市西城区百万庄大街 22 号　邮政编码：100037）	
责任编辑：吴　怡	责任校对：张惠兰
印　　刷：北京捷迅佳彩印刷有限公司	版　次：2025 年 1 月第 1 版第 5 次印刷
开　　本：186mm×240mm　1/16	印　张：26.75
书　　号：ISBN 978-7-111-60868-4	定　价：89.00 元

客服电话：（010）88361066　68326294

版权所有·侵权必究
封底无防伪标均为盗版

前言

最近几年，随着 DevOps 和以 Docker 为主的容器技术的发展，云原生应用架构和微服务变得流行起来。云原生包含的内容很多，如 DevOps、持续交付、微服务、敏捷等，本书关注的是其中的微服务。在大概三年前，我在互联网上查找关于微服务落地的方案，搜索到了 Spring 社区推出的 Spring Cloud 项目，在那个时候就开始关注 Spring Cloud，发现 Spring Cloud 基于 Spring Boot，引入依赖后开箱即用，使用非常方便。当时 Spring Cloud 中的组件数量和成熟度远不如现今，Spring Cloud 的版本为 Brixton。后来我在项目中尝试使用 Spring Cloud，主要用了 Spring Cloud Config 和 Spring Cloud Stream，使用过程中发现这两个组件在易用性、功能性等各方面都令人满意，慢慢地便在项目中铺开使用。

在应用 Spring Cloud 的过程中，我见证了它的不断完善和丰富。在其间也遇到了一些"坑"，通过源码分析才解决了一些问题。Spring Cloud 并没有重复造轮子，这些组件有些是 Spring Cloud 的全新项目，如 Spring Cloud Gateway、Spring Cloud Config 等，还有很多是基于业界现有的开源组件，如 Netflix 的合集 Netflix Ribbon 等。

在 2017 年下半年的时候，我开始对每个组件进行梳理，深入到每个组件的实现原理和源码。毕竟 Spring Cloud 中包含了众多组件，我断断续续花了半年时间把各个组件大概梳理了一遍，没想到这些积累成为了本书的写作基础。

本书详细介绍 Spring Cloud 相关组件及其在微服务架构中的应用。全书共 13 章，第 1 章介绍微服务架构相关的基本概念；第 2 章介绍 Spring Cloud 中包含的组件以及 Spring Cloud 约定的上下文；第 3 章介绍 Spring Cloud 的基础 Spring Boot，包括如何构建一个 Spring Boot 服务、Spring Boot 的配置等；第 4～13 章详细讲解 Spring Cloud 组件，包括 Eureka（服务注册与发现）、OpenFeign（声明式 RESTful 客户端）、Hystrix（断路器）、Ribbon（客户端负载均衡器）、Gateway（API 网关）、Config（配置中心）、Stream（消息驱动）、Bus（消息总线）、Security（认证与授权）、Sleuth（服务链路追踪）。本书的目标是深入到 Spring Cloud 组件实现的技术内幕，并介绍了进阶应用的思路，为读者提供使用 Spring Cloud 进行微服务架构实践的参考。

本书在介绍 Spring Cloud 中的重要组件时，从基础应用的案例着手，尽可能将这类组件的设计思路和实现原理讲清楚，以帮助读者加深理解，并结合源码讲解组件的实现原理，最后还介绍了组件的进阶功能与应用。本书适合具有一些 Java 基础的开发人员，特别适合正在尝试微服务实践并想要深入了解 Spring Cloud 各个组件原理的开发人员和架构师。书中的很多案例都提供了源代码，可以随时下载，下载地址为：

- github 地址：https://github.com/Advanced-SpringCloud/cloud-book
- gitee 地址：https://gitee.com/Advanced-SpringCloud/cloud-book

本书最终由三个人共同完成，具体分工如下：第 1、8、9、11、13 章由朱荣鑫编写，第 2、5、7、10 章由张天编写，第 3、4、6、12 章由黄迪璇编写，全书由朱荣鑫统稿。能够完成本书需要感谢很多人，丁二玉老师在本书的撰写过程中提供了很多内容组织方面的建议，花了很多休息时间帮助审稿，非常感谢丁老师的大力帮助；感谢笔者所在的公司，一个年轻而富有活力的公司，为我们提供了很好的平台，从而积累了很多微服务架构实践的经验；感谢机械工业出版社的吴怡编辑及其他工作人员，为本书投入了很多精力。由于时间有限，书中难免存在一些问题，请读者不吝赐教。

朱荣鑫
2018 年 5 月

Contents 目 录

前言

第1章 微服务架构介绍 ··········· 1
1.1 微服务架构的出现 ··········· 1
 1.1.1 单体应用架构 ··········· 1
 1.1.2 SOA 架构 ··········· 2
 1.1.3 微服务架构 ··········· 3
1.2 微服务架构的流派 ··········· 5
1.3 云原生与微服务 ··········· 9
1.4 本章小结 ··········· 12

第2章 Spring Cloud 总览 ··········· 13
2.1 Spring Cloud 架构 ··········· 13
2.2 Spring Cloud 特性 ··········· 16
 2.2.1 Spring Cloud Context：应用上下文 ··········· 16
 2.2.2 Spring Cloud Commons：公共抽象 ··········· 19
2.3 本章小结 ··········· 21

第3章 Spring Cloud 的基础：Spring Boot ··········· 22
3.1 Spring Boot 简介 ··········· 22
3.2 构建一个微服务 ··········· 24
3.3 Spring Boot 配置文件 ··········· 29
 3.3.1 默认配置文件 ··········· 29
 3.3.2 外部化配置 ··········· 29
 3.3.3 YAML ··········· 30
 3.3.4 自动载入外部属性到 Bean ··········· 30
 3.3.5 多 Profile ··········· 31
 3.3.6 Starter ··········· 32
 3.3.7 自制一个 Starter ··········· 32
 3.3.8 Actuator ··········· 36
3.4 本章小结 ··········· 38

第4章 服务注册与发现：Eureka ··········· 39
4.1 基础应用 ··········· 40
 4.1.1 Eureka 简介 ··········· 40
 4.1.2 搭建 Eureka 服务注册中心 ··········· 40
 4.1.3 搭建 Eureka 服务提供者 ··········· 42
 4.1.4 搭建 Eureka 服务调用者 ··········· 43
 4.1.5 Eureka 服务注册和发现 ··········· 44
 4.1.6 Consul 的简单应用 ··········· 46
4.2 服务发现原理 ··········· 48
4.3 Eureka Client 源码解析 ··········· 49
 4.3.1 读取应用自身配置信息 ··········· 50

4.3.2	服务发现客户端	52	5.3.1	Decoder 与 Encoder 的定制化 111
4.3.3	拉取注册表信息	56	5.3.2	请求/响应压缩 112
4.3.4	服务注册	61	5.4	本章小结 113
4.3.5	初始化定时任务	62		
4.3.6	服务下线	68		

第 5 章 声明式 RESTful 客户端：Spring Cloud OpenFeign ... 88

- 4.4 Eureka Server 源码解析 ... 70
  - 4.4.1 服务实例注册表 ... 70
  - 4.4.2 服务注册 ... 72
  - 4.4.3 接受服务心跳 ... 74
  - 4.4.4 服务剔除 ... 75
  - 4.4.5 服务下线 ... 77
  - 4.4.6 集群同步 ... 78
  - 4.4.7 获取注册表中服务实例信息 ... 82
- 4.5 进阶应用 ... 84
  - 4.5.1 Eureka Instance 和 Client 的元数据 ... 84
  - 4.5.2 状态页和健康检查页端口设置 ... 85
  - 4.5.3 区域与可用区 ... 85
  - 4.5.4 高可用性服务注册中心 ... 86
- 4.6 本章小结 ... 87

第 5 章 声明式 RESTful 客户端：Spring Cloud OpenFeign ... 88

- 5.1 基础应用 ... 88
  - 5.1.1 微服务之间的交互 ... 88
  - 5.1.2 OpenFeign 简介 ... 89
  - 5.1.3 代码示例 ... 89
- 5.2 源码分析 ... 91
  - 5.2.1 核心组件与概念 ... 91
  - 5.2.2 动态注册 BeanDefinition ... 92
  - 5.2.3 实例初始化 ... 98
  - 5.2.4 函数调用和网络请求 ... 107
- 5.3 进阶应用 ... 111

第 6 章 断路器：Hystrix ... 114

- 6.1 基础应用 ... 114
  - 6.1.1 RestTemplate 与 Hystrix ... 115
  - 6.1.2 OpenFeign 与 Hystrix ... 117
- 6.2 Hystrix 原理 ... 118
  - 6.2.1 服务雪崩 ... 118
  - 6.2.2 断路器 ... 119
  - 6.2.3 服务降级操作 ... 120
  - 6.2.4 资源隔离 ... 121
  - 6.2.5 Hystrix 实现思路 ... 122
- 6.3 源码解析 ... 123
  - 6.3.1 封装 HystrixCommand ... 123
  - 6.3.2 HystrixCommand 类结构 ... 129
  - 6.3.3 异步回调执行命令 ... 129
  - 6.3.4 异步执行命令和同步执行命令 ... 137
  - 6.3.5 断路器逻辑 ... 137
  - 6.3.6 资源隔离 ... 143
  - 6.3.7 请求超时监控 ... 148
  - 6.3.8 失败回滚逻辑 ... 150
- 6.4 进阶应用 ... 152
  - 6.4.1 异步与异步回调执行命令 ... 152
  - 6.4.2 继承 HystrixCommand ... 153
  - 6.4.3 请求合并 ... 157
- 6.5 本章小结 ... 161

第 7 章 客户端负载均衡器：Spring Cloud Netflix Ribbon ... 162

- 7.1 负载均衡 ... 162
- 7.2 基础应用 ... 163

7.3 源码分析 165
    7.3.1 配置和实例初始化 165
    7.3.2 与 OpenFeign 的集成 167
    7.3.3 负载均衡器 LoadBalancerClient 171
    7.3.4 ILoadBalancer 173
    7.3.5 负载均衡策略实现 177
7.4 进阶应用 184
    7.4.1 Ribbon API 184
    7.4.2 使用 Netty 发送网络请求 185
    7.4.3 只读数据库的负载均衡实现 186
7.5 本章小结 187

第 8 章 API 网关：Spring Cloud Gateway 189

8.1 Spring Cloud Gateway 介绍 189
8.2 基础应用 190
    8.2.1 用户服务 191
    8.2.2 网关服务 192
    8.2.3 客户端的访问 195
8.3 源码解析 195
    8.3.1 初始化配置 196
    8.3.2 网关处理器 197
    8.3.3 路由定义定位器 202
    8.3.4 路由定位器 205
    8.3.5 路由断言 208
    8.3.6 网关过滤器 216
    8.3.7 全局过滤器 227
    8.3.8 API 端点 234
8.4 应用进阶 235
    8.4.1 限流机制 235
    8.4.2 熔断降级 238
    8.4.3 网关重试过滤器 240
8.5 本章小结 241

第 9 章 配置中心：Spring Cloud Config 243

9.1 基础应用 244
    9.1.1 配置客户端 244
    9.1.2 配置仓库 245
    9.1.3 服务端 246
    9.1.4 配置验证 248
    9.1.5 配置动态更新 249
9.2 源码解析 250
    9.2.1 配置服务器 251
    9.2.2 配置客户端 261
9.3 应用进阶 267
    9.3.1 为 Config Server 配置多个 repo 268
    9.3.2 客户端覆写远端的配置属性 268
    9.3.3 属性覆盖 269
    9.3.4 安全保护 269
    9.3.5 加密解密 270
    9.3.6 快速响应失败与重试机制 272
9.4 本章小结 272

第 10 章 消息驱动：Spring Cloud Stream 274

10.1 消息队列 274
10.2 基础应用 276
    10.2.1 声明和绑定通道 276
    10.2.2 自定义通道 276
    10.2.3 接收消息 276
    10.2.4 配置 278
10.3 源码分析 278
    10.3.1 动态注册 BeanDefinition 279
    10.3.2 绑定服务 282
    10.3.3 获取绑定器 284
    10.3.4 绑定生产者 289

	10.3.5	消息发送的流程 291	12.1.2	JWT 336
	10.3.6	StreamListener 注解的处理 293	12.1.3	搭建授权服务器 338
	10.3.7	绑定消费者 298	12.1.4	配置资源服务器 341
	10.3.8	消息的接收 304	12.1.5	访问受限资源 344
10.4	进阶应用 306		12.2	整体架构 346
	10.4.1	Binder For RocketMQ 306	12.3	源码解析 348
	10.4.2	多实例 311		12.3.1 安全上下文 349
	10.4.3	分区 311		12.3.2 认证 350
10.5	本章小结 313			12.3.3 授权 357
				12.3.4 Spring Security 中的过滤器与拦截器 361

第 11 章 消息总线：Spring Cloud Bus 314

				12.3.5 授权服务器 372
11.1	基础应用 314			12.3.6 资源服务器 387
	11.1.1	配置服务器 315		12.3.7 令牌中继机制 394
	11.1.2	配置客户端 316	12.4	进阶应用 395
	11.1.3	结果验证 316		12.4.1 Spring Security 定制 395
11.2	源码解析 318			12.4.2 OAuth2 定制 399
	11.2.1	事件的定义与事件监听器 319		12.4.3 SSO 单点登录 403
	11.2.2	消息的订阅与发布 326	12.5	本章小结 406
	11.2.3	控制端点 328		

第 13 章 服务链路追踪：Spring Cloud Sleuth 407

11.3	应用进阶 329		13.1	链路监控组件简介 407
	11.3.1	在自定义的包中注册事件 329	13.2	基础应用 410
	11.3.2	自定义监听器 330		13.2.1 特性 411
	11.3.3	事件的发起者 331		13.2.2 项目准备 411
11.4	本章小结 332			13.2.3 Spring Cloud Sleuth 独立实现 414

第 12 章 认证与授权：Spring Cloud Security 333

				13.2.4 集成 Zipkin 414
12.1	基础应用 333		13.3	本章小结 420
	12.1.1	OAuth2 简介 334		

第 1 章 Chapter 1

微服务架构介绍

近年来，微服务架构一直是互联网技术圈的热点之一，越来越多的互联网应用都采用了微服务架构作为系统构建的基础，很多新技术和理念如 Docker、Kubernetes、DevOps、持续交付、Service Mesh 等也都在关注、支持和跟随微服务架构的发展。

本章将会概要性地介绍微服务架构：包括微服务架构是如何演进的，微服务架构的主要流派，当前主流的云原生应用与微服务之间的关系等。

1.1 微服务架构的出现

从单体应用架构发展到 SOA 架构，再到微服务架构，应用架构经历了多年的不断演进。微服务架构不是凭空产生的，而是技术发展的必然结果，分布式云平台的应用环境使得微服务代替单体应用成为互联网大型系统的架构选择。目前，虽然微服务架构还没有公认的技术标准和规范草案，但业界已经有了一些很有影响力的开源微服务架构解决方案，在进行微服务化开发或改造时可以进行相应的参考。

1.1.1 单体应用架构

与微服务架构对比的是传统的单体应用。Web 应用程序发展的早期，大部分 Web 工程是将所有的功能模块打包到一起部署和运行，例如 Java 应用程序打包为一个 war 包，其他语言（Ruby、Python 或者 C++）编写的应用程序也有类似的做法。单体应用的实现架构类似于图 1-1 中的电影售票系统。

这个电影售票系统采用分层架构，按照调用顺序，从上到下为表示层、业务层、数据访问（DAO）层、DB 层。表示层负责用户体验；业务层负责业务逻辑，包括电影、订单和用户三个模块；数据访问层负责 DB 层的数据存取，实现增删改查的功能。业务层定义了应用的业务逻辑，是整个应用的核心。在单体应用中，所有这些模块都集成在一起，这样的系统架构就叫做单体应用架构，或称为巨石型应用架构。单体应用是最早的应用形态，开发和部署都很简单。在中小型项目中使用单体应用架构，能体现出其优势，且单体应用的整体性能主要依赖于硬件资源和逻辑代码实现，应用架构自身不需要特别关注。

单体应用的集成非常简洁，IDE 集成开发环境（如 Eclipse）和其他工具都擅长开发一个简单应用；单体应用易于调试，由于一个应用包含所有功能，所以只需要简单运行此应用即可进行开发测试；单体应用易于部署，只需要把应用打包，拷贝到服务器端即可；通过负载均衡器，运行多个服务实例，单体应用可以轻松实现应用扩展。

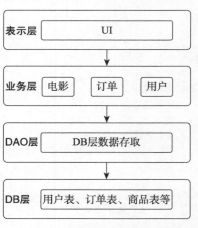

图 1-1　电影售票系统的架构图

但是，随着应用项目变得复杂、开发团队不断扩张之后，单体应用的不足和弊端将会变得很明显，主要有如下不足：

- **可靠性**：每个 bug 都可能会影响到整个应用的可靠性。因为所有模块都运行在一个进程中，任何一个模块中的一个 bug，比如内存泄露，都可能拖垮整个进程。
- **复杂性高**：单体应用巨大的代码库可能会令人望而生畏，特别是对团队新成员来说。应用难以理解和迭代，进而导致开发速度减慢。由于没有清晰的模块边界，模块化会逐渐消失。
- **持续部署困难**：巨大的单体应用本身就是频繁部署的一大障碍。为了更新一个组件，必须重新部署整个应用。这会中断那些可能与更改无关的后台任务（例如 Java 应用中的 Quartz 任务），同时可能引发其他问题。另外，未被更新的组件有可能无法正常启动。重新部署会增加风险，进而阻碍频繁更新。
- **扩展能力受限**：单体架构只能进行一维扩展。一方面，它可以通过运行多个应用服务实例来增加业务容量，实现扩展。但另一方面，不同的应用组件有不同的资源需求：有的是 CPU 密集型的，有的是内存密集型的。单体架构无法单独扩展每个组件。
- **阻碍技术创新**：单体应用往往使用统一的技术平台或方案解决所有问题，团队的每个成员都必须使用相同的开发语言和架构，想要引入新的框架或技术平台非常困难。单体架构迫使团队长期使用在开发初期选定的技术栈，比如选择了 JVM 的语言，此时以非 JVM 语言编写的组件就无法在该单体架构的应用中使用。

1.1.2　SOA 架构

面向服务的架构（SOA）是 Gartner 于 20 世纪 90 年代中期提出的。2002 年 12 月，Gartner 提出"面向服务的架构"是现代应用开发领域最重要的课题之一。

SOA 的核心主体是服务，其目标是通过服务的流程化来实现业务的灵活性。服务就像一堆"元器件"，这些元器件通过封装形成标准服务，它们有相同的接口和语义表达规则。但服务要组装成一个流程和应用，还需要有效的"管理"，包括如何注册服务、如何发现服务、如何包装服务的安全性和可靠性，这些就是 SOA 治理。SOA 治理是将 SOA 的一堆元器件进行有效组装。这是形成一个"产品"的关键，否则那些永远是一堆元器件，而无法形成一个有机整体。

完整的 SOA 架构由五大部分组成：基础设施服务、企业服务总线、关键服务组件、开发工具、管理工具等。

- **基础设施**：为整个 SOA 组件和框架提供一个可靠的运行环境，以及服务组件容器，它的核心组件是应用服务器等基础软件支撑设施，提供运行期完整、可靠的软件支撑。
- **企业服务总线**：提供可靠消息传输、服务接入、协议转换、数据格式转换、基于内容的路由等功能，屏蔽了服务的物理位置、协议和数据格式。
- **关键服务组件**：SOA 在各种业务服务组件的分类。
- **开发工具和管理工具**：提供完善的、可视化的服务开发和流程编排工具，包括服务的设计、开发、配置、部署、监控、重构等完整的 SOA 项目开发生命周期。

具体来说，就是在分布式的环境中，将各种功能都以服务的形式提供给最终用户或者其他服务。企业级应用的开发多采用面向服务的体系架构来满足灵活多变、可重用性的需求。它将应用程序的不同功能单元（称为服务）通过这些服务之间定义良好的接口和契约联系起来。接口采用中立的方式进行定义，它应该独立于实现服务的硬件平台、操作系统和编程语言。这使得构建在各种各样的系统中的服务可以以一种统一和通用的方式进行交互。

SOA 是在企业计算领域中提出的，目的是要将紧耦合的系统，划分为面向业务的、粗粒度、松耦合、无状态的服务。服务发布出来供其他服务调用，一组互相依赖的服务就构成了 SOA 架构的系统。基于这些基础的服务，可以将业务过程用类似 BPEL（业务流程执行语言）流程的方式编排起来，BPEL 反映的是业务处理的过程，这些过程对于业务人员更为直观。企业还需要一些服务治理的工具，比如服务注册库、监控管理等。在企业计算领域，如果不是交易系统的话，并发量都不是很大，所以大多数情况下，一台服务器就可以容纳许多的服务，这些服务采用统一的基础设施，可能都运行在一个应用服务器的进程中。虽然说是 SOA 架构，但还可能是单一的系统。

1.1.3 微服务架构

微服务最早是由 Martin Fowler 与 James Lewis 于 2014 年共同提出，需要了解细节的读者可以阅览 https://martinfowler.com/articles/microservices.html。其实 Martin 先生并没有给出明确的微服务定义，根据其描述，微服务的定义可以概括如下：微服务架构是一种使用一系列粒度较小的服务来开发单个应用的方式；每个服务运行在自己的进程中；服务间采用轻量级的方式进行通信（通常是 HTTP API）；这些服务是基于业务逻辑和范围，通过自动化部署的机制来独立部署的，并且服务的集中管理应该是最低限度的，即每个服务可以采用不同的编程语言编写，使用不同的数据存储技术。

如今，微服务架构已经不是一个新概念了，很多业界前沿互联网公司的实践表明，微服务是一种渐进式的演进架构，是企业应对业务复杂性，支持大规模持续创新行之有效的架构手段。

1. 组成

微服务架构是一种比较复杂、内涵丰富的架构模式，它包含很多支撑"微"服务的具体组件和概念，其中一些常用的组件及其概念如下：

- **服务注册与发现**：服务提供方将己方调用地址注册到服务注册中心，让服务调用方能够方便地找到自己；服务调用方从服务注册中心找到自己需要调用的服务的地址。
- **负载均衡**：服务提供方一般以多实例的形式提供服务，负载均衡功能能够让服务调用方连接到合适的服务节点。并且，服务节点选择的过程对服务调用方来说是透明的。
- **服务网关**：服务网关是服务调用的唯一入口，可以在这个组件中实现用户鉴权、动态路由、灰度发布、A/B 测试、负载限流等功能。
- **配置中心**：将本地化的配置信息（Properties、XML、YAML 等形式）注册到配置中心，实现程序包在开发、测试、生产环境中的无差别性，方便程序包的迁移。
- **集成框架**：微服务组件都以职责单一的程序包对外提供服务，集成框架以配置的形式将所有微服务组件（特别是管理端组件）集成到统一的界面框架下，让用户能够在统一的界面中使用系统。
- **调用链监控**：记录完成一次请求的先后衔接和调用关系，并将这种串行或并行的调用关系展示出来。在系统出错时，可以方便地找到出错点。
- **支撑平台**：系统微服务化后，各个业务模块经过拆分变得更加细化，系统的部署、运维、监控等都比单体应用架构更加复杂，这就需要将大部分的工作自动化。现在，Docker 等工具可以给微服务架构的部署带来较多的便利，例如持续集成、蓝绿发布、健康检查、性能健康等等。如果没有合适的支撑平台或工具，微服务架构就无法发挥它最大的功效。

2. 优点

微服务架构模式有很多优势可以有效解决单体应用扩大之后出现的大部分问题。首先，通过将巨大单体式应用分解为多个服务的方法解决了复杂性问题。在功能不变的情况下，应用分解为多个可管理的模块或服务。每个服务都有一个用 RPC 或者消息驱动 API 定义清楚的边界。微服务架构模式为采用单体式编码方式很难实现的功能提供了模块化的解决方案。由此，单个服务变得很容易开发、理解和维护。

其次，微服务架构模式使得团队并行开发得以推进，每个服务都可以由专门开发团队来开发。不同团队的开发者可以自由选择开发技术，提供 API 服务。这种自由意味着开发者不需要被迫使用之前采用的过时技术，他们可以选择最新的技术。甚至于，因为服务都是相对简单的，即使用新技术重写以前的代码也不是很困难的事情。

再次，微服务架构模式中每个微服务独立都是部署的。理想情况下，开发者不需要协调其他服务部署对本服务的影响。这种改变可以加快部署速度。UI 团队可以采用 AB 测试，快速地部署变化。微服务架构模式使得持续化部署成为可能。

最后，微服务架构模式使得每个服务易于独立扩展。

3. 挑战

微服务的一些想法是好的，但在实践中也会呈现出其复杂性，具体如下：

- **运维要求较高**。更多的服务意味着需要更多的运维投入。在单体架构中只需要保证一个应用的正常运行即可；而在微服务中，需要保证几十甚至几百个服务的正常运行与协作，这带来了巨大的挑战。
- **分布式固有的复杂性**。使用微服务构建的是分布式系统。对于一个分布式系统来说，系统容错、网络延迟、分布式事务等都会带来巨大的挑战。
- **接口调整成本高**。微服务之间通过接口进行通信。如果修改某个微服务的 API，可能所有使用了该接口的微服务都需要做调整。
- **重复劳动**。很多服务可能都会使用到相同的功能，而这个功能并没有达到分解为一个微服务的程度，这个时候，可能各个服务都会开发这一功能，导致代码重复。
- **可测试性的挑战**。在动态环境下，服务间的交互会产生非常微妙的行为，难以进行可视化及全面测试。

1.2 微服务架构的流派

常见的微服务架构方案有四种，分别是 ZeroC IceGrid、基于消息队列、Docker Swarm 和 Spring Cloud。下面分别介绍这四种方案。

1. ZeroC IceGrid

ZeroC IceGrid 是基于 RPC 框架 Ice 发展而来的一种微服务架构，Ice 不仅仅是一个 RPC 框架，它还为网络应用程序提供了一些补充服务。Ice 是一个全面的 RPC 框架，支持 C++、C#、Java、JavaScript、Python 等语言。IceGrid 具有定位、部署和管理 Ice 服务器的功能，具有良好的性能与分布式能力，下面具体介绍 IceGrid 的功能。

Ice 的 DNS。DNS 用于将域名信息映射到具体的 IP 地址，通过域名得到该域名对应的 IP 地址的过程叫做域名解析。IceGrid 为 Ice 提供了类似的服务：它允许 Ice 客户端通过简单的名称来查找 Ice 对象。Ice 客户端可以通过提供此对象的完整寻址信息来访问服务器中的 Ice 对象，例如 chatRoom1：ssl -h demo.zeroc.com -p 10000。这样的硬编码虽然很简单，但缺乏灵活性。因为需要为 Ice 服务器（本例中端口为 10000）选择一个固定的端口号，因此将 Ice 服务器移到不同的主机上需要更新其客户端。

IceGrid 提供了对这种寻址信息使用符号名称的选项，例如 chatRoom1@chatRoomHost。当 Ice 客户端尝试访问 chatRoomHost 中的对象时，它会要求 IceGrid 提供与此符号名称关联的实际地址。例如，IceGrid 返回 -h demo.zeroc.com -p 65431，客户端就可以直接并透明地连接到服务器。IceGrid 架构如图 1-2 所示。

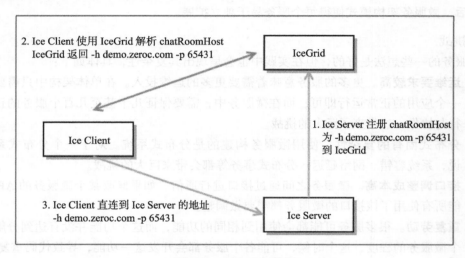

图 1-2　IceGrid 架构图

服务器部署。若直接通过 IP 地址+端口号的方式，当 Ice 客户端尝试连接到未运行的服务器时，连接将会失败。通过符号或间接寻址，IceGrid 有机会检查目标服务器是否正在运行，并在服务器未运行时启动或重新启动服务器。可以配置 IceGrid 以各种方式启动服务器：手动（通过管理命令）、按需（无论何时请求服务器）和 IceGrid 始终保持服务器运行。

服务器的复制。IceGrid 允许部署同一服务器的多个副本，并可以配置 IceGrid 将符号名称解析到此服务器副本的策略。可以在所有副本中对客户端进行负载平衡，或者使用主备配置，客户端只要保持可用状态，就使用主备配置。

管理和监控。IceGrid 的管理工具可以完全控制已部署的应用程序。诸如启动服务器或修改配置设置等活动只需单击鼠标即可。图 1-3 为 IceGrid 管理工具的界面。

IceGrid 当前最新的版本为 3.7.1，在 3.6 版本之后增加了容器化的运行方式。总的来说，IceGrid 作为微服务架构早期的实践方案，其流行时间并不是很久，当前国内选用这种微服务架构方案的公司非常少。

2. 基于消息队列

在微服务架构的定义中讲到，各个微服务之间使用"轻量级"的通信机制。所谓轻量级，是指通信协议与语言无关、与平台无关。微服务之间的通信方式有两种：同步和异步。

同步方式有 RPC，REST 等；除了标准的基于同步通信方式的微服务架构外，还有基于消息队列异步方式通信的微服务架构。

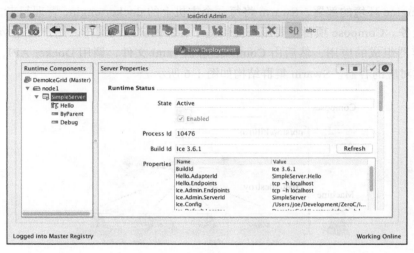

图 1-3 IceGrid Admin

在基于消息队列的微服务架构方式中，微服务之间采用发布消息与监听消息的方式来实现服务之间的交互。图 1-4 是一个简单的电商系统中商品服务、用户服务、订单服务和库存服务之间的交互示意图，可以看到消息中间件（MQ）是关键，它负责连通各个微服务，承担了整个系统互联互通的重任。

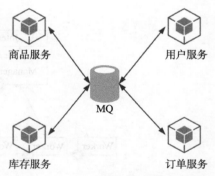

图 1-4 基于消息队列的微服务架构

基于消息队列的微服务架构是全异步通信模式的一种设计，各个组件之间没有直接的耦合关系，也不存在服务接口与服务调用的说法，服务之间通过消息来实现彼此的通信与业务流程的驱动。基于消息队列的微服务架构应用的案例并不多，更多地体现为一种与业务相关的设计经验，每个公司都有不同的实现方式，缺乏公认的设计思路与参考架构，也没有形成一个知名的开源平台。因此，如要实施这种微服务架构，需要项目组自己从零开始去设计实现一个微服务架构基础平台，这可能会造成项目的成本较高且风险较大，决策之前需要进行全盘思考与客观评价。

3. Docker Swarm

Swarm 项目是 Docker 公司发布的三剑客中的一员，用来提供容器集群服务，目的是更好地帮助用户管理多个 Docker Engine，方便用户使用。通过把多个 Docker Engine 聚集在一起，形成一个大的 Docker Engine，对外提供容器的集群服务。同时这个集群对外提供 Swarm API，用户可以像使用 Docker Engine 一样使用 Docker 集群。

如图 1-5 所示 Docker 三剑客包括：Machine、Compose 和 Swarm。通过 Machine 可以在不同云平台上创建包含 docker-engine 的主机。Machine 通过 driver 机制，支持多个平台的 docker-engine 环境的部署。Swarm 将每一个主机上的 docker-engine 管理起来，对外提供容器集群服务。Compose 项目主要用来提供基于容器的应用的编排。用户通过 yaml 文件描述由多个容器组成的应用，然后由 Compose 解析 yaml 文件，调用 Docker API，在 Swarm 集群上创建对应的容器。Swarm 集群结构如图 1-6 所示。

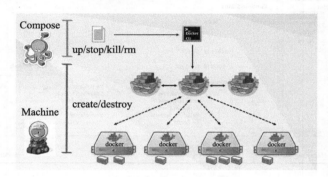

图 1-5　Docker 三剑客

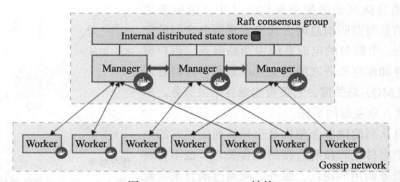

图 1-6　Docker Swarm 结构

从图 1-6 中我们看到一个 Swarm 集群中有两种角色的节点：
- Manager：负责集群的管理、集群状态的维持及将任务（Task）调度到工作节点上等。
- Worker：承载运行在 Swarm 集群中的容器实例，每个节点主动汇报其上运行的任务并维持同步状态。

Docker Swarm 对外提供 Docker API，自身轻量，学习成本、二次开发成本都比较低，是一个插件式框架。从功能上讲，Swarm 是类似于 Google 开源的 Kubernetes 微服务架构平台的一个产品。

4. Spring Cloud

Spring Cloud 是一个基于 Spring Boot 实现的云应用开发工具，是一系列框架的集合，当添加这些工具库到应用后会增强应用的行为。Spring Boot 秉持约定优于配置的思想，因此可以利用这些组件基本的默认行为来快速入门，并在需要的时候可以配置或扩展，以创建自定义解决方案。

Spring Cloud 利用 Spring Boot 的开发便利性，巧妙地简化了分布式系统基础设施的开发，如服务发现注册、配置中心、消息总线、负载均衡、断路器、数据监控等，都可以基于 Spring Boot 组件进行开发，做到一键启动和部署。Spring Cloud 并没有重复制造轮子，它只是将目前比较成熟、经得起实际考验、优秀的开源服务框架组合起来，通过 Spring Boot 进行封装，屏蔽掉复杂的配置和实现原理，最终给开发者留出了一套简单易懂、易部署和易维护的分布式系统开发工具包。

以下为 Spring Cloud 的核心功能：
- 分布式/版本化配置
- 服务注册和发现
- 服务路由
- 服务和服务之间的调用
- 负载均衡
- 断路器
- 分布式消息传递

还有很多基础的功能没有列出，每个功能对应 Spring Cloud 中的一个组件，包括 Spring Cloud Config、Spring Cloud Netflix（Eureka、Hystrix、Zuul、Archaius …）、Spring Cloud Bus 等组件。

1.3　云原生与微服务

提及云原生，首先需要了解一下 CNCF，即云原生计算基金会，2015 年由谷歌牵头成立，基金会成员目前已有一百多个企业与机构，包括亚马逊、微软、思科等巨头。目前 CNCF 所托管的应用已达 14 个，知名的项目有 Kubernetes、Prometheus、Envoy 等。

1. 云原生

CNCF 宪章中给出了云原生应用的三大特征，概括如下：
- **容器化封装**：以容器为基础，提高整体开发水平，形成代码和组件重用，简化云原生应用程序的维护。在容器中运行应用程序和进程，并作为应用程序部署的独立单元，实现高水平资源隔离。
- **动态管理**：通过集中式的编排调度系统来动态管理和调度。
- **面向微服务**：明确服务间的依赖，互相解耦。

云原生包含了一组应用的模式，用于帮助企业快速、持续、可靠、规模化地交付业务软件。如图 1-7 所示，云原生由微服务架构、DevOps 和以容器为代表的敏捷基础架构组成。

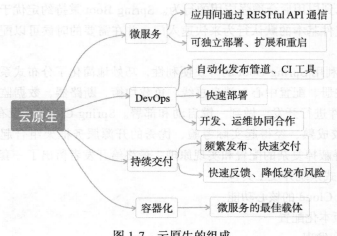

图 1-7　云原生的组成

2. 12 原则

12 原则（12-Factors）经常直译为 12 要素，12 原则由公有云 PaaS 的先驱 Heroku 于 2012 年提出（https://12factor.net/），目的是告诉开发者如何利用云平台提供的便利来开发更具可靠性和扩展性、更加易于维护的云原生应用。12 原则包括：

- 基准代码。
- 显式声明依赖关系。
- 在环境中存储配置。
- 把后端服务当作附加资源。
- 严格分离构建、发布和运行。
- 无状态进程。
- 通过端口绑定提供服务。
- 通过进程模型进行扩展。
- 快速启动和优雅终止。
- 开发环境与线上环境等价。
- 日志作为事件流。
- 管理进程。

另外还有补充的三点：API 声明管理、认证和授权、监控与告警。

12 原则提出来已有六年多，12 原则的有些细节可能已经不那么跟得上时代，也有人批评 12 原则的提出从一开始就有过于依赖 Heroku 自身特性的倾向。不过不管怎样，12 原则依旧是业界最为系统的云原生应用开发指南。

3. 容器化

Docker 是一个开源引擎，可以轻松地为任何应用创建一个轻量级的、可移植的、自给自足的容器。最近几年 Docker 容器化技术很火，在各种场合都能够听到关于 Docker 的分享。Docker 让开发工程师可以将他们的应用和依赖封装到一个可移植的容器中。Docker 根本的想法是创建软件程序可移植的轻量容器，让其可以在任何安装了 Docker 的机器上运行，而不用关心底层操作系统。

Docker 可以解决虚拟机能够解决的问题，同时也能够解决虚拟机由于资源要求过高而无法解决的问题。其优势包括：

- 隔离应用依赖。
- 创建应用镜像并进行复制。
- 创建容易分发的、即启即用的应用。
- 允许实例简单、快速地扩展。
- 测试应用并随后销毁它们。

虽然自动化运维工具可以降低环境搭建的复杂度，但仍然不能从根本上解决环境的问题。在看似稳定而成熟的场景下，结合使用 Docker 显然能带来更多的好处。

一旦拥抱了容器，这就需要一个编排框架来调度和管理容器。最常见的编排框架有 Kubernetes、Mesos、Docker Swarm。编排框架是容器平台的关键组成部分。

4. DevOps

DevOps 是软件开发人员（Dev）和 IT 运维技术人员（Ops）之间的合作，目标是自动执行软件交付和基础架构更改流程，使得构建、测试、发布软件能够更加地快捷和可靠。它创造了一种文化和环境，可在其中快速、频繁且更可靠地构建、测试和发布软件。通过 DevOps 流水线加上 Docker 容器，可以实现自动化工程管理，实现开发、测试环境的自动申请和构建。通过 Docker 标准化打包应用配置和环境，可以生成轻量容器镜像，并使用镜像方式迭代开发、测试、部署，加速应用上线。自动化运维在合理保障应用高可用的前提下，能够提升资源利用率，降低成本。

DevOps 的引入能对产品交付、测试、功能开发和维护起到意义深远的影响。在缺乏 DevOps 能力的组织中，开发与运营之间存在着信息"鸿沟"，例如运营人员要求更好的可靠性和安全性，开发人员则希望基础设施响应更快，而业务用户的需求则是更快地将更多的特性发布给最终用户使用。

5. 微服务

微服务将单体业务系统分解为多个可独立部署的服务。这个服务通常只关注某项业务，或者最小可提供业务价值的"原子"服务单元。

微服务架构有以下优势：

- 当人们将业务领域分解为可独立部署的环境时，能够将相关的变更周期解耦。只要变更限于单一有限的环境，并且服务继续履行其现有合约，那么这些变更可以独立于与其他业务来进行和部署。其结果是实现了更频繁和快速的部署，实现了持续的价值流动。
- 扩展更多的部署组件本身可以加快部署。在原本的单体应用中，由于沟通和协调的开销，在添加更多的人手时，往往会使软件开发流程变得更长。《人月神话》的作者弗雷德里克·布鲁克斯很多年前就教导我们，在软件项目的晚期增加更多的人力往往会使软件项目更加延期。但是在微服务架构中可以在有限的环境中构建更多的沙箱，而不是将所有的开发者都放在同一个沙箱中。由于学习业务领域和现有代码的负担减少，并且团队较小，因此添加到每个沙箱的新开发人员可以更快速地提高并使得工作变得更高效。
- 可以加快采用新技术的步伐。大型单体应用程序架构通常与对技术堆栈的长期保证有关。这些保证的存在是为了减轻采用新技术的风险。技术选型的错误在单体架构中的代价非常昂贵，因为这些错误可能会影响整个企业架构。如果可以在单个服务的范围内采用新技术，就将隔离并最大限度地降低风险，就像隔离和最小化运行时故障的风险一样。
- 微服务提供独立、高效的服务扩展。单体架构也可以扩展，但要求我们扩展所有组件，而不仅仅是那些负载较重的组件。当且仅当相关联的负载需要它时，才缩放微服务。

1.4 本章小结

单体模式开发速度很快，但是随着互联网用户的高速增长，已经变得不适应如此大用户量的应用。架构的演进从单体到 SOA，再到微服务架构，技术的发展遵循一定的规律，使得移动互联网的发展在技术上得到了支持。微服务架构是 SOA 思想的一种具体实践。微服务架构将早期的单体应用从数据存储开始垂直拆分成多个不同的服务，每个服务都能独立部署，独立维护，独立扩展，服务与服务间通过诸如 RESTful API 的方式互相调用。

微服务架构并不能解决所有问题，所以一定要慎重考虑何时采用微服务，只有业务复杂到一定程度的时候才适合采用微服务架构。微服务的理念和框架，使得它对团队的要求比较高，需要团队所有成员都要认可微服务的理念和治理框架，否则执行过程中会有很多问题。

第 2 章 Chapter 2

Spring Cloud 总览

在介绍微服务架构相关的概念之后，本章将会介绍 Spring Cloud 相关的概念。Spring Cloud 是一系列框架的有机集合，基于 Spring Boot 实现的云应用开发工具，为云原生应用开发中的服务发现与注册、熔断机制、服务路由、分布式配置中心、消息总线、负载均衡和链路监控等功能的实现提供了一种简单的开发方式。

本章将会对 Spring Cloud 架构及相关组件进行初步介绍，然后介绍 Spring Cloud 上下文和 Spring Cloud 的公共抽象，这部分内容可能会在多个组件中出现，所以此处的介绍并不依赖具体的组件。

2.1 Spring Cloud 架构

Spring 社区推出 Spring Cloud 框架与其自身的理念演变密切相关。Spring 是于 2003 年兴起的一个轻量级的 Java 开发框架，但是随着其不断地发展壮大，框架的代码规模越来越庞大，集成的项目越来越多，配置文件也变得越来越混乱，慢慢地背离最初的理念。如今，分布式系统和微服务架构等更多新的技术理念的陆续出现，催生了这样一种想法：Spring 社区需要一款框架来改善以前的开发模式并适应微服务开发环境。Spring Cloud 也就此应运而生。

1. 版本说明

Spring Cloud 不同于其他独立项目，它是拥有众多子项目的项目集合。其所有版本由版本名和版本号组成，如 Finchley M9 和 Edgware SR3。Finchley 和 Edgware 就是版本名，名字是以伦敦地铁站命名并且遵循字母顺序（第一个 release 版本为 Angel，第二个为 Brixton），而 M9 和 SR3 是版本号。M 是里程碑（milestone）的含义，M9 就是指第九个里程碑。SR（Service Releases）则代表着稳定版本，后边一般都会有一个递增的数字，例如 SR3 就是指第三个发布版本。图 2-1 为 Spring Cloud 包含的组件的详细列表，最新的版本为 Finchley RELEASE，基于 Spring Boot 2.0.3，Spring Boot 2.X 相对之前版本有较大的变动。本书的源码分析根据最新的 Finchley RELEASE 展开。

组件	Edgware.SR3	Finchley.RELEASE	Finchley.BUILD-SNAPSHOT
spring-cloud-aws	1.2.2.RELEASE	2.0.0.RELEASE	2.0.0.BUILD-SNAPSHOT
spring-cloud-bus	1.3.2.RELEASE	2.0.0.RELEASE	2.0.0.BUILD-SNAPSHOT
spring-cloud-cli	1.4.1.RELEASE	2.0.0.RELEASE	2.0.0.BUILD-SNAPSHOT
spring-cloud-commons	1.3.3.RELEASE	2.0.0.RELEASE	2.0.0.BUILD-SNAPSHOT
spring-cloud-contract	1.2.4.RELEASE	2.0.0.RELEASE	2.0.0.BUILD-SNAPSHOT
spring-cloud-config	1.4.3.RELEASE	2.0.0.RELEASE	2.0.0.BUILD-SNAPSHOT
spring-cloud-netflix	1.4.4.RELEASE	2.0.0.RELEASE	2.0.0.BUILD-SNAPSHOT
spring-cloud-security	1.2.2.RELEASE	2.0.0.RELEASE	2.0.0.BUILD-SNAPSHOT
spring-cloud-cloudfoundry	1.1.1.RELEASE	2.0.0.RELEASE	2.0.0.BUILD-SNAPSHOT
spring-cloud-consul	1.3.3.RELEASE	2.0.0.RELEASE	2.0.0.BUILD-SNAPSHOT
spring-cloud-sleuth	1.3.3.RELEASE	2.0.0.RELEASE	2.0.0.BUILD-SNAPSHOT
spring-cloud-stream	Ditmars.SR3	Elmhurst.RELEASE	Elmhurst.BUILD-SNAPSHOT
spring-cloud-zookeeper	1.2.1.RELEASE	2.0.0.RELEASE	2.0.0.BUILD-SNAPSHOT
spring-boot	1.5.10.RELEASE	2.0.2.RELEASE	2.0.0.BUILD-SNAPSHOT
spring-cloud-task	1.2.2.RELEASE	2.0.0.RELEASE	2.0.0.RELEASE
spring-cloud-vault	1.1.0.RELEASE	2.0.0.RELEASE	2.0.0.BUILD-SNAPSHOT
spring-cloud-gateway	1.0.1.RELEASE	2.0.0.RELEASE	2.0.0.BUILD-SNAPSHOT
spring-cloud-openfeign		2.0.0.RELEASE	2.0.0.BUILD-SNAPSHOT

图 2-1　Spring Cloud 组件及其对应的版本

2. Spring Cloud 组成

Spring Cloud 是在 Spring Boot 的基础上构建的，用于简化分布式系统构建的工具集，为开发人员提供快速建立分布式系统中一些常见的组件。Spring Cloud 技术体系如图 2-2 所示。

Spring Cloud 包含的组件众多，各个组件都有各自不同的特色和优点，为用户提供丰富的选择：

- ❑ 服务注册与发现组件：Eureka、Zookeeper 和 Consul 等。本书将会重点讲解 Eureka，Eureka 是一个 REST 风格的服务注册与发现的基础服务组件。
- ❑ 服务调用组件：Hystrix、Ribbon 和 OpenFeign；其中 Hystrix 能够使系统在出现依赖服务失效的情况下，通过隔离系统依赖服务的方式，防止服务级联失败，同时提供失败回滚机制，使系统能够更快地从异常中恢复；Ribbon 用于提供客户端的软件负载均衡算法，还提供了一系列完善的配置项如连接超时、重试等；OpenFeign 是一个声明式 RESTful 网络请求客户端，它使编写 Web 服务客户端变得更加方便和快捷。

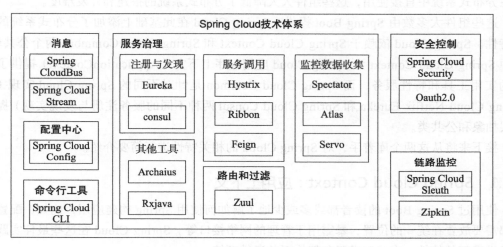

图 2-2　Spring Cloud 技术体系

- 路由和过滤组件：包括 Zuul 和 Spring Cloud Gateway。Spring Cloud Gateway 提供了一个构建在 Spring 生态之上的 API 网关，其旨在提供一种简单而有效的途径来发送 API，并为他们提供横切关注点，如：安全性、监控指标和弹性。
- 配置中心组件：Spring Cloud Config 实现了配置集中管理、动态刷新等配置中心的功能。配置通过 Git 或者简单文件来存储，支持加解密。
- 消息组件：Spring Cloud Stream 和 Spring Cloud Bus。Spring Cloud Stream 对于分布式消息的各种需求进行了抽象，包括发布订阅、分组消费和消息分区等功能，实现了微服务之间的异步通信。Spring Cloud Bus 主要提供了服务间的事件通信（如刷新配置）。
- 安全控制组件：Spring Cloud Security 基于 OAuth2.0 开放网络的安全标准，提供了微服务环境下的单点登录、资源授权和令牌管理等功能。
- 链路监控组件：Spring Cloud Sleuth 提供了全自动、可配置的数据埋点，以收集微服务调用链路上的性能数据，并可以结合 Zipkin 进行数据存储、统计和展示。

除了上述组件之外，Spring Cloud 还提供了命令行工具 Spring Cloud Cli 和集群工具 Spring Cloud Cluster。Spring Cloud Cli 提供了以命令行和脚本的方式来管理微服务及 Spring Cloud 组件的方式，Spring Cloud Cluster 提供了集群选主、分布式锁和一次性令牌等分布式集群需要的技术组件。

2.2 Spring Cloud 特性

云原生应用程序开发风格鼓励在持续交付和价值驱动开发上采取最佳的实践策略，Spring Cloud 提供了多种方式来促进云原生开发风格。Spring Cloud 提供了一系列组件，可以在分布式系统中直接使用，这些组件大大降低了分布式系统的搭建和开发难度。

这些组件大多数由 Spring Boot 提供，Spring Cloud 在此基础上添加了分布式系统的相关特性。Spring Cloud 依赖于 Spring Cloud Context 和 Spring Cloud Commons 两个公共库，其中 Spring Cloud Context 为 Spring Cloud 应用程序上下文（ApplicationContext）提供了大量的实用工具和特性服务，而 Spring Cloud Common 是针对不同的 Spring Cloud 实现（如 Spring Cloud Netflix Eureka 和 Spring Cloud Consul 两种不同的服务注册与发现实现）提供上层抽象和公共类。

接下来将从这两个库着手，对 Spring Cloud 的相关特性进行简要介绍。

2.2.1 Spring Cloud Context：应用上下文

使用过 Spring Boot 的读者都或多或少地了解如何使用 Spring 构建应用，比如，配置一般都需要放置在统一的位置，暴露用于管理的网络接口等。Spring Cloud 在这些最佳实践之上提供了新的特性，来适配微服务架构下的运维环境。

1. Bootstrap 上下文

除了应用上下文配置（application.yml 或者 application.properties）之外，Spring Cloud 应用程序还额外提供与 Bootstrap 上下文配置相关的应用属性。Bootstrap 上下文对于主程序来说是一个父级上下文，它支持从外部资源中加载配置文件，和解密本地外部配置文件中的属性。Bootstrap 上下文和应用上下文将共享一个环境（Environment），这是所有 Spring 应用程序的外部属性来源。一般来讲，Bootstrap 上下文中的属性优先级较高，所以它们不能被本地配置所覆盖。

Bootstrap 上下文使用与主程序不同的规则来加载外部配置。因此 bootstrap.yml 用于为 Bootstrap 上下文加载外部配置，区别于应用上下文的 application.yml 或者 application.properties。一个简单的 bootstrap.yml 的例子如下所示：

```
spring:
    application:
        name: my-application
    cloud:
        config:
            uri: ${CONFIG_SERVER:http://localhost:8080}
```

如果想要禁止 Bootstrap 引导过程，可以在 bootstrap.yml 中设置，如下所示：

```
spring:
    cloud:
```

```
bootstrap:
    enabled: false
```

2. 应用上下文层级

Spring 的上下文有一个特性：子级上下文将从父级中继承属性源和配置文件。如果通过 SpringApplication 或者 SpringApplicationBuilder 来构建应用程序上下文，那么 Bootstrap 上下文将会成为该应用程序上下文的父级上下文。

在 Bootstrap 上下文中扫描到的非空的 PropertySourceLocators 会以高优先级添加到 CompositePropertySource 中。如果通过 bootstrap.yml 来配置 Bootstrap 上下文，且在设定好父级上下文的情况下，bootstrap.yml 中的属性会添加到子级的上下文。它们的优先级低于 application.yml 和其他添加到子级中作为创建 Spring Boot 应用的属性源。

基于属性源的排序规则，Bootstrap 上下文中的属性优先，但是需要注意这些属性并不包含任何来自 bootstrap.yml 的数据。bootstrap.yml 中的属性具备非常低的优先级，因此可以作为默认值。

可以简单地将父级上下文设置为应用上下文来扩展上下文的层次结构。Bootstrap 上下文将会是最高级别上下文的父级。每一个在层次结构中的上下文都有它自己的 Bootstrap 属性源（可能为空），来避免无意中将父级上下文中的属性传递到它的后代中。层次结构中的每一个上下文原则上应该拥有自己不同的 spring.application.name，以便在有配置中心的时候也能有不同的远程属性源。来自子级上下文的属性可以覆盖父级中的具有相同名称和属性源名称的属性。

3. 修改 Bootstrap 配置文件的位置

bootstrap.yml 的位置可以通过在配置属性中设置 spring.cloud.bootstrap.name（默认是 bootstrap）或者 spring.cloud.bootstrap.location 来修改。

4. 重载远程属性

通过 Bootstrap 上下文添加到应用程序的属性源通常是远程的，例如来自配置中心，通常本地的配置文件不能覆盖这些远程属性源。一般来说过，启动命令行参数的优先级高于远程配置，可以通过设定启动命令行参数的方式覆盖远程配置。

如果想使用应用程序的系统属性或者配置文件覆盖远程属性，那么远程属性源必须设置为 spring.cloud.config.allowOverride=true（这个配置在本地设置不会生效）。在远程属性源中设定上述配置后，就可以通过更为细粒度的设置来控制远程属性是否能被重载，具体配置如下所示。

```
spring:
    cloud:
        config:
            overrideNone: true #本地属性覆盖所有的远程属性源
            overrideSystemProperties: false#仅覆盖远程属性源中的系统属性和环境变量
```

5. 自定义 Bootstrap 配置

自定义 Bootstrap 配置过程与 Spring Boot 自动配置运行原理类似，具体操作是在 /META-INF/spring.factories 文件中添加 org.springframework.cloud.bootstrap.BootstrapConfiguration 配置项。配置项的值是一系列用来创建 Context 的 @Configuration 配置类，配置类之间以逗号分隔。这些类可以为应用上下文提供 Bean 实例，配置类可以通过标记 @Order 来控制 Bean 实例初始化序列。如下例所示，在引导过程中添加了一个 LogAutoConfiguration 的配置类，为应用程序添加日志相关的 Bean 实例：

```
org.springframework.cloud.bootstrap.BootstrapConfiguration=\
    com.demo.starter.config.LogAutoConfiguration
```

6. 自定义 Bootstrap 属性源

默认的 Bootstrap 外部配置属性源是 Spring Cloud Config Server，即使用配置中心加载外部属性。但是用户也可以通过将 PropertySourceLocator 类型的 Bean 实例添加到 Bootstrap 上下文（在 spring.factories 添加对应的配置类）来添加额外的属性来源。通过这种方法可以从不同的服务器或者数据库中加载额外的属性，如下所示：

```
@Configuration
public class CustomPropertySourceLocator implements PropertySourceLocator {
    @Override
    public PropertySource<?> locate(Environment environment) {
        return new MapPropertySource("customProperty",
            Collections.<String, Object>singletonMap("property.from.sample.custom.
                source", "worked as intended"));
    }
}
```

上述代码中传入的 Environment 参数用于创建应用上下文，它具有 Spring Boot 提供的属性源，可以使用它们来加载特定的属性源（例如重新设置 spring.application.name）。可以在 META-INF/spring.factories 文件中添加如下记录来配置属性源：

```
org.springframework.cloud.bootstrap.BootstrapConfiguration=\
    sample.custom.CustomPropertySourceLocator
```

上述配置令应用程序可以使用 CustomPropertySourceLocator 作为其属性源。

7. Environment 变化

Config Client 应用程序会监听 EnvironmentChangeEvent 事件，当监听到一个 EnvironmentChangeEvent 时，它将持有一个被改变的键值对列表，应用程序使用这些值来：

- 重新绑定所有的 @ConfigurationProperties 的 Bean 实例，更新本地的配置属性。
- 为在 logging.level.* 的所有属性设置日志的等级。

一般来讲，Config Client 默认不会使用轮询方法来监听 Environment 中的改变。在 Spring Cloud 中，Spring Cloud Config Server 使用 Spring Cloud Bus 将 EnvironmentChangeEvent 广播

到所有的 Config Client 中，通知它们 Environment 发生变化。

EnvironmentChangeEvent 是一个事件类，用于在 Environment 发生修改时发布事件。开发者可以通过访问 /configprops 端点（常规的 Spring Boot Actuator 端点）来验证这些更改是否绑定到 @ConfigurationProperties 的 Bean 实例上。例如一个 DataSource 的最大连接数量在运行时被改变了（DataSource 默认由 Spring Boot 创建，属于 @ConfigurationProperties 的 Bean）并且动态增加容量，可以通过查看 Config Client 应用程序的 /configprops 端点来验证 DataSource 的最大连接池数量是否发生变化。

8. 刷新范围

一个被标记为 @RefreshScope 的 Spring Bean 实例在配置发生变更时可以重新进行初始化，即动态刷新配置，这是为了解决状态 Bean 实例只能在初始化的时候才能进行属性注入的问题。

被 @RefreshScope 修饰的 Bean 实例是懒加载的，即当它们被使用的时候才会进行初始化（方法被调用的时候），想要在下次方法调用前强制重新初始化一个 Bean 实例，只需要将它的缓存失效即可。

RefreshScope 是上下文中的一个 Bean 实例，它有一个公共方法 refreshAll，该方法可以通过清除目标缓存来刷新作用域中的所有 Bean 实例。RefreshScope 也有一个 refresh 方法来按照名字刷新单个 Bean。

2.2.2 Spring Cloud Commons：公共抽象

Spring Cloud 将服务发现、负载均衡和断路器等通用模型封装在一个公共抽象中，可以被所有的 Spring Cloud 客户端使用，不依赖于具体的实现（例如服务发现就有 Eureka 和 Consul 等不同的实现），这些公共抽象位于 Spring Cloud Commons 项目中。

1. @EnableDiscoveryClient 注解

@EnableDiscoveryClient 注解用于在 META-INF/spring.factories 文件中查找 DiscoveryClient（DiscoveryClient 为服务发现功能抽象类）的实现。spring.factories 文件的 org.springframework.cloud.client.discovery.EnableDiscoveryClient 配置项可以指定 DiscoveryClient 的实现类。DiscoveryClient 目前的实现有 Spring Cloud Netflix Eureka、Spring Cloud Consul Discovery 和 Spring Cloud Zookeeper Discovery。

DiscoveryClient 的实现类会自动将本地的 Spring Boot 服务注册到远程服务发现中心。可以通过在 @EnableDiscoveryClient 中设置 autoRegister=false 来禁止自动注册行为。

在 Finchley 版本的 Spring Cloud 中，不需要显式使用 @EnableDiscoveryClient 来开启客户端的服务注册与发现功能。只要在类路径中，有 DiscoveryClient 的实现就能使 Spring Cloud 应用注册到服务发现中心。

2. 服务注册（ServiceRegistry）

Spring Cloud Commons 的 ServiceRegister 接口提供 register（服务注册）和 deregister（服务下线）方法，使开发者可以自定义注册服务的逻辑，如下所示：

```
public interface ServiceRegistry<R extends Registration> {
    void register(R registration);
    void deregister(R registration);
}
```

每一个 ServiceRegistry 的实现都拥有自己的注册表实现，如 Eureka、Consul 等。

3. RestTemplate 的负载均衡

创建 RestTemplate 实例的时候，使用 @LoadBalanced 注解可以将 RestTemplate 自动配置为使用负载均衡的状态。@LoadBalanced 将使用 Ribbon 为 RestTemplate 执行负载均衡策略。

创建负载均衡的 RestTemplate 不再能通过自动配置来创建，必须通过配置类创建，具体实例如下所示：

```
@Configuration
public class MyConfiguration {

    @LoadBalanced
    @Bean
    RestTemplate restTemplate() {
        return new RestTemplate();
    }
}

public class MyApplication {
    @Autowired
    private RestTemplate restTemplate;

    public String getMyApplicationName() {
        //使用restTemplate访问my-application微服务的/name接口
        String name = restTemplate.getForObject("http://my-application/name",
            String.class);
        return name;
    }
}
```

URI 需要使用服务名来指定需要访问应用服务，Ribbon 客户端将通过服务名从服务发现应用处获取具体的服务地址来创建一个完整的网络地址，以实现网络调用。

4. RestTemplate 的失败重试

负载均衡的 RestTemplate 可以添加失败重试机制。默认情况下，失败重试机制是关闭的，启用方式是将 Spring Retry 添加到应用程序的类路径中。还可以设置 spring.cloud.loadbalancer.retry.enabled=false 禁止类路径中 Spring retry 的重试逻辑。

如果想要添加一个或者多个 RetryListener 到重试请求中，可以创建一个类型为 LoadBalancedRetryListenerFactory 的 Bean，用来返回将要用于重试机制的 RetryListener 的列表，如下代码所示：

```
@Configuration
public class RryListenerConfiguration {
    @Bean
    LoadBalancedRetryListenerFactory retryListenerFactory() {
        return new LoadBalancedRetryListenerFactory() {
            @Override
            public RetryListener[] createRetryListeners(String service) {
                return new RetryListener[]{new RetryListener() {
                    @Override
                    // 重试开始前的工作
                    public <T, E extends Throwable> boolean open(RetryContext
                        context, RetryCallback<T, E> callback) {
                        return true;
                    }
                    // 重试结束后的工作
                    @Override
                    public <T, E extends Throwable> void close(RetryContext
                        context, RetryCallback<T, E> callback, Throwable throwable) {
                    }
                    // 重试出错后的工作
                    @Override
                    public <T, E extends Throwable> void onError(RetryContext
                        context, RetryCallback<T, E> callback, Throwable throwable) {
                    }
                }};
            }
        };
    }
}
```

其中，自定义配置类中定义了生成 LoadBalancedRetryListenerFactory 实例的 @Bean 方法，该工厂类的 createRetryListeners 方法会生成一个 RetryListener 实例，用于进行网络请求的重试。

2.3 本章小结

Spring Cloud 并不能与微服务或者微服务架构划上等号，不能误认为使用了 Spring Cloud 的应用服务就是微服务。微服务架构是一种架构的理念，重点是微服务的设计原则，从理论上为具体的技术落地提供了指导思想。Spring Cloud 是一个基于 Spring Boot 实现的服务治理工具包，关注全局的服务治理框架。目前来说，Spring Cloud 仍是 Java 世界中微服务实践的最佳落地方案。

Chapter 3 第 3 章

Spring Cloud 的基础：Spring Boot

工欲善其事，必先利其器。在对 Spring Cloud 各部分组件进行具体介绍之前，我们会对 Spring Cloud 微服务的基础 Spring Boot 进行介绍。Spring Boot 是 Spring 一套快速配置开发的脚手架，可以基于 Spring Boot 快速集成开发单个 Spring 应用。Spring Cloud 是基于 Spring Boot 实现的云应用开发工具，很大一部分实现依赖于 Spring Boot。可以说 Spring Boot 是整个 Spring Cloud 微服务架构的服务基础。

本章第一小节对 Spring Boot 功能和特点进行概述，包括 Spring Boot 中的核心特性；第二小节将搭建一个 Spring Boot 项目来快速了解基本开发流程；在第三小节中总结了 Spring Boot 的配置文件及应用。

3.1 Spring Boot 简介

Spring 框架功能很强大，但是就算是一个很简单的项目，开发者也需要进行大量的配置工作，因此在 Spring 4.0 之后出现了 Spring Boot 框架，它的作用很简单，就是帮助开发者自动配置 Spring 的相关依赖。Spring Boot 是 Pivotal 团队于 2013 年推出的全新项目，主要用来简化 Spring 开发框架的开发、配置、调试和部署工作，同时在项目内集成了大量易于使用且实用的基础框架。Spring Boot 使用了特殊的方式来进行初始化配置，这使得开发者不需要额外定义样板化的配置。

1. Spring Boot 2.0

2018 年 3 月初，Spring Boot 2.0 正式发布。该版本历经了 17 个月，是继 4 年前 Spring Boot 1.0 之后的第一个主要版本，也是第一个正式支持 Spring Framework 5.0 的发行版本。

Spring Boot 2.0 相对之前的 1.x 发生了以下的变化：
- 不再支持 JDK1.6 和 1.7，JDK 最低要求 1.8+，并支持 1.9。
- 支持 Spring webflux/webflux.fn 响应式的 Web 编程。
- 提供 Spring Data Cassandra、MongoDB、Couchbase 和 Redis 的响应式自动配置及 Starter POM。
- 支持嵌入式的 Netty。
- TLS 配置和 HTTP/2 的支持：Tomcat、Reactor Netty、Undertow 和 Jetty。
- 全新的体系结构，支持 Spring MVC、WebFlux 和 Jersey。
- Spring Boot 2 针对 Quartz 调度器提供了支持。可以加入 spring-boot-starter-quartz 的 Starter 依赖来启用。
- 极大简化了安全配置。

需要注意的是，许多配置属性在 Spring Boot 2.0 中已经重命名或被删除，为了方便从 1.x 升级，Spring Boot 发布了一个新的 spring-boot-properties-migrator 模块。只要将其作为依赖添加到项目中，它不仅会分析应用程序的环境并在启动时打印诊断信息，还会在运行阶段为项目临时将属性迁移至新的配置方式。

2. Spring Boot 与 Spring Cloud

Spring Cloud 基于 Spring Boot 框架开发应用，为微服务开发中的架构问题提供了一整套的解决方案：如服务注册与发现、服务消费、服务容错、API 网关、分布式调用追踪和分布式配置管理等。

Spring Cloud 与 Spring Boot 的联系如下：
- Spring Boot 是 Spring 的一套快速配置脚手架，可以基于 Spring Boot 快速开发单个服务，Spring Cloud 是一个基于 Spring Boot 实现的云应用开发工具。
- Spring Boot 专注于快速、方便集成单个服务，Spring Cloud 是关注全局的服务治理框架。
- Spring Boot 使用了约定优先于配置的理念，有很多集成方案已经设置好，减少了用户的配置，Spring Cloud 很大一部分是基于 Spring Boot 实现的。
- Spring Boot 可以离开 Spring Cloud 独立使用开发项目，但是 Spring Cloud 离不开 Spring Boot，属于依赖与被依赖的关系。

3. Spring Boot 核心特性

从本质上来讲，Spring Boot 是一个框架中的框架，它专注于框架的整合，让基础的框架能够更好地集成使用。它简化了集成过程中的模板化配置，提供了本应该由开发者自身去实现的 Spring Bean 配置，从而使开发者能够从繁琐的配置中解放出来，更专注于应用程序的业务逻辑。

Spring Boot 具备以下特性：

- SpringApplication：提供一种简便的方式来引导启动 Spring 应用程序，通过 main 的方式启动应用。
- 外部化配置（External Configuration）：通过外部化配置的方式，可以让开发者自定义相关配置以使相同的应用运行于不同的环境中。
- Profiles：Spring Profiles 可以将配置文件隔离成不同的模块，并且使这些模块中的配置只在特定的环境中生效。
- 日志（Logging）：Spring Boot 完善的日志系统更利于开发者调试和监控应用程序。
- MVC：Spring Web MVC 框架，使得开发者得心应手地搭建 Web 后端应用。
- 嵌入式容器（Embedded Containers）：支持内置的 Tomcat、Jetty 和 Undertow 服务器，使得应用程序通过内嵌的服务器容器一键启动。
- SQL：提供了使用 SQL 数据库的广泛支持，从直接通过 JDBC（JdbcTemplate）访问数据库到通过对象关系映射框架（如 Hibernate）访问数据库等等。
- NoSQL：集成了大量的框架来提供使用 NoSQL 数据库的技术，支持 Redis、MongoDB、Neo4j、Elasticsearch、Solr Cassandra、Couchbase 和 LDAP 等诸多 NoSQL 数据库，这些 NoSQL 数据库的使用都可以由 Spring Boot 提供相关的自动配置实现。
- 消息系统（Message）：提供了对消息系统的广泛支持，包括 JMS、RabbitMQ、Kafka 和 WebSocket 等。
- 测试（Testing）：提供了一系列实用工具和注解，以帮助开发者测试应用程序。
- 自动配置（Auto-Configuration）：自动配置不仅减少了 Spring 框架中本该由开发者自主实现 Bean 相关配置，也使得开发者具备根据自己的业务需要为应用程序定义各种特定 Bean 的能力。
- 监控（Monitoring）：Spring Boot 的监控能够使开发者更好地掌控应用程序的运行状态。

在下面 Spring Boot 的构建中，将对其中的部分特性进行较为详细的讲解，对于其他未涉及的特性，读者们可以通过阅读官方文档或者其他资料进行了解。

3.2 构建一个微服务

下面通过构建简单的 RESTful 应用，了解 Spring Boot 的基本开发流程，体验其简单、易用的特性。创建 Spring Boot 项目很简单，如果使用 STS 或 IDEA 的话，新建 Spring Starter 项目（Spring Initializer）即可。也可以从 https://start.spring.io/ 直接创建项目，下面分别介绍这两种方式。

1. IDEA 生成

使用 IDEA 生成项目的主要过程如下：

1)Spring Initializer 创建项目。

IDEA 新建项目有多种方式,在图 3-1 中选择 Spring Initializer 的方式创建项目。

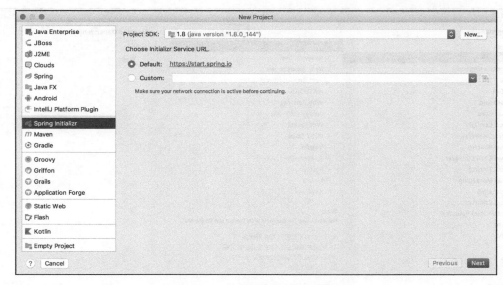

图 3-1 Spring Initializer 创建项目

2)设置项目的基本信息。

如图 3-2 所示,可以设置创建项目的包名、Group 的 Id、Artifact 的 Id 和 Java 的版本等信息。

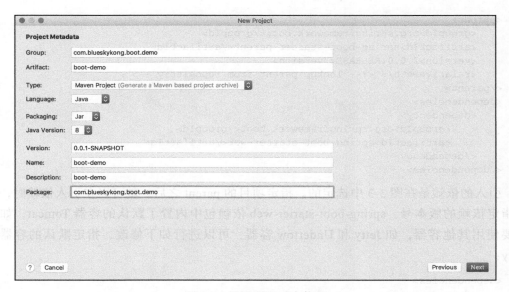

图 3-2 设置项目基本信息

3）添加依赖。

图 3-3 用于添加 Spring Boot 的依赖。

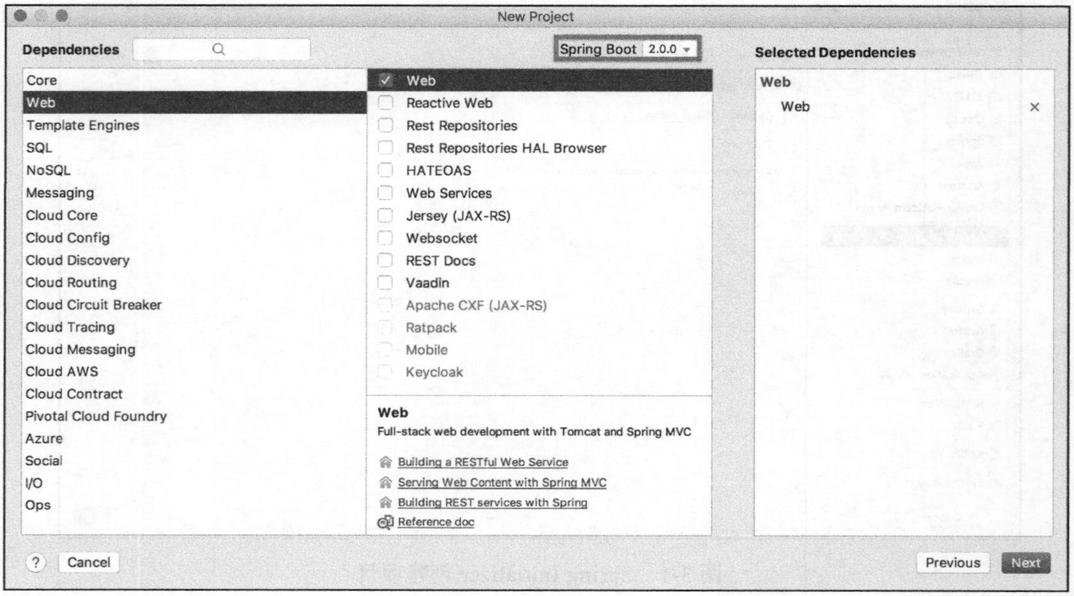

图 3-3　添加依赖

首先是 Spring Boot 的版本，图中使用了 2.0.0 正式版。并且添加了 Web 开发的依赖，这里可以添加的依赖很多，如 spring-boot-data、security、lombok 等等。添加依赖如下：

```
<parent>
    <groupId>org.springframework.boot</groupId>
    <artifactId>spring-boot-starter-parent</artifactId>
    <version>2.0.0.RELEASE</version>
    <relativePath/> <!-- lookup parent from repository -->
</parent>
<dependencies>
    <dependency>
        <groupId>org.springframework.boot</groupId>
        <artifactId>spring-boot-starter-web</artifactId>
    </dependency>
</dependencies>
```

引入的依赖是在图 3-3 中选定的。指定项目的 parent 之后，就可以在引入依赖时，不再指定依赖的版本号。spring-boot-starter-web 依赖包中内置了默认的容器 Tomcat，如果想要使用其他容器，如 Jetty 和 Undertow 容器，可以进行如下修改，指定默认的容器为 Jetty：

```
<dependencies>
    <dependency>
```

```xml
            <groupId>org.springframework.boot</groupId>
            <artifactId>spring-boot-starter-web</artifactId>
            <exclusions>
                <exclusion>
                    <groupId>org.springframework.boot</groupId>
                    <artifactId>spring-boot-starter-tomcat</artifactId>
                </exclusion>
            </exclusions>
        </dependency>
        <dependency>
            <groupId>org.springframework.boot</groupId>
            <artifactId>spring-boot-starter-jetty</artifactId>
        </dependency>
<dependencies>
```

4）启动类与控制类。

启动类和控制类的代码如下所示：

```
@SpringBootApplication
@RestController
public class Chapter3BootDemoApplication {
    public static void main(String[] args) {
        SpringApplication.run(Chapter3BootDemoApplication.class, args);
    }
    @GetMapping("/test")
    public String test() {
        return "this is a demo boot.";
    }
}
```

SpringMVC 中使用了 3 个注解作用于 Chapter3BootDemoApplication 类，分别是 @Configuration（2.0.0 版本中添加了 @SpringBootConfiguration 注解来代替 Spring 的标准配置注解 @Configuration）、@EnableAutoConfiguration 和 @ComponentScan。SpringBoot 提供了一个统一的注解 @SpringBootApplication，默认属性下等于上述 3 个注解。

@RestController 组合了 @Controller 和 @ResponseBody 注解，表明该类可以处理 HTTP 请求，并且返回 JSON 类型的响应。Spring Initializer 会自动为应用生成对应的启动类，一般以 *Application 方式进行命名。

在启动类中增加控制类的端点，暴露出 /test 的端点。同时在 application.properties 中设置服务器启动的端点，如下所示：

```
server.port=8000
```

服务器会使用内置的 Tomcat 容器进行启动，服务器端口为 8000。这样一个简单的 Spring Boot Web 应用就写好了，正常访问接口 http://localhost:8000/test 即可。

2. Initial 生成

如果不想使用 IDEA 的话，也可以在 Spring 官方网站 https://start.spring.io/ 创建项目，

再将创建好的项目下载到本地,解压之后导入到 IDEA 中。

1)创建项目。如图 3-4 所示,填写 Group 的 Id、Artifact 的 Id 和项目依赖。添加项目依赖时,根据输入的关键字,会有下拉框选择提示。填好这些信息,就可以生成对应的项目。生成的项目会自动下载。

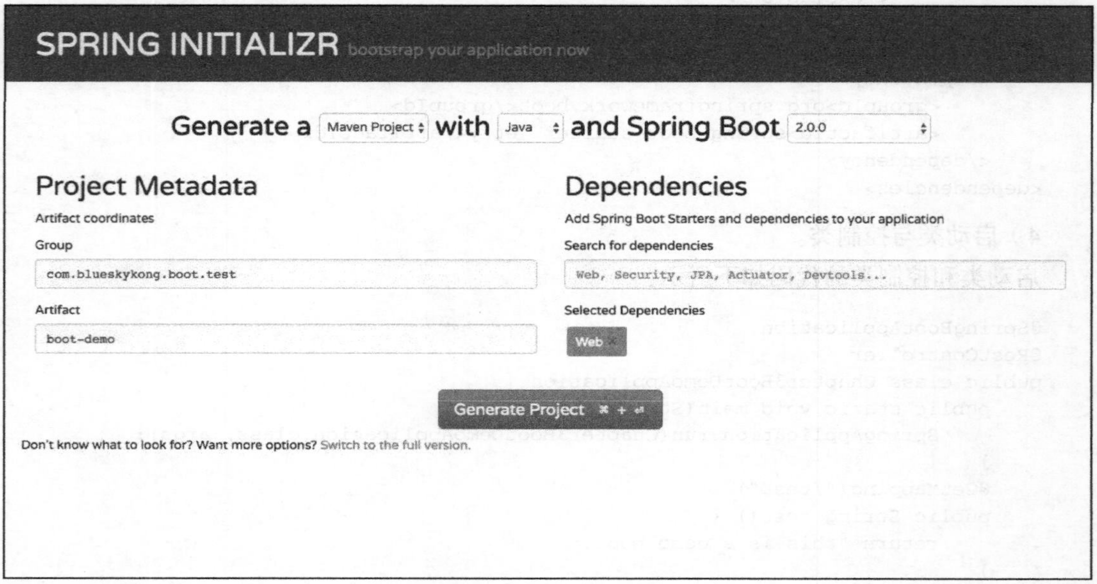

图 3-4 创建项目

2)解压并导入项目。解压下载好的项目之后,会发现项目结构和 IDEA 中生成的一样,如图 3-5 所示,因为 IDEA 中调用的 API 接口是 Spring 官方的项目生成器接口。

图 3-5 解压后的项目结构

3.3 Spring Boot 配置文件

3.3.1 默认配置文件

在创建 Spring Boot 项目时，会默认在 resource 文件夹下创建 application.propertities 的属性文件。在上面的例子中，我们在 application.propertities 中加上了服务启动端口的配置，代码如下所示：

```
server.port=8000
```

属性文件是最常见的管理配置属性的方式。Spring Boot 提供的 SpringApplication 类会搜索并加载 application.properties 文件来获取配置属性值。SpringApplication 类会在下面位置搜索该文件：

- 当前目录的"/config"子目录
- 当前目录
- classpath 中的"/config"包
- classpath

上面的顺序也表示了该位置上包含的属性文件的优先级。优先级按照从高到低的顺序排列。可以通过 spring.config.name 配置属性来指定不同的属性文件名称。还可以通过 spring.config.location 来添加额外的属性文件的搜索路径。如果 Spring Boot 在优先级更高的位置找到了配置，那么它就会忽略低优先级的配置。

3.3.2 外部化配置

Spring Boot 所提供的配置优先级顺序比较复杂。按照优先级从高到低的顺序，具体的列表如下所示：

1）命令行参数。SpringApplication 类默认会把命令行参数转化成应用中可以使用的配置参数。

2）通过 System#getProperties 方法获取的 Java 系统参数。

3）操作系统环境变量。使用 Docker 启动时，经常会设置系统变量。

4）从 java:comp/env 得到的 JNDI 属性。

5）通过 RandomValuePropertySource 生成的 random.* 属性。

6）应用 jar 文件之外的属性文件，如通过 spring.config.location 参数指定的属性文件。

7）应用 jar 文件内部的属性文件，这是常用的方式。

8）在应用配置 Java 类（包含 @Configuration 注解的 Java 类）中通过 @PropertySource 注解声明的属性文件。

9）通过 SpringApplication#setDefaultProperties 方法声明的默认属性。

有些系统会涉及一些数据库或其他第三方账户等信息，出于安全考虑，这些信息并不会提前配置在项目中。对于这种情况，可以在运行程序的时候通过参数指定一个外部配置文件的方式来解决。

以 demo.jar 为例，方法如下：

```
java -jar demo.jar --spring.config.location=/opt/config/application.properties
```

这里指定的外部文件名无固定要求，如果需要通过 bash 脚本来加载不同环境的外部配置文件，文件的命名可以参考 application-{env}.properties。

3.3.3 YAML

YAML 是 JSON 的超集，是一种非常方便的格式，它一般用于指定分层配置数据。SpringApplication 类能够自动支持 YAML，将其作为 Properties 属性文件的替代者。SpringFramework 提供了两个可用于加载 YAML 文档的类：YamlPropertiesFactoryBean 将 YAML 加载为 Properties；YamlMapFactoryBean 将 YAML 加载为 Map。如下面的属性配置：

```
env[0]=dev
env[1]=prod
dev.port=8080
```

相应的 YAML 的配置如下：

```
env:
    - dev
    - pro
dev:
    port: 8080
```

从上面的对比可以看出 YAML 利于开发者阅读，更易于实施和使用。

3.3.4 自动载入外部属性到 Bean

当 @EnableConfigurationProperties 注解应用到 @Configuration 修饰的配置类时，任何被 @ConfigurationProperties 注解的 Bean 将自动由 Environment 配置。这种风格的配置特别适合与 SpringApplication 的外部 YAML 配置配合使用。

Spring Boot 使用相对宽松的规则来将环境属性绑定到使用 @ConfigurationProperties 注解的 Bean 中，所以不需要环境属性名称和 Bean 属性名称完全匹配。Java 中的属性变量命名使用驼峰法，例如当属性名定义为 maxLength 时，在配置文件中 max-length 和 MAXLENGTH 都将会被映射到 maxLength 属性。我们看几个实例应用。

属性实体类代码如下：

```
@ConfigurationProperties(prefix = "sms")
public class SMS {
    private int retryLimitationMinutes;
    private int validityMinute;
    private final List<String> types = new ArrayList<>();

    //...getter,setter
```

YAML 配置文件代码如下：

```yaml
sms:
    retry-limitation-minutes: 1
    validity-minute: 3
    type:
       - register
       - login
```

在 @ConfigurationProperties 中指明 prefix，定义实体的映射规则，可以简化配置。这种配置方式支持复杂的属性类型，如数组、Map 和子对象等。

3.3.5 多 Profile

Spring Boot 使用 Profile 为不同环境提供不同配置，可以和全局 Profile 配合使用。

Spring 通过配置 spring.profiles.active 指定激活某个具体的 Profile。除了使用 spring.profiles.active 激活一个或者多个 Profile 之外，还可以用 spring.profiles.include 来叠加 Profile。如下所示：

```
spring.profiles.include: prod,dev
```

一个多 Profile 的 application.yml 配置如下所示：

```yaml
spring:
    profiles:
         active: dev
---
#开发环境配置
spring:
    profiles: dev
server:
    port: 8080
---
#测试环境配置
spring:
    profiles: test
server:
    port: 8081
---
#生产环境配置
spring:
    profiles: prod
server:
    port: 8082
```

上述 application.yml 文件分为四部分，使用一组（---）来作为分隔符。第一部分，通用配置部分，表示三个环境都通用的属性，默认激活了 dev 的 Profile；后面三部分分别表示不同的环境，指定了不同的端口（port）。

Spring Boot 应用通常会被打包成 jar 包部署到服务器中。在启动 jar 应用时，可以加上参数 --spring.profiles.active=test 指定应用加载哪个环境的配置。在 IDEA 中也可以直接指定应用启动时激活哪个 Profile。

3.3.6 Starter

Spring Boot 项目的快速发展与流行，很大程度依赖于 Starter 的出现。Starter 方便了 Spring 各项依赖的集成，通过 Starter，可以在 Spring Boot 中获取到所需相关技术的一站式支持（依赖、相关的自动配置文件和相关的 Bean），而无需通过实例代码和复制粘贴来获取依赖。例如当需要 Spring 中的 Web 支持时，可以通过引入 spring-boot-starter-web 这个 Starter 依赖，它将自动为项目配置一个内嵌的 Tomcat 以及开启 Spring WebMvc 的功能，下面是一些常用的 Starter：

- spring-boot-starter：核心 Starter，包含自动配置的支持、日志以及 YAML 解析等。
- spring-boot-starter-aop：提供 Spring AOP 和 AspectJ 的面向切面的编程支持。
- spring-boot-starter-data-jpa：提供 Spring Data JPA 支持（由 Hibernate 提供底层支持）。
- spring-boot-starter-data-mongodb：提供 Spring Data MongoDB 和 MongoDB 支持。
- spring-boot-starter-jdbc：提供 JDBC 支持（由 Tomcat JDBC 连接池提供支持）。
- spring-boot-starter-jersey：提供使用 JAX-RS 和 Jersey 构建 RESTful 风格的 Web 应用的支持。
- spring-boot-starter-web：提供使用 Spring MVC 构建 Web（包含 RESTful）应用的支持，使用 Tomcat 作为默认嵌入式容器。
- spring-boot-starter-webflux：提供使用 Spring Framework 的 Reactive Web 构建 WebFlux 应用的支持。
- spring-boot-starter-actuator：Spring Boot 的 Actuator 支持，其提供了生产就绪功能，帮助开发者监控和管理应用。

更多的 Starter 可以通过官方文档查阅。Starter 大大简化了开发者对 Spring 相关依赖的配置，让他们能够将更多的精力放置在业务开发上，提高生产效率。

3.3.7 自制一个 Starter

下面通过自定义一个 Starter 来理解 Starter 的运行机制。自定义 Starter 的名字叫做 filter-log-starter，它的作用非常简单：为应用添加一个过滤器，在控制台中打印每次访问的 URI。

1. 创建 Starter 项目和导入依赖

首先创建一个新的 Spring Boot 项目，导入以下依赖：

```
<dependencies>
```

```xml
<dependency>
    <groupId>org.springframework.boot</groupId>
    <artifactId>spring-boot-starter-web</artifactId>
</dependency>
</dependencies>
```

因为过滤器属于 Serlvet 包中的类，所以需要引入 Spring WebMvc 相关依赖。

2. 定义过滤器

定义 LogFilter 地址日志过滤器，具体代码如下：

```java
public class LogFilter implements Filter{
    private Logger logger = LoggerFactory.getLogger(LogFilter.class);
    @Override
    public void init(FilterConfig filterConfig) throws ServletException {
        logger.info("logFilter init...");
    }
    @Override
    public void doFilter(ServletRequest servletRequest, ServletResponse servletResponse,
        FilterChain filterChain) throws IOException, ServletException {
        // 从request中获取到访问的地址，并在控制台中打印出来
        HttpServletRequest request = (HttpServletRequest) servletRequest;
        logger.info("uri {} is working.", request.getRequestURI());
        filterChain.doFilter(servletRequest, servletResponse);
    }
    @Override
    public void destroy() {
        logger.info("logFilter destroy...");
    }
}
```

定义 LogFilterRegistrationBean 用于将 LogFilter 过滤器封装成 Spring Bean，具体代码如下：

```java
public class LogFilterRegistrationBean extends FilterRegistrationBean<LogFilter>{
    public LogFilterRegistrationBean(){
        super();
        this.setFilter(new LogFilter()); //添加LogFilter过滤器
        this.addUrlPatterns("/*"); // 匹配所有路径
        this.setName("LogFilter"); // 定义过滤器名
        this.setOrder(1); // 设置优先级
    }
}
```

3. 定义自动配置类

接着，定义一个自动配置类将 LogFilterRegistrationBean 注入到 Spring 的上下文中，具体代码如下：

```java
@Configuration
@ConditionalOnClass({LogFilterRegistrationBean.class, LogFilter.class})
public class LogFilterAutoConfiguration {
```

```
@Bean
@ConditionalOnMissingBean(LogFilterRegistrationBean.class)
public LogFilterRegistrationBean logFilterRegistrationBean(){
    return new LogFilterRegistrationBean();
}
```

}

@Configuration 通常与 @Bean 相配合，使用这两个注解可以创建一个简单的 Spring 配置类，代替相应的 xml 配置文件。@ConditionalOnClass 声明只有当某个或某些 class 位于类路径上，才会实例化一个 Bean。如上述代码中只有当 LogFilterRegistrationBean 和 LogFilter 的 class 在类路径上，LogFilterAutoConfiguration 配置类才会生效。

添加 @Bean 注解的方法将返回一个对象，该对象会被注册为 Spring 上下文中的 Bean。@ConditionalOnMissingBean 声明仅仅在当前 Spring 上下文中不存在某个对象时，才会实例化一个 Bean。上述代码中，当 LogFilterRegistrationBean 不存在于 Spring 上下文时，才会创建 LogFilterRegistrationBean 的 Bean 并注入到 Spring 上下文中。

4. 定义使自动配置类生效的注解

然后我们需要一个注解使 LogFilterAutoConfiguration 配置类生效。因为 Starter 是通过 jar 包的方式引入项目中，对应的 classes 并不在项目的 Spring 扫描范围内，所以无法自动引入项目的 Spring 管理中。对此需要用额外的方式将 LogFilterAutoConfiguration 引入到项目的 Spring 管理中，如通过注解的方式将配置类引入项目的 Spring 扫描范围内。

定义 EnableLogFilter 引入 LogFilterAutoConfiguration 配置类到项目的 Spring 扫描范围内，具体代码如下：

```
@Target({ElementType.TYPE})
@Retention(RetentionPolicy.RUNTIME)
@Import(LogFilterAutoConfiguration.class) //引入LogFilterAutoConfiguration配置类
public @interface EnableLogFilter {
}
```

除了在注解中直接通过 @Import 引入对应的配置类外，还可以通过 ImportSelector 的方式从 resources/META-INF/spring.factories 中读入需要加载的自动配置类。

首先需要在 META-INF 中定义 spring.factories，代码如下所示：

```
com.xuan.chapter3.starter.annotation.EnableLogFilter=\
    com.xuan.chapter3.starter.config.LogFilterAutoConfiguration
```

声明 EnableLogFilter 注解将使 LogFilterAutoConfiguration 自动配置类生效。

定义 EnableLogFilterImportSelector 用于从 spring.factories 获取需要加载的自动配置类。具体代码如下：

```
public class EnableLogFilterImportSelector
        implements DeferredImportSelector, BeanClassLoaderAware, EnvironmentAware {
```

```java
private static final Logger logger = LoggerFactory.getLogger(EnableLogFi
    lterImportSelector.class);
private Class annotationClass = EnableLogFilter.class;
...
@Override
public String[] selectImports(AnnotationMetadata metadata) {
    // 是否生效,默认为true
    if (!isEnabled()) {
        return new String[0];
    }
    // 获取注解中的属性
    AnnotationAttributes attributes = AnnotationAttributes.fromMap(
        metadata.getAnnotationAttributes(this.annotationClass.getName(),
            true));
    Assert.notNull(attributes, "can not be null…");
    // 从spring.factories中获取所有通过EnableLogFilter注解引入的自动配置类,并
        进行去重操作
    List<String> factories = new ArrayList<>(new LinkedHashSet<>(Sprin
        gFactoriesLoader.loadFactoryNames(this.annotationClass, this.
        beanClassLoader)));
    if (factories.isEmpty() && !hasDefaultFactory()) {
        throw new IllegalStateException("…");
    }
    if (factories.size() > 1) {
        logger.warn("More than one implementation ");
    }
    return factories.toArray(new String[factories.size()]);
}
...
}
```

通过覆盖 selectImports 方法,使用 SpringFactoriesLoader 从 spring.factories 中获取通过 EnableLogFilter 注解引入的自动配置类,获取的自动配置类是在 spring.factories 中配置的 LogFilterAutoConfiguration。EnableLogFilterImportSelector 可以扩展成一个通用的配置加载类,从而实现在 spring.factories 中通过不同的注解引入不同的自动配置类的功能。

修改 EnableLogFilter 注解,代码如下所示:

```java
@Target({ElementType.TYPE})
@Retention(RetentionPolicy.RUNTIME)
@Import(EnableLogFilterImportSelector.class) // 引入自动配置加载类
public @interface EnableLogFilter {
}
```

5. 使用 Starter

最后我们使用一下 Starter,在 boot-demo 项目中添加 Starter 的依赖,如下所示:

```xml
<dependency>
    <groupId>com.xuan</groupId>
    <artifactId>chapter3-filter-log-starter</artifactId>
    <version>0.0.1-SNAPSHOT</version>
```

```
</dependency>
```

在启动类中添加 @EnableLogFilter 注解，修改后代码如下所示：

```
@SpringBootApplication
@RestController
@EnableLogFilter
public class Chapter3BootDemoApplication {
    public static void main(String[] args) {
        SpringApplication.run(Chapter3BootDemoApplication.class, args);
    }

    @GetMapping("/test")
    public String test() {
        return "this is a demo boot.";
    }
}
```

启动项目，访问 http://localhost:8000/test 接口，将会在控制台中看到如下访问日志：

```
2018-04-09 21:30:36.045  INFO 32080 --- [nio-8008-exec-1] c.x.chapter1.starter.
   filter.LogFilter     : uri /test is working.
```

3.3.8 Actuator

Spring Boot 提供了一系列额外的特性来帮助开发者监控和管理生产环境中的 Spring Boot 应用。通过引入 spring-boot-actuator 模块，审计、健康检查和度量收集等监控功能都将会自动配置到 Spring Boot 应用中，最简单的方式是直接引入相关模块的 Starter，如下所示：

```
<dependency>
    <groupId>org.springframework.boot</groupId>
    <artifactId>spring-boot-starter-actuator</artifactId>
</dependency>
```

Actuator 的端点能够使开发者监控应用和与应用进行交互，Spring Boot 中包含了一系列内置的端点，同时也允许自定义端点。每一个端点都能够独立决定开启还是关闭。为了可以进行远程调用，这些端点需要暴露到 HTTP 中或者 JMX 中。当采用 HTTP 的方式时，这些端点通常会配置在 /actuator 的前缀之后，例如 health 端点的地址为 /actuator/health（这是 Spring Boot 2.0 之后的改动，之前的版本中端点地址不包含 /actuator 前缀）。表 3-1 列出了 Actuator 提供的相关端点。

表 3-1 Actuator 提供的端点

ID	描述	默认是否开启	JMX 是否默认暴露	HTTP 是否默认暴露
auditevents	公开当前应用的审批事件信息	√	√	×
beans	展示应用中的所有 Spring Bean	√	√	×

（续）

ID	描述	默认是否开启	JMX 是否默认暴露	HTTP 是否默认暴露
conditions	显示在配置和自动配置类中进行评估的条件以及它们为什么匹配或者不匹配的原因	√	√	×
configprops	展示整理过的所有配置类属性 (@ConfigurationProperties) 的列表	√	√	×
env	暴露 Spring ConfigurableEnvironment 的属性	√	√	×
flyway	展示已经被应用的任何 flyway 数据库迁移	√	√	×
health	显示应用的健康信息	√	√	√
httptrace	展示 HTTP 跟踪信息。默认情况下，展示最后 100 个 HTTP 请求响应的交换	√	√	×
info	展示主观的应用信息，由开发者配置	√	√	√
loggers	展示和修改应用中的 loggers 配置	√	√	×
liquibase	展示已应用的 liquibase 数据库迁移	√	√	×
metrics	展示当前应用的度量信息	√	√	×
mappings	展示整理过的所有的 @RequestMapping 路径	√	√	×
scheduledtasks	展示应用中的定时任务	√	√	×
sessions	允许从 Spring 会话支持的会话存储库中检索和删除会话。在使用 Spring Session 支持的 reactive Web 应用时不可用	√	√	×
shutdown	优雅地关闭应用	×	√	×
threaddump	执行线程转存	√	√	×

开启端点可以通过 management.endpoint.<id>.enabled 的方式进行配置，通过 id 对特定的端点进行操作，如下所示：

```
management:
    endpoint:
        shutdown:
            enabled: true
```

通过上面的配置，开启了 shutdown 端点。可以通过 mangement.endpoints.enabled-by-default=true 统一开启和关闭端点。

即使端点已经开启，大多数端点是通过 JMX 的方式暴露出去的，通过 HTTP 暴露出来的端点默认仅有 /health 和 /info，可以通过 management.endpoints.<method>.exposure.exclude 和 management.endpoints.<method>.exposure.include 的配置，来指定 JMX 和 HTTP

需要暴露的端点，如下所示：

```
management:
    endpoints:
        jmx:
            exposure:
                exclude:
                include: *
        web:
            exposure:
                exclude:
                include: info, health, beans
```

上面的配置中，JMX 暴露所有的端点（*表示通配符，用来选择所有的端点），HTTP 仅暴露 /info、/health、/beans 三个端点。

3.4 本章小结

Spring Boot 是伴随着 Spring4.0 诞生的。从字面理解，Boot 是引导的意思，因此 Spring Boot 能帮助开发者快速搭建 Spring 框架、快速启动一个 Web 容器。Spring Boot 继承了原有 Spring 框架的优秀基因，并简化了使用 Spring 的过程。

Spring Boot 的哲学是约定大于配置，避免了很多的配置。可以基于 Spring Boot 快速开发单个服务。本章简单介绍了 Spring Boot 的相关特性和用法，为后面章节具体讲解 Spring Cloud 组件做准备。

第 4 章　服务注册与发现：Eureka

在传统的单体应用中，组件之间的调用通过有规范约束的接口进行，从而实现不同模块间良好协作。在微服务架构中，原本的"巨石"应用按照业务被分割成相对独立的、提供特定功能的微服务，每一个微服务都可以通过集群或者其他方式进行动态扩展，每一个微服务实例的网络地址都可能动态变化，这使得原本通过硬编码地址的调用方式失去了作用。微服务架构中，服务地址的动态变化和数量变动，迫切需要系统建立一个中心化的组件对各个微服务实例信息进行登记和管理，同时让各个微服务实例之间能够互相发现，从而达到互相调用的结果。

通常来说服务注册与发现包括两部分，一个是服务器端，另一个是客户端。Server 是一个公共服务，为 Client 提供服务注册和发现的功能，维护注册到自身的 Client 的相关信息，同时提供接口给 Client 获取注册表中其他服务的信息，使得动态变化的 Client 能够进行服务间的相互调用。Client 将自己的服务信息通过一定的方式登记到 Server 上，并在正常范围内维护自己信息一致性，方便其他服务发现自己，同时可以通过 Server 获取到自己依赖的其他服务信息，完成服务调用。

Spring Cloud Netflix Eureka 是 Spring Cloud 提供用于服务发现和注册的基础组件，是搭建 Spring Cloud 微服务架构的前提之一。Eureka 作为一个开箱即用的基础组件，屏蔽了底层 Server 和 Client 交互的细节，使得开发者能够将精力更多地放在业务逻辑上，加快微服务架构的实施和项目的开发。

本章中，第一小节将对 Eureka 进行一个综合性的概述，同时搭建一个 Eureka 应用的简单例子，以演示运行机制；第二小节将从宏观的角度对 Eureka 的整体架构进行概述，对 Eureka 各组件和组件间的行为进行介绍；第三小节将从源码的角度对 Eureka Client 的运行原理进行解析，分析它与 Eureka Server 之间的交互行为；第四小节将对 Euerka Server

的源码进行讲解,分析其如何在微服务架构中发挥服务注册中心的作用;在第五小节将对 Eureka 中的配置属性和高级特性进行讲解。

4.1 基础应用

4.1.1 Eureka 简介

Eureka 这个词来源于古希腊语,意为"我找到了!我发现了!"。据传,阿基米德在洗澡时发现浮力原理,高兴得来不及穿上衣服,跑到街上大喊:"Eureka!"。

在 Netflix 中,Eureka 是一个 RESTful 风格的服务注册与发现的基础服务组件。Eureka 由两部分组成,一个是 Eureka Server,提供服务注册和发现功能,即我们上面所说的服务器端;另一个是 Eureka Client,它简化了客户端与服务端之间的交互。Eureka Client 会定时将自己的信息注册到 Eureka Server 中,并从 Server 中发现其他服务。Eureka Client 中内置一个负载均衡器,用来进行基本的负载均衡。

本章内容基于 Spring Cloud 的 Finchley.RELEASE 版本,spring-cloud-netflix 版本基于 2.0.0.RELEASE,Eureka 版本基于 v1.9.2。

下面我们将通过搭建一个简单的 Eureka 例子来了解 Eureka 的运作原理。

4.1.2 搭建 Eureka 服务注册中心

可以搭建包含 Eurake Server 依赖的 Spring Boot 项目。主要依赖如下:

```xml
<dependency> <!--eureka-client相关依赖-->
    <groupId>org.springframework.cloud</groupId>
    <artifactId>spring-cloud-starter-netflix-eureka-server</artifactId>
</dependency>
<dependency> <!--actuator相关依赖-->
    <groupId>org.springframework.boot</groupId>
    <artifactId>spring-boot-starter-actuator</artifactId>
</dependency>
<dependency> <!--Spring Web MVC相关依赖-->
    <groupId>org.springframework.boot</groupId>
    <artifactId>spring-boot-starter-web</artifactId>
</dependency>
```

在启动类中添加注解 @EnableEurekaServer,代码如下所示:

```java
@SpringBootApplication
//会为项目自动配置必须的配置类,标识该服务为注册中心
@EnableEurekaServer
public class Chapter4EurekaServerApplication {
    public static void main(String[] args) {
        SpringApplication.run(Chapter4EurekaServerApplication.class, args);
    }
}
```

在 application.yml 配置文件中添加以下配置，配置注册中心的端口和标识自己为 Eureka Server：

```yaml
server:
    port: 8761
eureka:
    instance:
        hostname: standalone
        instance-id: ${spring.application.name}:${vcap.application.instance_id:$
            {spring.application.instance_id:${random.value}}}
    client:
        register-with-eureka: false  # 表明该服务不会向Eureka Server注册自己的信息
        fetch-registry: false  # 表明该服务不会向Eureka Server获取注册信息
        service-url:    # Eureka Server注册中心的地址，用于client与server进行交流
            defaultZone: http://${eureka.instance.hostname}:${server.port}/eureka/
spring:
    application:
        name: eureka-service
```

InstanceId 是 Eureka 服务的唯一标记，主要用于区分同一服务集群的不同实例。一般来讲，一个 Eureka 服务实例默认注册的 InstanceId 是它的主机名（即一个主机只有一个服务）。但是这样会引发一个问题，一台主机不能启动多个属于同一服务的服务实例。为了解决这种情况，spring-cloud-netflix-eureka 提供了一个合理的实现，如上面代码中的 InstanceId 设置样式。通过设置 random.value 可以使得每一个服务实例的 InstanceId 独一无二，从而可以唯一标记它自身。

Eureka Server 既可以独立部署，也可以集群部署。在集群部署的情况下，Eureka Server 间会进行注册表信息同步的操作，这时被同步注册表信息的 Eureka Server 将会被其他同步注册表信息的 Eureka Server 称为 peer。

请注意，上述配置中的 service-url 指向的注册中心为实例本身。通常来讲，一个 Eureka Server 也是一个 Eureka Client，它会尝试注册自己，所以需要至少一个注册中心的 URL 来定位对等点 peer。如果不提供这样一个注册端点，注册中心也能工作，但是会在日志中打印无法向 peer 注册自己的信息。在独立（Standalone）Eureka Server 的模式下，Eureka Server 一般会关闭作为客户端注册自己的行为。

Eureka Server 与 Eureka Client 之间的联系主要通过心跳的方式实现。心跳（Heartbeat）即 Eureka Client 定时向 Eureka Server 汇报本服务实例当前的状态，维护本服务实例在注册表中租约的有效性。

Eureka Server 需要随时维持最新的服务实例信息，所以在注册表中的每个服务实例都需要定期发送心跳到 Server 中以使自己的注册保持最新的状态（数据一般直接保存在内存中）。为了避免 Eureka Client 在每次服务间调用都向注册中心请求依赖服务实例的信息，Eureka Client 将定时从 Eureka Server 中拉取注册表中的信息，并将这些信息缓存到本地，用于服务发现。

启动 Eureka Server 后，应用会有一个主页面用来展示当前注册表中的服务实例信息并同时暴露一些基于 HTTP 协议的端点在 /eureka 路径下，这些端点将由 Eureka Client 用于注册自身、获取注册表信息以及发送心跳等。

4.1.3 搭建 Eureka 服务提供者

可以搭建包含 Eurake Client 依赖的 Spring Boot 项目。主要依赖有：

```xml
<dependency> <!--eureka-client相关依赖-->
    <groupId>org.springframework.cloud</groupId>
    <artifactId>spring-cloud-starter-netflix-eureka-client</artifactId>
</dependency>
<dependency>
    <groupId>org.springframework.boot</groupId>
    <artifactId>spring-boot-starter-web</artifactId>
</dependency>
```

启动类如下：

```java
@SpringBootApplication
public class Chapter4EurekaClientServiceApplication
{
    public static void main(String[] args) {
        SpringApplication.run(Chapter4EurekaClientApplication.class, args);
    }
}
```

在 Spring Cloud 的 Finchley 版本中，只要引入 spring-cloud-starter-netflix-eureka-client 的依赖，应用就会自动注册到 Eureka Server，但是需要在配置文件中添加 Eureka Server 的地址。在 application.yml 添加以下配置：

```yaml
eureka:
  instance:
    hostname: client
    instance-id: ${spring.application.name}:${vcap.application.instance_id:${spring.application.instance_id:${random.value}}}
  client:
    service-url: # Eureka Server注册中心的地址，用于Client与Server进行交流
      defaultZone: http://localhost:8761/eureka/
server:
  port: 8760
spring:
  application:
    name: eureka-client-service
```

为服务提供者添加一个提供服务的接口，代码如下：

```java
@RestController
public class SayHelloController {
```

```
@RequestMapping(value = "/hello/{name}")
public String sayHello(@PathVarivable("name") String name){
    return "Hello, ".concat(name).concat("!");
    }
}
```

上述接口将会向请求者返回打招呼的响应信息。

4.1.4 搭建 Eureka 服务调用者

可以搭建包含 Eurake Client 依赖的 Spring Boot 项目。主要依赖有：

```
<dependency> <!--eureka-client相关依赖-->
    <groupId>org.springframework.cloud</groupId>
    <artifactId>spring-cloud-starter-netflix-eureka-client</artifactId>
</dependency>
<dependency>
    <groupId>org.springframework.boot</groupId>
    <artifactId>spring-boot-starter-web</artifactId>
</dependency>
```

启动类代码如下：

```
@SpringBootApplication
public class Chapter4EurekaClientApplication {
    public static void main(String[] args) {
        SpringApplication.run(Chapter4EurekaClientApplication.class, args);
    }
}
```

在 application.yml 添加 eureka-client 相关配置，代码如下所示：

```
eureka:
    instance:
        hostname: client
        instance-id: ${spring.application.name}:${vcap.application.instance_id:${spring.application.instance_id:${random.value}}}

    client:
        service-url: // Eureka Server注册中心的地址，用于client与server进行交流
            defaultZone: http://localhost:8761/eureka/

server:
    port: 8765
spring:
    application:
        name: eureka-client
```

添加一个 AskController 向 eureka-client-service 请求 sayHello 的服务。通过使用可以进行负载均衡的 RestTemplate 向 eureka-client-service 发起打招呼的请求，并直接返回对应的响应结果。具体代码如下所示：

```
@RestController
```

```
@Configuration
public class AskController {
    // 注入本地服务名
    @Value("${spring.application.name}")
    private String name;
    @Autowired
    RestTemplate restTemplate;
    @RequestMapping(value = "/ask")
    public String ask(){
        // 从eureka-client-service服务提供者中请求sayHello服务
        String askHelloFromService = restTemplate.getForEntity("http://EUREKA-
            CLIENT-SERVICE/hello/{name}", String.class, name).getBody();
        return askHelloFromService;
    }
    // 注入一个可以进行负载均衡的RestTemple用于服务间调用
    @Bean
    @LoadBalanced
    public RestTemplate restTemplate(){
        return new RestTemplate();
    }
}
```

4.1.5　Eureka 服务注册和发现

搭建好上述三个 Eureka 应用后，依次启动三个应用。

1. Eureka Server 主页

访问 Eureka Server 的主页 http://localhost:8761，可以看到图 4-1 所示的界面。

从图中可以看到以下信息：

- 展示当前注册到 Eureka Server 上的服务实例信息。
- 展示 Eureka Server 运行环境的通用信息。
- 展示 Eureka Server 实例的信息。

2. 服务间调用

访问 http://localhost:8762/ask，eureka-client 将调用 eureka-client-service 的 sayHello 服务，向 eureka-client-service 传递服务名，等待 eureka-client-service 返回请求响应，响应结果如下：

```
Hello, eureka-client!
```

RestTemplate 将根据服务名 eureka-client-service 通过预先从 eureka-service 缓存到本地的注册表中获取到 eureka-client-service 服务的具体地址，从而发起服务间调用。

3. 与服务注册中心交换信息

DiscoveryClient 来源于 spring-cloud-client-discovery，是 Spring Cloud 中定义用来服务发现的公共接口，在 Spring Cloud 的各类服务发现组件中（如 Netflix Eureka 或 Consul）都有相

应的实现。它提供从服务注册中心根据 serviceId 获取到对应服务实例信息的能力。当一个服务实例拥有 DiscoveryClient 的具体实现时，就可以从服务注册中心中发现其他的服务实例。

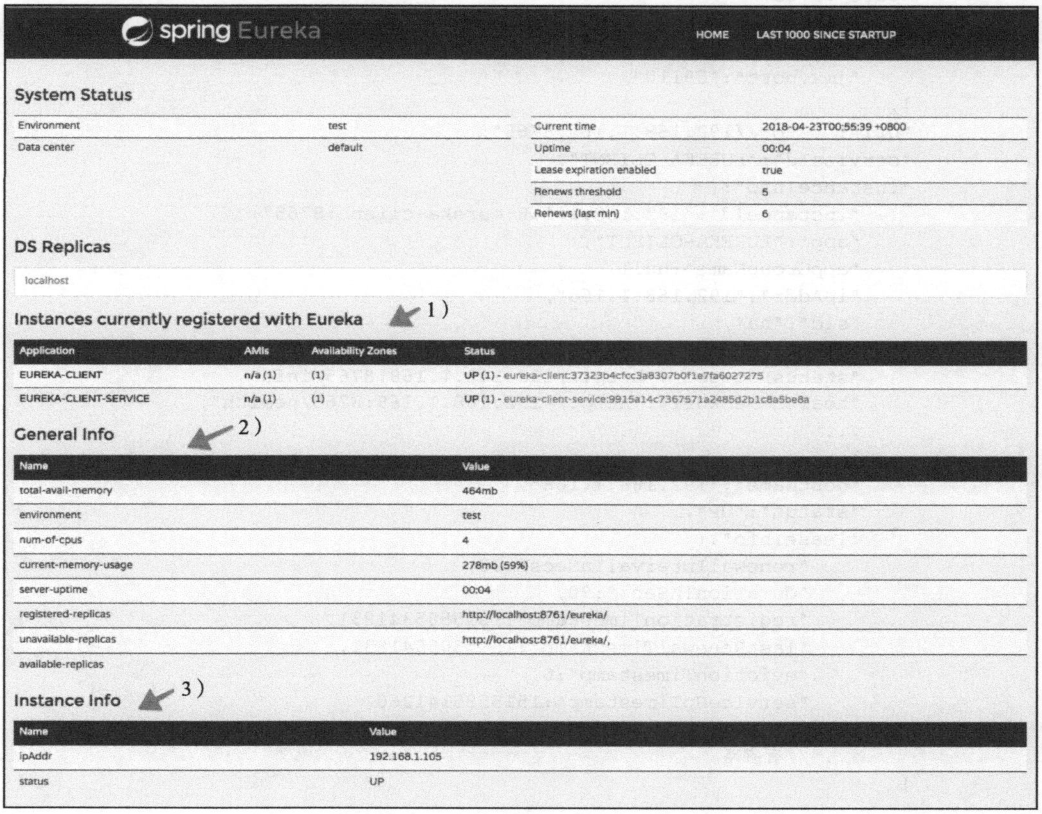

图 4-1　Eureka Server 主页

在 Eureka Client 中注入 DiscoveryClient，并从 Eureka Server 获取服务实例的信息。在 chapter4-eureka-client 添加一个 ServiceInstanceRestController 的 controller，如下所示：

```
@RestController
public class ServiceInstanceRestController {
    @Autowired
    private DiscoveryClient discoveryClient;
    @RequestMapping("/service-instances/{applicationName}")
    public List<ServiceInstance> serviceInstancesByApplicationName(
            @PathVariable String applicationName) {
        return this.discoveryClient.getInstances(applicationName);
    }
}
```

启动应用后，访问地址 http://localhost:8765/service-instances/eureka-client，获取应用名为 eureka-client（服务本身）的服务实例元数据，结果如下所示：

```json
[
    {
        "host":"192.168.1.168",
        "port":8765,
        "metadata":{
            "management.port":"8765",
            "jmx.port":"59110"
        },
        "uri":"http://192.168.1.168:8765",
        "serviceId":"EUREKA-CLIENT",
        "instanceInfo":{
            "instanceId":"192.168.1.168:eureka-client:8765",
            "app":"EUREKA-CLIENT",
            "appGroupName":null,
            "ipAddr":"192.168.1.168",
            "sid":"na",
            "homePageUrl":"http://192.168.1.168:8765/",
            "statusPageUrl":"http://192.168.1.168:8765/info",
            "healthCheckUrl":"http://192.168.1.168:8765/health",
            ...
        },
        "hostName":"192.168.1.168",
        "status":"UP",
        "leaseInfo":{
            "renewalIntervalInSecs":30,
            "durationInSecs":90,
            "registrationTimestamp":1515585341831,
            "lastRenewalTimestamp":1515585341831,
            "evictionTimestamp":0,
            "serviceUpTimestamp":1515585341260
        },
        ... //有省略
    }
]
```

Eureka 中标准元数据有主机名、IP 地址、端口号、状态页 url 和健康检查 url 等，这些元数据都会保存在 Eureka Server 的注册表中，Eureka Client 根据服务名读取这些元数据，来发现和调用其他服务实例。元数据可以自定义以适应特定的业务场景，这些内容将在下面章节进行讲解。

4.1.6　Consul 的简单应用

除了 Eureka，Spring Cloud 还提供了 Consul 服务注册与发现的支持实现。

Consul 是由 HashiCorp 基于 Go 语言开发的服务软件，支持多数据中心、分布式和高可用的服务注册和发现。它采用 Raft 算法保证服务的一致性，支持健康检查。

1. 安装和启动 Consul

笔者以自身的 mac 系统为例，通过 HomeBrew 的方式安装 Consul，命令为：

```
brew install consul
```

笔者安装的 Consul 版本为 v1.0.6。检查 Consul 安装结果,如下所示:

```
$ consul
Usage: consul [--version] [--help] <command> [<args>]

Available commands are:
    agent           Runs a Consul agent
    ....
```

通过 -dev 的方式启动单节点 Consul(dev 方式的启动不适用于生产环境,因为不会持久化任何状态,这里只是为了快速便捷地启动单节点的 Consul),如下所示:

```
$ consul agent -dev
==> Starting Consul agent...
==> Consul agent running!
           Version: 'v1.0.6'
           Node ID: '13214b1d-a505-f397-4bed-5b96aa8d5731'
         Node name: 'xuans-iMac.local'
        Datacenter: 'dc1' (Segment: '<all>')
            Server: true (Bootstrap: false)
       Client Addr: [127.0.0.1] (HTTP: 8500, HTTPS: -1, DNS: 8600)
      Cluster Addr: 127.0.0.1 (LAN: 8301, WAN: 8302)
           Encrypt: Gossip: false, TLS-Outgoing: false, TLS-Incoming: false
```

2. 搭建 Consul 客户端的 Spring Boot 服务

可以搭建包含 Consul 依赖的 Spring Boot 项目。主要依赖如下所示:

```xml
<dependency> <!--Spring WebMvc相关依赖-->
    <groupId>org.springframework.boot</groupId>
    <artifactId>spring-boot-starter-web</artifactId>
</dependency>
<dependency> <!-- consul客户端相关依赖-->
    <groupId>org.springframework.cloud</groupId>
    <artifactId>spring-cloud-starter-consul-discovery</artifactId>
</dependency>
<dependency> <!-- actuator相关依赖-->
    <groupId>org.springframework.boot</groupId>
    <artifactId>spring-boot-starter-actuator</artifactId>
</dependency>
```

在 application.ym 文件中添加 Consul 客户端的相关配置,如下所示:

```
spring:
  application:
    name: demo-consul
  cloud:
    consul:
      discovery:
        instance-id: ${spring.application.name}:${vcap.application.
            instance_id:${spring.application.instance_id:${random.value}}}
            #服务实例ID
```

```
          service-name: demo-consul # 服务名
        host: localhost # consul服务器hostname
        port: 8500 # consul服务器端口号
```

3. 查看 Consul 主界面

如图 4-2 所示，启动 consul-client 服务，访问 localhost:8500，将会发现 consul-client 已注册到 Consul 中。

图 4-2 consul 主界面

4.2 服务发现原理

目前 SpringCloud 的 Finchley 版本采用的 Eureka 版本为 1.9，本部分的架构图主要基于 Eureka v1 进行介绍。图 4-3 为 Eureka 架构图。

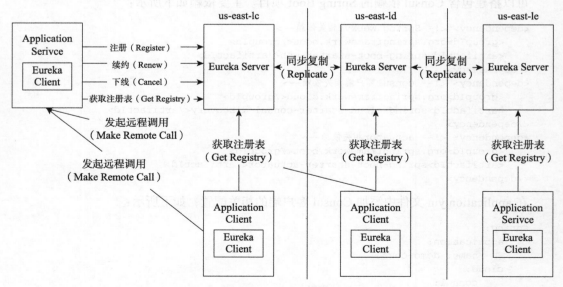

图 4-3 Eureka 架构图

1. Region 与 Availability Zone

Eureka 最初设计的目的是 AWS（亚马逊网络服务系统）中用于部署分布式系统，所以首先对 AWS 上的区域（Regin）和可用区（Availability Zone）进行简单的介绍。
- 区域：AWS 根据地理位置把某个地区的基础设施服务集合称为一个区域，区域之间相对独立。在架构图上，us-east-1c、us-east-1d、us-east-1e 表示 AWS 中的三个设施服务区域，这些区域中分别部署了一个 Eureka 集群。
- 可用区：AWS 的每个区域都是由多个可用区组成的，而一个可用区一般都是由多个数据中心（简单理解成一个原子服务设施）组成的。可用区与可用区之间是相互独立的，有独立的网络和供电等，保证了应用程序的高可用性。在上述的架构图中，一个可用区中可能部署了多个 Eureka，一个区域中有多个可用区，这些 Eureka 共同组成了一个 Eureka 集群。

2. 组件与行为

下面介绍组件与行为：
- Application Service：是一个 Eureka Client，扮演服务提供者的角色，提供业务服务，向 Eureka Server 注册和更新自己的信息，同时能从 Eureka Server 注册表中获取到其他服务的信息。
- Eureka Server：扮演服务注册中心的角色，提供服务注册和发现的功能。每个 Eureka Cient 向 Eureka Server 注册自己的信息，也可以通过 Eureka Server 获取到其他服务的信息达到发现和调用其他服务的目的。
- Application Client：是一个 Eureka Client，扮演了服务消费者的角色，通过 Eureka Server 获取注册到其上其他服务的信息，从而根据信息找到所需的服务发起远程调用。
- Replicate：Eureka Server 之间注册表信息的同步复制，使 Eureka Server 集群中不同注册表中服务实例信息保持一致。
- Make Remote Call：服务之间的远程调用。
- Register：注册服务实例，Client 端向 Server 端注册自身的元数据以供服务发现。
- Renew：续约，通过发送心跳到 Server 以维持和更新注册表中服务实例元数据的有效性。当在一定时长内，Server 没有收到 Client 的心跳信息，将默认服务下线，会把服务实例的信息从注册表中删除。
- Cancel：服务下线，Client 在关闭时主动向 Server 注销服务实例元数据，这时 Client 的服务实例数据将从 Server 的注册表中删除。
- Get Registry：获取注册表，Client 向 Server 请求注册表信息，用于服务发现，从而发起服务间远程调用。

4.3　Eureka Client 源码解析

Eureka Client 为了简化开发人员的开发工作，将很多与 Eureka Server 交互的工作隐藏起来，自主完成。在应用的不同运行阶段在后台完成工作如图 4-4 所示。

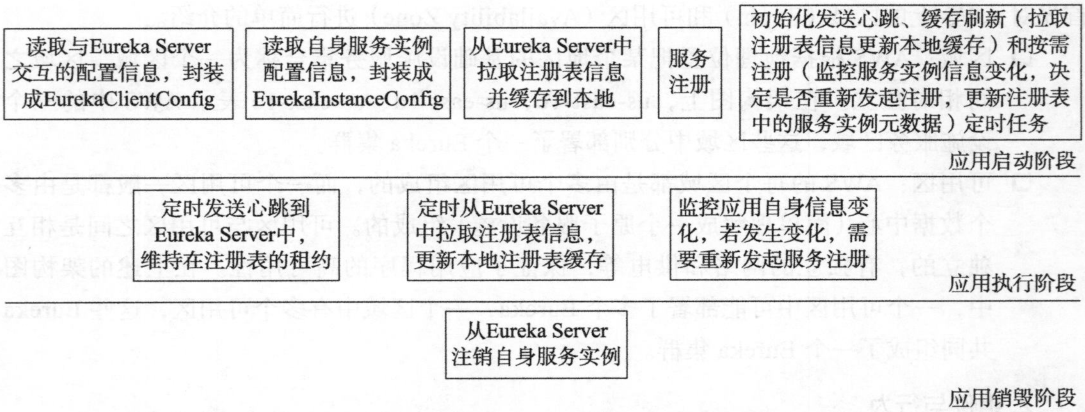

```
┌─────────────┐ ┌─────────────┐ ┌─────────────┐ ┌────┐ ┌─────────────────┐
│读取与Eureka   │ │读取自身服务  │ │从Eureka Server│ │服务 │ │初始化发送心跳、  │
│Server        │ │实例配置信息, │ │中拉取注册表  │ │注册 │ │缓存刷新(拉取    │
│交互的配置信息,│ │封装成        │ │信息并缓存到  │ │    │ │注册表信息更新本地│
│封装成        │ │EurekaInstance│ │本地          │ │    │ │缓存)和按需注册  │
│EurekaClient  │ │Config        │ │              │ │    │ │(监控服务实例信息│
│Config        │ │              │ │              │ │    │ │变化,决定是否重新│
│              │ │              │ │              │ │    │ │发起注册,更新注册│
│              │ │              │ │              │ │    │ │表中的服务实例元 │
│              │ │              │ │              │ │    │ │数据)定时任务    │
└─────────────┘ └─────────────┘ └─────────────┘ └────┘ └─────────────────┘
```

应用启动阶段

```
┌──────────────┐  ┌──────────────┐  ┌──────────────┐
│定时发送心跳到 │  │定时从Eureka   │  │监控应用自身信息│
│Eureka Server中,│ │Server中拉取   │  │变化,若发生变化,│
│维持在注册表的 │  │注册表信息,更新│  │需要重新发起服务│
│租约           │  │本地注册表缓存 │  │注册           │
└──────────────┘  └──────────────┘  └──────────────┘
```

应用执行阶段

```
                  ┌──────────────┐
                  │从Eureka Server│
                  │注销自身服务实例│
                  └──────────────┘
```

应用销毁阶段

图 4-4　Eureka Client 的工作

为了跟踪 Eureka 的运行机制，读者可以通过打开 Spring Boot 的 Debug 模式来查看更多的输出日志，如下所示：

```
logging:
    level:
        org.springframework: DEBUG
```

Eukeka Client 通过 Starter 的方式引入依赖，Spring Boot 将会为项目使用以下的自动配置类：

- EurekaClientAutoConfiguration：Eureke Client 自动配置类，负责 Eureka Client 中关键 Beans 的配置和初始化，如 ApplicationInfoManager 和 EurekaClientConfig 等。
- RibbonEurekaAutoConfiguration：Ribbon 负载均衡相关配置。
- EurekaDiscoveryClientConfiguration：配置自动注册和应用的健康检查器。

4.3.1 读取应用自身配置信息

通过 EurekaDiscoveryClientConfiguration 配置类，Spring Boot 帮助 Eureka Client 完成很多必要 Bean 的属性读取和配置，表 4-1 列出了 EurekaDiscoveryClientConfiguration 中的属性读取和配置类。

表 4-1　EurekaDiscoveryClientConfiguration 中属性读取和配置类

类　名	作用和介绍
EurekaClientConfig	封装 Eureka Client 与 Eureka Server 交互所需要的配置信息。Spring Cloud 为其提供了一个默认配置类的 EurekaClientConfigBean，可以在配置文件中通过前缀 eureka.client+ 属性名进行属性覆盖
ApplicationInfoManager	作为应用信息管理器，管理服务实例的信息类 InstanceInfo 和服务实例的配置信息类 EurekaInstanceConfig
InstanceInfo	封装将被发送到 Eureka Server 进行服务注册的服务实例元数据。它在 Eureka Server 的注册表中代表一个服务实例，其他服务实例可以通过 InstanceInfo 了解该服务实例的相关信息从而发起服务请求

(续)

类　　名	作用和介绍
EurekaInstanceConfig	封装 Eureka Client 自身服务实例的配置信息，主要用于构建 InstanceInfo 通常这些信息在配置文件中的 eureka.instance 前缀下进行设置，Spring Cloud 通过 EurekaInstanceConfigBean 配置类提供了默认配置
DiscoveryClient	Spring Cloud 中定义用来服务发现的客户端接口

下面我们对 Spring Cloud 中的服务发现客户端 DiscoveryClient 进行进一步的介绍，它是客户端进行服务发现的核心接口。

DiscoveryClient 是 Spring Cloud 中用来进行服务发现的顶级接口，在 Netflix Eureka 或者 Consul 中都有相应的具体实现类，在 4.1 节基础应用中有所介绍，该接口提供的方法如下：

```
//DiscoveryClient.java
public interface DiscoveryClient {
    // 获取实现类的描述
    String description();
    // 通过服务Id获取服务实例的信息
    List<ServiceInstance> getInstances(String serviceId);
    // 获取所有的服务实例Id
    List<String> getServices();
...
}
```

其在 Eureka 方面的实现类结构如图 4-5 所示。

EurekaDiscoveryClient 继承了 DiscoveryClient 接口，但是通过查看 EurekaDiscoveryClient 中代码，会发现它是通过组合 EurekaClien 类实现接口的功能，如下为 getInstance 方法的实现：

```
//EurekaDiscoveryClient.java
@Override
public List<ServiceInstance> getInstances (String serviceId) {
    List<InstanceInfo> infos = this.eurekaClient.getInstancesByVipAddress(service
        Id, false);
    List<ServiceInstance> instances = new ArrayList<>();
    for (InstanceInfo info : infos) {
        instances.add(new EurekaServiceInstance(info));
    }
    return instances;
}
```

EurekaClient 来自于 com.netflix.discovery 包中，其默认实现为 com.netflix.discovery. DiscoveryClient，属于 eureka-client 的源代码，它提供了 Eureka Client 注册到 Server 上、续租、下线以及获取 Server 中注册表信息等诸多关键功能。Spring Cloud 通过组合方式调用了

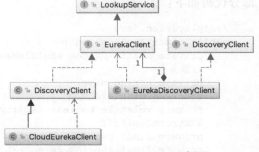

图 4-5　DiscoveryClient 类图

Eureka 中的服务发现方法，关于 EurekaClient 的详细代码分析将放在下面的章节进行介绍。

4.3.2 服务发现客户端

为了对 Eureka Client 的执行原理进行讲解，首先需要对服务发现客户端 com.netflix.discover.DiscoveryClient 职能以及相关类进行讲解，它负责了与 Eureka Server 交互的关键逻辑。

1. DiscoveryClient 职责

DiscoveryClient 是 Eureka Client 的核心类，包括与 Eureka Server 交互的关键逻辑，具备了以下职能：

- 注册服务实例到 Eureka Server 中；
- 发送心跳更新与 Eureka Server 的租约；
- 在服务关闭时从 Eureka Server 中取消租约，服务下线；
- 查询在 Eureka Server 中注册的服务实例列表。

2. DiscoveryClient 类结构

DiscoverClient 的核心类图如图 4-6 所示。

DiscoveryClient 继承了 LookupService 接口，LookupService 作用是发现活跃的服务实例，主要方法如下：

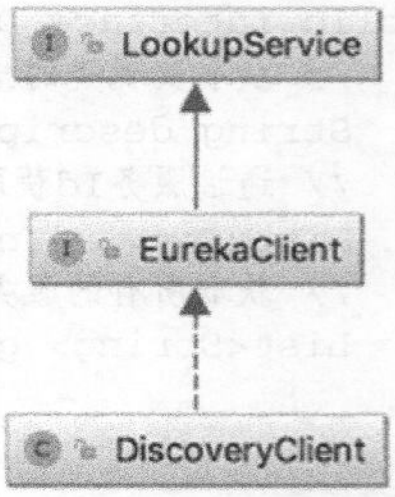

```
//LookupService.java
public interface LookupService<T> {
    //根据服务实例注册的appName来获取封装有相同appName的服务实例信息容器
    Application getApplication(String appName);
    //返回当前注册表中所有的服务实例信息
    Applications getApplications();
    //根据服务实例的id获取服务实例信息
    List<InstanceInfo> getInstancesById(String id);
    ...
}
```

图 4-6　DiscoveryClient 核心类图

Application 持有服务实例信息列表，它可以理解成同一个服务的集群信息，这些服务实例都挂在同一个服务名 appName 下。InstanceInfo 代表一个服务实例信息。Application 部分代码如下：

```
//Application.java
public class Application {
    private static Random shuffleRandom = new Random();
    //服务名
    private String name;
    @XStreamOmitField
    private volatile boolean isDirty = false;
    @XStreamImplicit
    private final Set<InstanceInfo> instances;
    private final AtomicReference<List<InstanceInfo>> shuffledInstances;
    private final Map<String, InstanceInfo> instancesMap;
```

```
    ...
}
```

为了保证原子性操作，Application 中对 InstanceInfo 的操作都是同步操作。

Applications 是注册表中所有服务实例信息的集合，里面的操作大多也是同步操作。

EurekaClient 继承了 LookupService 接口，为 DiscoveryClient 提供了一个上层接口，目的是方便从 Eureka 1.x 到 Eureka 2.x（已停止开发）的升级过渡。EurekaClient 接口属于比较稳定的接口，即使在下一阶段也会被保留。

EurekaClient 在 LookupService 的基础上扩充了更多的接口，提供了更丰富的获取服务实例的方式，主要有：

- 提供了多种方式获取 InstanceInfo，例如根据区域、Eureka Server 地址等获取。
- 提供了本地客户端（所处的区域、可用区等）的数据，这部分与 AWS 密切相关。
- 提供了为客户端注册和获取健康检查处理器的能力。

除去查询相关的接口，我们主要关注 EurekaClient 中以下两个接口，代码如下所示：

```
// EurekaClient.java
//为Eureka Client注册健康检查处理器
public void registerHealthCheck(HealthCheckHandler healthCheckHandler);
//为Eureka Client注册一个EurekaEventListener(事件监听器)
// 监听Client服务实例信息的更新
public void registerEventListener(EurekaEventListener eventListener);
```

Eureka Server 一般通过心跳（heartbeats）来识别一个实例的状态。Eureka Client 中存在一个定时任务定时通过 HealthCheckHandler 检测当前 Client 的状态，如果 Client 的状态发生改变，将会触发新的注册事件，更新 Eureka Server 的注册表中该服务实例的相关信息。HealthCheckHandler 的代码如下所示：

```
// HealthCheckHandler.java
public interface HealthCheckHandler {
    InstanceInfo.InstanceStatus getStatus(InstanceInfo.InstanceStatus currentStatus);
}
```

HealthCheckHandler 接口的代码如上所示，其在 spring-cloud-netflix-eureka-client 中的实现类为 EurekaHealthCheckHandler，主要组合了 spring-boot-actuator 中的 HealthAggregator 和 HealthIndicator，以实现对 Spring Boot 应用的状态检测。

Eureka 中的事件模式属于观察者模式，事件监听器将监听 Client 的服务实例信息变化，触发对应的处理事件，图 4-7 为 Eureka 事件的类图：

3. DiscoveryClient 构造函数

在 DiscoveryClient 构造函数中，Eureka Client 会执行从 Eureka Server 中拉取注册表信息、服务注册、初始化发送心跳、缓存刷新（重新拉取注册表信息）和按需注册定时任务等操作，可以说 DiscoveryClient 的构造函数贯穿了 Eureka Client 启动阶段的各项工作。DiscoveryClient 的构造函数传入的参数如下所示：

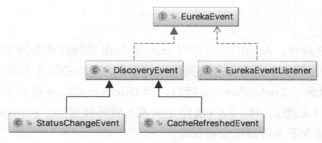

图 4-7 EurekaEvent 类图

```
//DiscoveryClient.java
DiscoveryClient(ApplicationInfoManager applicationInfoManager, EurekaClientConfig
    config, AbstractDiscoveryClientOptionalArgs args, Provider<BackupRegistry>
    backupRegistryProvider)
```

ApplicationInfoManager 和 **EurekaClientConfig** 在前面内容中已经做了介绍,一个是应用信息管理器,另一个是封装了 Client 与 Server 交互配置信息的类。

AbstractDiscoveryClientOptionalArgs 是用于注入一些可选参数,以及一些 jersey1 和 jersey2 通用的过滤器。而 BackupRegistry 充当了备份注册中心的职责,当 Eureka Client 无法从任何一个 Eureka Server 中获取注册表信息时,BackupRegistry 将被调用以获取注册表信息。默认的实现是 NotImplementedRegistryImpl,即没有实现。

在构造方法中,忽略掉构造方法中大部分的赋值操作,我们逐步了解了配置类中的属性会对 DiscoveryClient 的行为造成什么影响。DiscoveryClient 构造函数中的部分代码如下所示:

```
// DiscoveryClient.java
if (config.shouldFetchRegistry()) {
    this.registryStalenessMonitor = new ThresholdLevelsMetric(this, METRIC_
        REGISTRY_PREFIX + "lastUpdateSec_", new long[]{15L, 30L, 60L, 120L, 240L, 480L});
} else {
    this.registryStalenessMonitor = ThresholdLevelsMetric.NO_OP_METRIC;
}
if (config.shouldRegisterWithEureka()) {
    this.heartbeatStalenessMonitor = new ThresholdLevelsMetric(this, METRIC_
        REGISTRATION_PREFIX + "lastHeartbeatSec_", new long[]{15L, 30L, 60L,
        120L, 240L, 480L});
} else {
    this.heartbeatStalenessMonitor = ThresholdLevelsMetric.NO_OP_METRIC;
}
```

config#shouldFetchRegistry(对应配置为 eureka.client.fetch-register)为 true 表示 Eureka Client 将从 Eureka Server 中拉取注册表信息。config#shouldRegisterWithEureka(对应配置为 eureka.client.register-with-eureka)为 true 表示 Eureka Client 将注册到 Eureka Server 中。如果上述的两个配置均为 false,那么 Discovery 的初始化将直接结束,表示该客户端既不进行服务注册也不进行服务发现。

接着定义一个基于线程池的定时器线程池 ScheduledExecutorService，线程池大小为 2，一个线程用于发送心跳，另一个线程用于缓存刷新，同时定义了发送心跳和缓存刷新线程池，代码如下所示：

```
//DiscoveryClient.java
scheduler = Executors.newScheduledThreadPool(2, new ThreadFactoryBuilder()
            .setNameFormat("DiscoveryClient-%d").setDaemon(true).build());
    heartbeatExecutor = new ThreadPoolExecutor(...);
    cacheRefreshExecutor = new ThreadPoolExecutor(...);
```

之后，初始化 Eureka Client 与 Eureka Server 进行 HTTP 交互的 Jersey 客户端，将 AbstractDiscoveryClientOptionalArgs 中的属性用来构建 EurekaTransport，如下所示：

```
// DiscoveryClient.java
eurekaTransport = new EurekaTransport();
scheduleServerEndpointTask(eurekaTransport, args);
```

EurekaTransport 是 DiscoveryClient 中的一个内部类，其内封装了 DiscoveryClient 与 Eureka Server 进行 HTTP 调用的 Jersey 客户端。

再接着从 Eureka Server 中拉取注册表信息，代码如下所示：

```
// DiscoveryClient.java
if (clientConfig.shouldFetchRegistry() && !fetchRegistry(false)) {
    fetchRegistryFromBackup();
}
```

如果 EurekaClientConfig#shouldFetchRegistry 为 true 时，fetchRegistry 方法将会被调用。在 Eureka Client 向 Eureka Server 注册前，需要先从 Eureka Server 拉取注册表中的信息，这是服务发现的前提。通过将 Eureka Server 中的注册表信息缓存到本地，就可以就近获取其他服务的相关信息，减少与 Eureka Server 的网络通信。

拉取完 Eureka Server 中的注册表信息后，将对服务实例进行注册，代码如下所示：

```
// DiscoveryClient.java
if (this.preRegistrationHandler != null) {
    this.preRegistrationHandler.beforeRegistration();
}
if (clientConfig.shouldRegisterWithEureka() && clientConfig.shouldEnforceRegistr
    ationAtInit()) {
    try {
        // 发起服务注册
        if (!register() ) {
            // 注册失败，抛出异常
            throw new IllegalStateException("Registration error at startup. Invalid
                server response.");
        }
    } catch (Throwable th) {
        throw new IllegalStateException(th);
    }
}
    initScheduledTasks(); // 初始化定时任务
```

在服务注册之前会进行注册预处理，Eureka 没有对此提供默认实现。构造函数的最后将初始化并启动发送心跳、缓存刷新和按需注册等定时任务。

最后总结一下，在 DiscoveryClient 的构造函数中，主要依次做了以下的事情：

1）相关配置的赋值，类似 ApplicationInfoManager、EurekaClientConfig 等。
2）备份注册中心的初始化，默认没有实现。
3）拉取 Eureka Server 注册表中的信息。
4）注册前的预处理。
5）向 Eureka Server 注册自身。
6）初始化心跳定时任务、缓存刷新和按需注册等定时任务。

4.3.3 拉取注册表信息

在 DiscoveryClient 的构造函数中，调用了 DiscoveryClient#fetchRegistry 方法从 Eureka Server 中拉取注册表信息，方法执行如下所示：

```java
//DiscoveryClient.java
private boolean fetchRegistry(boolean forceFullRegistryFetch) {
    Stopwatch tracer = FETCH_REGISTRY_TIMER.start();
    try {
        // 如果增量式拉取被禁止，或者Applications为null，进行全量拉取
        Applications applications = getApplications();
        if (clientConfig.shouldDisableDelta()
            || (!Strings.isNullOrEmpty(clientConfig.getRegistryRefreshSingleVipAddress()))
            || forceFullRegistryFetch
            || (applications == null)
            || (applications.getRegisteredApplications().size() == 0)
            || (applications.getVersion() == -1))
        {
            ...
            // 全量拉取注册表信息
            getAndStoreFullRegistry();
        } else {
        // 增量拉取注册表信息
        getAndUpdateDelta(applications);
        }
            // 计算应用集合一致性哈希码
            applications.setAppsHashCode(applications.getReconcileHashCode());
            // 打印注册表上所有服务实例的总数量
            logTotalInstances();
    } catch (Throwable e) {
        return false;
    } finally {
        if (tracer != null) {
            tracer.stop();
        }
    }
    // 在更新远程实例状态之前推送缓存刷新事件，但是Eureka中并没有提供默认的事件监听器
    onCacheRefreshed();
```

```
        // 基于缓存中被刷新的数据更新远程实例状态
        updateInstanceRemoteStatus();
        // 注册表拉取成功, 返回true
        return true;
    }
```

一般来讲，在 Eureka 客户端，除了第一次拉取注册表信息，之后的信息拉取都会尝试只进行增量拉取（第一次拉取注册表信息为全量拉取），下面将分别介绍拉取注册表信息的两种实现，全量拉取注册表信息 DiscoveryClient#getAndStoreFullRegistry 和增量式拉取注册表信息 DiscoveryClient#getAndUpdateDelta。

1. 全量拉取注册表信息

一般只有在第一次拉取的时候，才会进行注册表信息的全量拉取，主要在 DiscoveryClient#getAndStoreFullRegistry 方法中进行。代码如下所示：

```java
// DiscoveryClient.java
private void getAndStoreFullRegistry() throws Throwable {
    // 获取拉取的注册表的版本，防止拉取版本落后(由其他的线程引起)
    long currentUpdateGeneration = fetchRegistryGeneration.get();
    Applications apps = null;
    EurekaHttpResponse<Applications> httpResponse = clientConfig.getRegistryRefreshSingleVipAddress()
        == null
        ? eurekaTransport.queryClient.getApplications(remoteRegionsRef.get()):
        eurekaTransport.queryClient.getVip(clientConfig.getRegistryRefreshSingleVipAddress(),
            remoteRegionsRef.get());
    // 获取成功
    if (httpResponse.getStatusCode() == Status.OK.getStatusCode()) {
        apps = httpResponse.getEntity();
    }
    if (apps == null) {
        // 日志
        // 检查fetchRegistryGeneration的更新版本是否发生改变，无改变的话说明本次拉取是最新的
    } else if (fetchRegistryGeneration.compareAndSet(currentUpdateGeneration,
        currentUpdateGeneration + 1)) {
        // 从apps中筛选出状态为UP的实例，同时打乱实例的顺序，防止同一个服务的不同实例在启动时
        //   接受流量
        localRegionApps.set(this.filterAndShuffle(apps));
    } else {
        // 日志
    }
}
```

全量拉取将从 Eureka Server 中拉取注册表中所有的服务实例信息（封装在 Applications 中），并经过处理后替换掉本地注册表缓存 Applications。

通过跟踪调用链，在 AbstractJerseyEurekaHttpClient#getApplicationsInternal 方法中发现了相关的请求 url，接口地址为 /eureka/apps，如图 4-8 所示。

该接口位于 Eureka Server 中，可以直接访问，用于获取当前 Eureka Server 中持有的所有注册表信息。

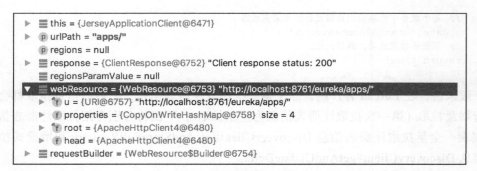

图 4-8　全量拉取注册表信息

　　getAndStoreFullRegistry 方法可能被多个线程同时调用，导致新拉取的注册表被旧的注册表覆盖（有可能出现先拉取注册表信息的线程在覆盖 apps 时被阻塞，被后拉取注册表信息的线程抢先设置了 apps，被阻塞的线程恢复后再次设置了 apps，导致 apps 数据版本落后），产生脏数据，对此，Eureka 通过类型为 AtomicLong 的 currentUpdateGeneration 对 apps 的更新版本进行跟踪。如果更新版本不一致，说明本次拉取注册表信息已过时，不需要缓存到本地。拉取到注册表信息之后会对获取到的 apps 进行筛选，只保留状态为 UP 的服务实例信息。

2. 增量式拉取注册表信息

　　增量式的拉取方式，一般发生在第一次拉取注册表信息之后，拉取的信息定义为从某一段时间之后发生的所有变更信息，通常来讲是 3 分钟之内注册表的信息变化。在获取到更新的 delta 后，会根据 delta 中的增量更新对本地的数据进行更新。与 getAndStoreFullRegistry 方法一样，也通过 fetchRegistryGeneration 对更新的版本进行控制。增量式拉取是为了维护 Eureka Client 本地的注册表信息与 Eureka Server 注册表信息的一致性，防止数据过久而失效，采用增量式拉取的方式减少了拉取注册表信息的通信量。Client 中有一个注册表缓存刷新定时器专门负责维护两者之间信息的同步性。但是当增量式拉取出现意外时，定时器将执行全量拉取以更新本地缓存的注册表信息。具体代码如下所示：

```
// DiscoveryClient.java
private void getAndUpdateDelta(Applications applications) throws Throwable {
    long currentUpdateGeneration = fetchRegistryGeneration.get();
    Applications delta = null;
    EurekaHttpResponse<Applications> httpResponse = eurekaTransport.queryClient.
        getDelta(remoteRegionsRef.get());
    if (httpResponse.getStatusCode() == Status.OK.getStatusCode()) {
        delta = httpResponse.getEntity();
    }
    // 获取增量拉取失败
    if (delta == null) {
        // 进行全量拉取
```

```
            getAndStoreFullRegistry();
        } else if (fetchRegistryGeneration.compareAndSet(currentUpdateGeneration,
            currentUpdateGeneration + 1)) {
            String reconcileHashCode = "";
                if (fetchRegistryUpdateLock.tryLock()) {
                    try {
                        // 更新本地缓存
                        updateDelta(delta);
                        // 计算应用集合一致性哈希码
                        reconcileHashCode = getReconcileHashCode(applications);
                    } finally {
                        fetchRegistryUpdateLock.unlock();
                    }
                }
                // 比较应用集合一致性哈希码，如果不一致将认为本次增量式拉取数据已脏，将发起全量拉
                   取更新本地注册表信息
                if (!reconcileHashCode.equals(delta.getAppsHashCode()) || clientConfig.
                    shouldLogDeltaDiff()) {
                    reconcileAndLogDifference(delta, reconcileHashCode);
                }
            }
        ...
}
```

同理，在相同的位置也发现了增量式更新的 url，/eureka/app/delta，可以直接访问，如图 4-9 所示。

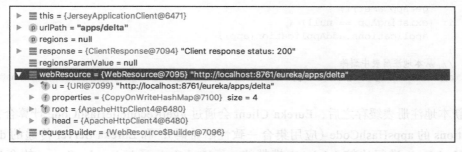

图 4-9 增量式拉取注册表信息

由于更新的过程过于漫长，时间成本为 O（N^2），所以通过同步代码块防止多个线程同时进行更新，污染数据。

在根据从 Eureka Server 拉取的 delta 信息更新本地缓存的时候，Eureka 定义了 ActionType 来标记变更状态，代码位于 InstanceInfo 类中，如下所示：

```
// InstanceInfo.java
public enum ActionType {
    ADDED,      // 添加Eureka Server
    MODIFIED,   // 在Euerka Server中的信息发生改变
    DELETED     // 被从Eureka Server中剔除
}
```

根据 InstanceInfo#ActionType 的不同，对 delta 中的 InstanceInfo 采取不同的操作，其中 ADDED 和 MODIFIED 状态变更的服务实例信息将添加到本地注册表，DELETED 状态变更的服务实例将从本地注册表中删除。具体代码如下所示：

```
// DiscoveryClient.java
// 变更类型为ADDED
if (ActionType.ADDED.equals(instance.getActionType())) {
    Application existingApp = applications.getRegisteredApplications(instance.
        getAppName());
    if (existingApp == null) {
        applications.addApplication(app);
    }
    // 添加到本地注册表中
    applications.getRegisteredApplications(instance.getAppName()).addInstance(instance);
// 变更类型为MODIFIED
} else if (ActionType.MODIFIED.equals(instance.getActionType())) {
    Application existingApp = applications.getRegisteredApplications(instance.
        getAppName());
    if (existingApp == null) {
        applications.addApplication(app);
    }
    // 添加到本地注册表中
    applications.getRegisteredApplications(instance.getAppName()).addInstance(instance);
// 变更类型为DELETE
} else if (ActionType.DELETED.equals(instance.getActionType())) {
    Application existingApp = applications.getRegisteredApplications(instance.
        getAppName());
    if (existingApp == null) {
        applications.addApplication(app);
    }
    // 从本地注册表中删除
    applications.getRegisteredApplications(instance.getAppName()).removeInstance(instance);
}
```

更新本地注册表缓存之后，Eureka Client 会通过 #getReconcileHashCode 计算合并后的 Applications 的 appsHashCode（应用集合一致性哈希码），和 Eureka Server 传递的 delta 上的 appsHashCode 进行比较（delta 中携带的 appsHashCode 通过 Eureka Server 的全量注册表计算得出），比对客户端和服务端上注册表的差异。如果哈希值不一致的话将再调用一次 getAndStoreFullRegistry 获取全量数据保证 Eureka Client 与 Eureka Server 之间注册表数据的一致。代码如下所示：

```
// DiscoveryClient.java
if (!reconcileHashCode.equals(delta.getAppsHashCode()) || clientConfig.
    shouldLogDeltaDiff()) {
    reconcileAndLogDifference(delta, reconcileHashCode);
}
```

reconcileAndLogDifference 方法中将会执行拉取全量注册表信息操作。
appsHashCode 的一般表示方式为：

```
appsHashCode = ${status}_${count}_
```

它通过将应用状态和数量拼接成字符串，表示了当前注册表中服务实例状态的统计信息。举个简单的例子，有 10 个应用实例的状态为 UP，有 5 个应用实例状态为 DOWN，其他状态的数量为 0（不进行表示），那么 appsHashCode 的形式将是：

```
appsHashCode = UP_10_DOWN_5_
```

4.3.4 服务注册

在拉取完 Eureka Server 中的注册表信息并将其缓存在本地后，Eureka Client 将向 Eureka Server 注册自身服务实例元数据，主要逻辑位于 Discovery#register 方法中。register 方法代码如下所示：

```
boolean register() throws Throwable {
    EurekaHttpResponse<Void> httpResponse;
    try {
        httpResponse = eurekaTransport.registrationClient.register(instanceInfo);
    } catch (Exception e) {
        throw e;
    }
    ...
    // 注册成功
    return httpResponse.getStatusCode() == 204;
}
```

Eureka Client 会将自身服务实例元数据（封装在 InstanceInfo 中）发送到 Eureka Server 中请求服务注册，当 Eureka Server 返回 204 状态码时，说明服务注册成功。

跟踪到 AbstractJerseyEurekaHttpClient#register 方法中，可以发现服务注册调用的接口以及传递的参数，如图 4-10 所示。

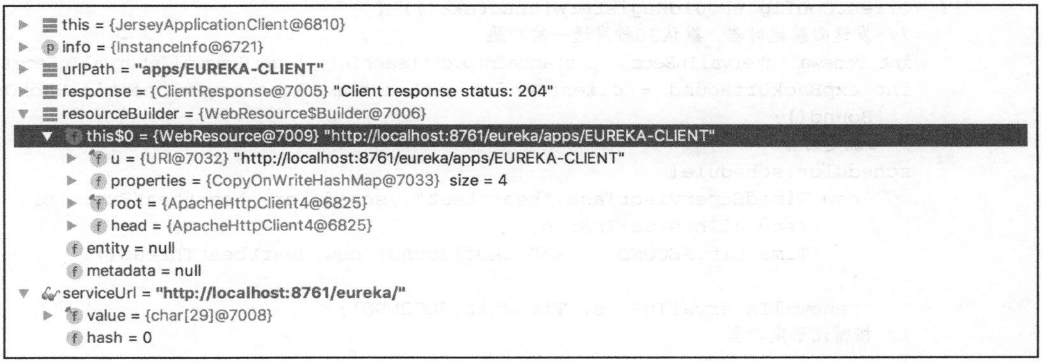

```
▶ ▦ this = {JerseyApplicationClient@6810}
▶ ⓟ info = {InstanceInfo@6721}
▶ ▦ urlPath = "apps/EUREKA-CLIENT"
▶ ▦ response = {ClientResponse@7005} "Client response status: 204"
▼ ▦ resourceBuilder = {WebResource$Builder@7006}
  ▼ ▦ this$0 = {WebResource@7009} "http://localhost:8761/eureka/apps/EUREKA-CLIENT"
    ▶ ⓕ u = {URI@7032} "http://localhost:8761/eureka/apps/EUREKA-CLIENT"
    ▶ ⓕ properties = {CopyOnWriteHashMap@7033} size = 4
    ▶ ⓕ root = {ApacheHttpClient4@6825}
    ▶ ⓕ head = {ApacheHttpClient4@6825}
    ⓕ entity = null
    ⓕ metadata = null
  ▼ ⓐ serviceUrl = "http://localhost:8761/eureka/"
    ▶ ⓕ value = {char[29]@7008}
    ⓕ hash = 0
```

图 4-10　服务注册

注册接口地址为 apps/${APP_NAME}，传递参数为 InstanceInfo，如果服务器返回 204 状态，则表明注册成功。

4.3.5 初始化定时任务

很明显，服务注册应该是一个持续的过程，Eureka Client 通过定时发送心跳的方式与 Eureka Server 进行通信，维持自己在 Server 注册表上的租约。同时 Eureka Server 注册表中的服务实例信息是动态变化的，为了保持 Eureka Client 与 Eureka Server 的注册表信息的一致性，Eureka Client 需要定时向 Eureka Server 拉取注册表信息并更新本地缓存。为了监控 Eureka Client 应用信息和状态的变化，Eureka Client 设置了一个按需注册定时器，定时检查应用信息或者状态的变化，并在发生变化时向 Eureka Server 重新注册，避免注册表中的本服务实例信息不可用。

在 DiscoveryClient#initScheduledTasks 方法中初始化了三个定时器任务，一个用于向 Eureka Server 拉取注册表信息刷新本地缓存；一个用于向 Eureka Server 发送心跳；一个用于进行按需注册的操作。代码如下所示：

```java
// DiscoveryClient.java
private void initScheduledTasks() {
    if (clientConfig.shouldFetchRegistry()) {
        // 注册表缓存刷新定时器
        // 获取配置文件中刷新间隔，默认为30s，可以通过eureka.client.registry-fetch-
           interval-seconds进行设置
        int registryFetchIntervalSeconds = clientConfig.getRegistryFetchInterval
           Seconds();
        int expBackOffBound = clientConfig.getCacheRefreshExecutorExponentialBac
           kOffBound(); scheduler.schedule(
            new TimedSupervisorTask("cacheRefresh", scheduler, cacheRefreshExecutor,
                registryFetchIntervalSeconds, TimeUnit.SECONDS, expBackOffBound,
                    new CacheRefreshThread()
            ),
            registryFetchIntervalSeconds, TimeUnit.SECONDS);
    }
    if (clientConfig.shouldRegisterWithEureka()) {
        // 发送心跳定时器，默认30秒发送一次心跳
        int renewalIntervalInSecs = instanceInfo.getLeaseInfo().getRenewalIntervalInSecs();
        int expBackOffBound = clientConfig.getHeartbeatExecutorExponentialBackOff
           Bound();
        // 心跳定时器
        scheduler.schedule(
            new TimedSupervisorTask("heartbeat", scheduler, heartbeatExecutor,
                renewalIntervalInSecs,
                TimeUnit.SECONDS, expBackOffBound, new HeartbeatThread()
            ),
            renewalIntervalInSecs, TimeUnit.SECONDS);
        // 按需注册定时器
        ...
    }
}
```

1. 缓存刷新定时任务与发送心跳定时任务

在 DiscoveryClient #initScheduledTasks 方法中，通过 ScheduledExecutorService#schedule

的方式提交缓存刷新任务和发送心跳任务,任务执行的方式为延时执行并且不循环,这两个任务的定时循环逻辑由 TimedSupervisorTask 提供实现。TimedSupervisorTask 继承了 TimerTask,提供执行定时任务的功能。它在 run 方法中定义执行定时任务的逻辑。具体代码如下所示:

```java
//TimedSupervisorTask.java
public class TimedSupervisorTask extends TimerTask {
    ...
    public void run() {
        Future future = null;
        try {
            // 执行任务
            future = executor.submit(task);
            threadPoolLevelGauge.set((long) executor.getActiveCount());
            // 等待任务执行结果
            future.get(timeoutMillis, TimeUnit.MILLISECONDS);
            // 执行完成,设置下次任务执行频率(时间间隔)
            delay.set(timeoutMillis);
            threadPoolLevelGauge.set((long) executor.getActiveCount());
        } catch (TimeoutException e) {
            // 执行任务超时
            timeoutCounter.increment();
            // 设置下次任务执行频率(时间间隔)
            long currentDelay = delay.get();
            long newDelay = Math.min(maxDelay, currentDelay * 2);
            delay.compareAndSet(currentDelay, newDelay);
        } catch (RejectedExecutionException e) {
            // 执行任务被拒绝
            // 统计被拒绝次数
            rejectedCounter.increment();
        } catch (Throwable e) {
            // 其他的异常
            // 统计异常次数
            throwableCounter.increment();
        } finally {
            // 取消未结束的任务
            if (future != null) {
                future.cancel(true);
            }
            // 如果定时任务服务未关闭,定义下一次任务
            if (!scheduler.isShutdown()) {
                scheduler.schedule(this, delay.get(), TimeUnit.MILLISECONDS);
            }
        }
    }
}
```

run 方法中存在以下的任务调度过程:

1)scheduler 初始化并延迟执行 TimedSupervisorTask;

2)TimedSupervisorTask 将 task 提 交 executor 中 执 行,task 和 executor 在 初 始 化

TimedSupervisorTask 时传入：
- 若 task 正常执行，TimedSupervisorTask 将自己提交到 scheduler，延迟 delay 时间后再次执行；
- 若 task 执行超时，计算新的 delay 时间（不超过 maxDelay），TimedSupervisorTask 将自己提交到 scheduler，延迟 delay 时间后再次执行；

执行流程图如图 4-11 所示。

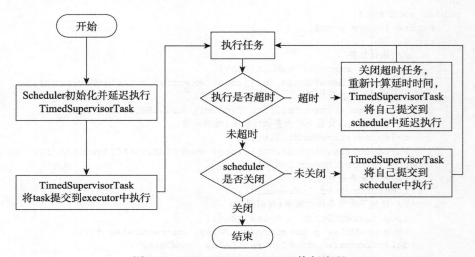

图 4-11 TimedSupervisorTask 执行流程

TimedSupervisorTask 通过这种不断循环提交任务的方式，完成定时执行任务的要求。

在 DiscoveryClient#initScheduledTasks 方法中，提交缓存刷新定时任务的线程任务为 CacheRefreshThread，提交发送心跳定时任务的线程为 HeartbeatThread。CacheRefreshThread 继承了 Runnable 接口，代码如下所示：

```
// DiscoveryClient.java
class CacheRefreshThread implements Runnable {
    public void run() {
        refreshRegistry();
    }
}
void refreshRegistry(){
    ...//判断远程Region是否改变（即Eureka Server地址是否发生变化），决定进行全量拉取还是增量式拉取
    boolean success = fetchRegistry(remoteRegionsModified);
    ...//打印更新注册表缓存后的变化
}
```

CacheRefreshThread 线程任务将委托 DiscoveryClient#fetchRegistry 方法进行缓存刷新的具体操作。

HeartbeatThread 同样继承了 Runnable 接口，该任务的作用是向 Eureka Server 发送心跳

请求,维持 Eureka Client 在注册表中的租约。代码如下所示:

```java
// DiscoveryClient.java
private class HeartbeatThread implements Runnable {
    public void run() {
        if (renew()) {
lastSuccessfulHeartbeatTimestamp = System.currentTimeMillis();
        }
    }
}
```

HeartbeatThread 主要逻辑代码位于 #renew 方法中,代码如下所示:

```java
//DiscovClient.java
boolean renew() {
    EurekaHttpResponse<InstanceInfo> httpResponse;
    try {
        // 调用HTTP发送心跳到Eureka Server中维持租约
        httpResponse = eurekaTransport.registrationClient.sendHeartBeat(instanceInfo.
            getAppName(), instanceInfo.getId(), instanceInfo, null);
        // Eureka Server中不存在该应用实例,
        if (httpResponse.getStatusCode() == 404) {
            REREGISTER_COUNTER.increment();
            // 重新注册
            return register();
        }
        // 续约成功
        return httpResponse.getStatusCode() == 200;
    } catch (Throwable e) {
        return false;
    }
}
```

Eureka Server 会根据续租提交的 appName 与 instanceInfoId 来更新注册表中的服务实例的租约。当注册表中不存在该服务实例时,将返回 404 状态码,发送心跳请求的 Eureka Client 在接收到 404 状态后将会重新发起注册;如果续约成功,将会返回 200 状态码。

跟踪到 AbstractJerseyEurekaHttpClient#sendHeartBeat 方法中,可以发现服务续租调用的接口以及传递的参数,如图 4-12 所示。

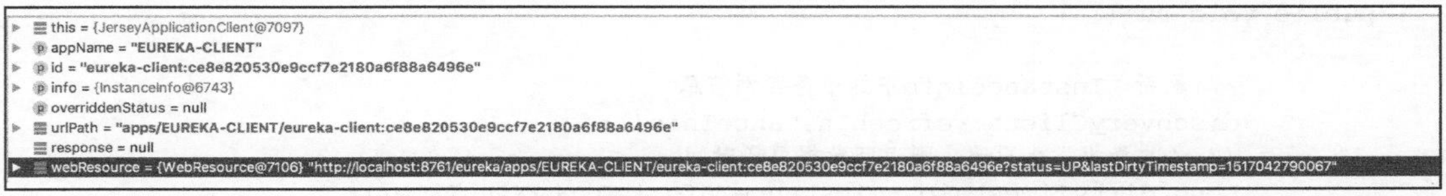

图 4-12 发送心跳

续租的接口地址为 apps/${APP_NAME}/${INSTANCE_INFO_ID},HTTP 方法为 put,参数主要有 status(当前服务的状态)、lastDirtyTimestamp(上次数据变化时间)以及 overriddenStatus。

2. 按需注册定时任务

按需注册定时任务的作用是当 Eureka Client 中的 InstanceInfo 或者 status 发生变化时，重新向 Eureka Server 发起注册请求，更新注册表中的服务实例信息，保证 Eureka Server 注册表中服务实例信息有效和可用。按需注册定时任务的代码如下：

```java
// DiscoveryClient.java
// 定时检查刷新服务实例信息，检查是否有变化，是否需要重新注册
instanceInfoReplicator = new InstanceInfoReplicator(
        this, instanceInfo, clientConfig.getInstanceInfoReplicationIntervalSeconds(), 2);
    // 监控应用的status变化，发生变化即可发起重新注册
    statusChangeListener = new ApplicationInfoManager.StatusChangeListener() {
        @Override
        public String getId() {
            return "statusChangeListener";
        }
        @Override
        public void notify(StatusChangeEvent statusChangeEvent) {
            ...
            instanceInfoReplicator.onDemandUpdate();
        }
    };
    if (clientConfig.shouldOnDemandUpdateStatusChange()) {
        // 注册应用状态改变监控器
        applicationInfoManager.registerStatusChangeListener(statusChangeListener);
    }
    // 启动定时按需注册定时任务
    instanceInfoReplicator.start(clientConfig.
        getInitialInstanceInfoReplicationIntervalSeconds());
}
```

按需注册分为两部分，一部分是定义了一个定时任务，定时刷新服务实例的信息和检查应用状态的变化，在服务实例信息发生改变的情况下向 Eureka Server 重新发起注册操作；另一部分是注册了状态改变监控器，在应用状态发生变化时，刷新服务实例信息，在服务实例信息发生改变的情况下向 Eureka Server 重新发起注册操作。InstanceInfoReplicator 中的定时任务逻辑位于 #run 方法中，如下所示：

```java
// InstanceInfoReplicator.java
public void run() {
    try {
        // 刷新了InstanceInfo中的服务实例信息
        discoveryClient.refreshInstanceInfo();
        // 如果数据发生更改，则返回数据更新时间
        Long dirtyTimestamp = instanceInfo.isDirtyWithTime();
        if (dirtyTimestamp != null) {
            // 注册服务实例
            discoveryClient.register();
            // 重置更新状态
            instanceInfo.unsetIsDirty(dirtyTimestamp);
        }
    } catch (Throwable t) {
```

```
        logger.warn("There was a problem with the instance info replicator", t);
    } finally {
        // 执行下一个延时任务
        Future next = scheduler.schedule(this, replicationIntervalSeconds,
            TimeUnit.SECONDS);
        scheduledPeriodicRef.set(next);
    }
}
```

DiscoveryClient 中刷新本地服务实例信息和检查服务状态变化的代码如下：

```
// DiscoveryClient.java
void refreshInstanceInfo() {
    // 刷新服务实例信息
    applicationInfoManager.refreshDataCenterInfoIfRequired();
    // 更新租约信息
    applicationInfoManager.refreshLeaseInfoIfRequired();

    InstanceStatus status;
    try {
        // 调用healthCheckHandler检查服务实例的状态变化
        status = getHealthCheckHandler().getStatus(instanceInfo.getStatus());
    } catch (Exception e) {
        status = InstanceStatus.DOWN;
    }

    if (null != status) {
        applicationInfoManager.setInstanceStatus(status);
    }
}
```

run 方法首先调用了 discoveryClient#refreshInstanceInfo 方法刷新当前的服务实例信息，查看当前服务实例信息和服务状态是否发生变化，如果当前服务实例信息或者服务状态发生变化将向 Eureka Server 重新发起服务注册操作。最后声明了下一个延时任务，用于再次调用 run 方法，继续检查服务实例信息和服务状态的变化，在服务实例信息发生变化的情况下重新发起注册。

如果 Eureka Client 的状态发生变化（在 Spring Boot 通过 Actuator 对服务状态进行监控，具体实现为 EurekaHealthCheckHandler），注册在 ApplicationInfoManager 的状态改变监控器将会被触发，从而调用 InstanceInfoReplicator#onDemandUpdate 方法，检查服务实例信息和服务状态的变化，可能会引发按需注册任务。代码如下所示：

```
//InstanceInfoReplicator.java
public boolean onDemandUpdate() {
    // 控制流量，当超过限制时，不能进行按需更新
    if (rateLimiter.acquire(burstSize, allowedRatePerMinute)) {
        scheduler.submit(new Runnable() {
            @Override
            public void run() {
                Future latestPeriodic = scheduledPeriodicRef.get();
                // 取消上次#run任务
```

```
            if (latestPeriodic != null && !latestPeriodic.isDone()) {
                latestPeriodic.cancel(false);
            }
            InstanceInfoReplicator.this.run();
        }
    });
    return true;
} else {
    return false;
}
}
```

InstanceInfoReplicator#onDemandUpdate 方法调用 InstanceInfoReplicator#run 方法检查服务实例信息和服务状态的变化，并在服务实例信息发生变化的情况下向 Eureka Server 发起重新注册的请求。为了防止重复执行 run 方法，onDemandUpdate 方法还会取消执行上次已提交且未完成的 run 方法，执行最新的按需注册任务。

按需注册定时任务的处理流程如图 4-13 所示。

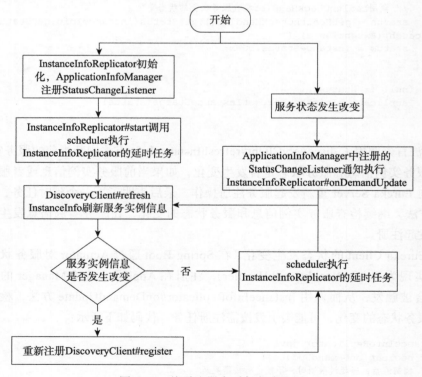

图 4-13 按需注册定时任务流程图

4.3.6 服务下线

一般情况下，应用服务在关闭的时候，Eureka Client 会主动向 Eureka Server 注销自身在注册表中的信息。DiscoveryClient 中对象销毁前执行的清理方法如下所示：

```
// DiscoveryClient.java
@PreDestroy
@Override
public synchronized void shutdown() {
    // 同步方法
    if (isShutdown.compareAndSet(false, true)) {
        // 原子操作，确保只会执行一次
        if (statusChangeListener != null && applicationInfoManager != null) {
            // 注销状态监听器
applicationInfoManager.unregisterStatusChangeListener(statusChangeListener.getId());
        }
        // 取消定时任务
        cancelScheduledTasks();
        if (applicationInfoManager != null && clientConfig.shouldRegisterWithEureka()) {
            // 服务下线
            applicationInfoManager.setInstanceStatus(InstanceStatus.DOWN);
            unregister();
        }
        // 关闭Jersy客户端
        if (eurekaTransport != null) {
            eurekaTransport.shutdown();
        }
        // 关闭相关Monitor
        heartbeatStalenessMonitor.shutdown();
        registryStalenessMonitor.shutdown();
    }
}
```

在销毁 DiscoveryClient 之前，会进行一系列清理工作，包括注销 ApplicationInfoManager 中的 StatusChangeListener、取消定时任务、服务下线和关闭 Jersey 客户端等。我们主要关注 unregister 服务下线方法，其实现代码如下所示：

```
void unregister() {
    // It can be null if shouldRegisterWithEureka == false
    if(eurekaTransport != null && eurekaTransport.registrationClient != null) {
        try {
            EurekaHttpResponse<Void> httpResponse = eurekaTransport.registrationClient.
                cancel(instanceInfo.getAppName(), instanceInfo.getId());
        } catch (Exception e) {
            ...
        }
    }
}
```

跟踪到 AbstractJerseyEurekaHttpClient#cancel 方法中，可以发现服务下线调用的接口以及传递的参数，代码如下所示：

```
@Override
public EurekaHttpResponse<Void> cancel(String appName, String id) {
    String urlPath = "apps/" + appName + '/' + id;
```

```
        ClientResponse response = null;
        try {
            Builder resourceBuilder = jerseyClient.resource(serviceUrl).path(urlPath).
                getRequestBuilder();
            addExtraHeaders(resourceBuilder);
            response = resourceBuilder.delete(ClientResponse.class);
            return anEurekaHttpResponse(response.getStatus()).headers(headersOf
                (response)).build();
        } finally {
            if (response != null) {
                response.close();
            }
        }
    }
```

服务下线的接口地址为 apps/${APP_NAME}/${INSTANCE_INFO_ID}，传递参数为服务名和服务实例 id，HTTP 方法为 delete。

4.4 Eureka Server 源码解析

Eureka Server 作为一个开箱即用的服务注册中心，提供了以下的功能，用以满足与 Eureka Client 交互的需求：

- 服务注册
- 接受服务心跳
- 服务剔除
- 服务下线
- 集群同步
- 获取注册表中服务实例信息

需要注意的是，Eureka Server 同时也是一个 Eureka Client，在不禁止 Eureka Server 的客户端行为时，它会向它配置文件中的其他 Eureka Server 进行拉取注册表、服务注册和发送心跳等操作。

4.4.1 服务实例注册表

InstanceRegistry 是 Eureka Server 中注册表管理的核心接口。它的类结构如图 4-14 所示。图中出现了两个 InstanceRegistry，最下面的 InstanceRegistry 对 Eureka Server 的注册表实现类 PeerAwareInstanceRegistryImpl 进行了继承和扩展，使其适配 Spring Cloud 的使用环境，主要实现由 PeerAwareInstanceRegistryImpl 提供。

上层的 InstanceRegistry 是 Eureka Server 注册表的最核心接口，其职责是在内存中管理注册到 Eureka Server 中的服务实例信息。

InstanceRegistry 分别继承了 LeaseManager 和 LookupService 接口。LeaseManager 接口的

功能是对注册到 Eureka Server 中的服务实例租约进行管理。而 LookupService 提供对服务实例进行检索的功能，在 Eureka Client 的源码解析中已进行介绍，在此不对其接口进行展示。

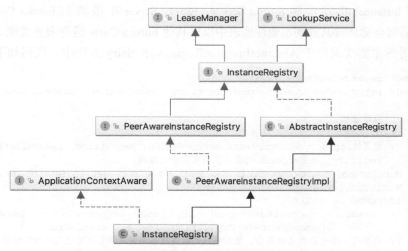

图 4-14 InstanceRegistry 类图

LeaseManager 接口提供的方法代码如下所示：

```
//LeaseManager.java
public interface LeaseManager<T> {
    void register(T r, int leaseDuration, boolean isReplication);
    boolean cancel(String appName, String id, boolean isReplication);
    boolean renew(String appName, String id, boolean isReplication);
    void evict();
}
```

LeaseManager 接口作用是对注册到 Eureka Server 中的服务实例租约进行管理，分别有服务注册、服务下线、服务租约更新以及服务剔除等操作。

LeaseManager 中管理的对象是 Lease，Lease 代表一个 Eureka Client 服务实例信息的租约，它提供了对其内持有的类的时间有效性操作。Lease 持有的类是代表服务实例信息的 InstanceInfo。Lease 中定义了租约的操作类型，分别是注册、下线、更新，同时提供了对租约中时间属性的各项操作。租约默认有效时长（duration）为 90 秒。

InstanceRegistry 在继承 LeaseManager 和 LookupService 接口的基础上，还添加了一些特有的方法，可以更为简单地管理服务实例租约和查询注册表中的服务实例信息。可以通过 AbstractInstanceRegistry 查看 InstanceRegistry 接口方法的具体实现。

PeerAwareInstanceRegistry 继承了 InstanceRegistry 接口，在其基础上添加了 Eureka Server 集群同步的操作，其实现类 PeerAwareInstanceRegistryImpl 继承了 AbstractInstanceRegistry 的实现，在对本地注册表操作的基础上添加了对其 peer 节点的同步复制操作，使得 Eureka Server 集群中的注册表信息保持一致。

4.4.2 服务注册

我们了解到 Eureka Client 在发起服务注册时会将自身的服务实例元数据封装在 InstanceInfo 中,然后将 InstanceInfo 发送到 Eureka Server。Eureka Server 在接收到 Eureka Client 发送的 InstanceInfo 后将会尝试将其放到本地注册表中以供其他 Eureka Client 进行服务发现。

服务注册的主要实现位于 AbstractInstanceRegistry#registry 方法中,代码如下所示:

```java
//AbstractInstanceRegistry.java
public void register(InstanceInfo registrant, int leaseDuration, boolean isReplication) {
    try {
        // 获取读锁
        read.lock();
        // 这里的registry是ConcurrentHashMap<String, Map<String, Lease<InstanceInfo>>>
        //   registry,根据appName对服务实例集群进行分类
        Map<String, Lease<InstanceInfo>> gMap = registry.get(registrant.getAppName());
        REGISTER.increment(isReplication);
        if (gMap == null) {
            final ConcurrentHashMap<String, Lease<InstanceInfo>> gNewMap = new
                    ConcurrentHashMap<String, Lease<InstanceInfo>>();
        // 这里有一个比较严谨的操作,防止在添加新的服务实例集群租约时把已有的其他线程添加的集群
        //   租约覆盖掉,如果存在该键值,直接返回已存在的值;否则添加该键值对,返回null
            gMap = registry.putIfAbsent(registrant.getAppName(), gNewMap);
            if (gMap == null) {
                gMap = gNewMap;
            }
        }
        //根据instanceId获取实例的租约
        Lease<InstanceInfo> existingLease = gMap.get(registrant.getId());
        if (existingLease != null && (existingLease.getHolder() != null)) {
            Long existingLastDirtyTimestamp = existingLease.getHolder().getLastDirtyTimestamp();
            Long registrationLastDirtyTimestamp = registrant.getLastDirtyTimestamp();
            // 如果该实例的租约已经存在,比较最后更新时间戳的大小,取最大值的注册信息为有效
            if (existingLastDirtyTimestamp > registrationLastDirtyTimestamp) {
                registrant = existingLease.getHolder();
            }
```

在 register 中,服务实例的 InstanceInfo 保存在 Lease 中,Lease 在 AbstractInstanceRegistry 中统一通过 ConcurrentHashMap 保存在内存中。在服务注册过程中,会先获取一个读锁,防止其他线程对 registry 注册表进行数据操作,避免数据的不一致。然后从 resgitry 查询对应的 InstanceInfo 租约是否已经存在注册表中,根据 appName 划分服务集群,使用 InstanceId 唯一标记服务实例。如果租约存在,比较两个租约中的 InstanceInfo 的最后更新时间 lastDirtyTimestamp,保留时间戳大的服务实例信息 InstanceInfo。如果租约不存在,意味这是一次全新的服务注册,将会进行自我保护的统计,创建新的租约保存 InstanceInfo。接着将租约放到 resgitry 注册表中。

之后将进行一系列缓存操作并根据覆盖状态规则设置服务实例的状态,缓存操作包括将 InstanceInfo 加入用于统计 Eureka Client 增量式获取注册表信息的 recentlyChangedQueue

和失效 responseCache 中对应的缓存。最后设置服务实例租约的上线时间用于计算租约的有效时间,释放读锁并完成服务注册。代码如下所示:

```java
// AbstractInstanceRegistry.java
        } else {
            // 如果租约不存在,这是一个新的注册实例
            synchronized (lock) {
                if (this.expectedNumberOfRenewsPerMin > 0) {
                // 自我保护机制
                    this.expectedNumberOfRenewsPerMin = this.expectedNumberOfRenewsPerMin + 2;
                    this.numberOfRenewsPerMinThreshold =
                    (int) (this.expectedNumberOfRenewsPerMin * serverConfig.
                        getRenewalPercentThreshold());
                }
            }
        }
        // 创建新的租约
        Lease<InstanceInfo> lease = new Lease<InstanceInfo>(registrant, leaseDuration);
        if (existingLease != null) {
            // 如果租约存在,继承租约的服务上线初始时间
            lease.setServiceUpTimestamp(existingLease.getServiceUpTimestamp());
        }
        // 保存租约
        gMap.put(registrant.getId(), lease);
        // 添加最近注册队列
        // private final CircularQueue<Pair<Long, String>> recentRegisteredQueue
        // 用来统计最近注册服务实例的数据
        synchronized (recentRegisteredQueue) {
            recentRegisteredQueue.add(new Pair<Long, String>(
                System.currentTimeMillis(),
                registrant.getAppName() + "(" + registrant.getId() + ")"));
        }
        ...
        // 根据覆盖状态规则得到服务实例的最终状态,并设置服务实例的当前状态
        InstanceStatus overriddenInstanceStatus = getOverriddenInstanceStatus(re
            gistrant, existingLease, isReplication);
        registrant.setStatusWithoutDirty(overriddenInstanceStatus);
        // 如果服务实例状态为UP,设置租约的服务上线时间,只有第一次设置有效
        if (InstanceStatus.UP.equals(registrant.getStatus())) {
            lease.serviceUp();
        }
        registrant.setActionType(ActionType.ADDED);
        // 添加最近租约变更记录队列,标识ActionType为ADDED
        // 这将用于Eureka Client增量式获取注册表信息
        // private ConcurrentLinkedQueue<RecentlyChangedItem>
        recentlyChangedQueue
        recentlyChangedQueue.add(new RecentlyChangedItem(lease));
        // 设置服务实例信息更新时间
        registrant.setLastUpdatedTimestamp();
        // 设置response缓存过期,这将用于Eureka Client全量获取注册表信息
        invalidateCache(registrant.getAppName(), registrant.getVIPAddress(), registrant.
            getSecureVipAddress());
```

```
        } finally {
            // 释放锁
            read.unlock();
        }
}
```

在register方法中有诸多的同步操作,为了防止数据被错误地覆盖,有兴趣的读者可以细细研究一下,在此不再展开讲述。

4.4.3 接受服务心跳

在Eureka Client完成服务注册之后,它需要定时向Eureka Server发送心跳请求(默认30秒一次),维持自己在Eureka Server中租约的有效性。

Eureka Server处理心跳请求的核心逻辑位于AbstractInstanceRegistry#renew方法中。renew方法是对Eureka Client位于注册表中的租约的续租操作,不像register方法需要服务实例信息,仅根据服务实例的服务名和服务实例id即可更新对应租约的有效时间。具体代码如下所示:

```
// AbstractInstanceRegistry.java
public boolean renew(String appName, String id, boolean isReplication) {
    RENEW.increment(isReplication);
    // 根据appName获取服务集群的租约集合
    Map<String, Lease<InstanceInfo>> gMap = registry.get(appName);
    Lease<InstanceInfo> leaseToRenew = null;
    if (gMap != null) {
    leaseToRenew = gMap.get(id);
    }
    // 租约不存在,直接返回false
    if (leaseToRenew == null) {
        RENEW_NOT_FOUND.increment(isReplication);
        return false;
    } else {
        InstanceInfo instanceInfo = leaseToRenew.getHolder();
        if (instanceInfo != null) {
            // 根据覆盖状态规则得到服务实例的最终状态
            InstanceStatus overriddenInstanceStatus = this.getOverriddenInstance
                Status(instanceInfo, leaseToRenew, isReplication);
            if (overriddenInstanceStatus == InstanceStatus.UNKNOWN) {
                // 如果得到的服务实例最后状态是UNKNOWN,取消续约
                RENEW_NOT_FOUND.increment(isReplication);
                return false;
            }
            if (!instanceInfo.getStatus().equals(overriddenInstanceStatus)) {
                instanceInfo.setStatus(overriddenInstanceStatus);
            }
        }
        // 统计每分钟续租的次数,用于自我保护
        renewsLastMin.increment();
        // 更新租约中的有效时间
        leaseToRenew.renew();
```

```
        return true;
    }
}
```

在 #renew 方法中，不关注 InstanceInfo，仅关注于租约本身以及租约的服务实例状态。如果根据服务实例的 appName 和 instanceInfoId 查询出服务实例的租约，并且根据 #getOverriddenInstanceStatus 方法得到的 instanceStatus 不为 InstanceStatus.UNKNOWN，那么更新租约中的有效时间，即更新租约 Lease 中的 lastUpdateTimestamp，达到续约的目的；如果租约不存在，那么返回续租失败的结果。

4.4.4 服务剔除

如果 Eureka Client 在注册后，既没有续约，也没有下线（服务崩溃或者网络异常等原因），那么服务的状态就处于不可知的状态，不能保证能够从该服务实例中获取到回馈，所以需要服务剔除 AbstractInstanceRegistry#evict 方法定时清理这些不稳定的服务，该方法会批量将注册表中所有过期租约剔除。实现代码如下所示：

```
// AbstractInstanceRegistry.java
@Override
public void evict() {
    evict(0l);
}
public void evict(long additionalLeaseMs) {
    // 自我保护相关，如果出现该状态，不允许剔除服务
    if (!isLeaseExpirationEnabled()) {
        return;
    }
    // 遍历注册表register，一次性获取所有的过期租约
    List<Lease<InstanceInfo>> expiredLeases = new ArrayList<>();
    for (Entry<String, Map<String, Lease<InstanceInfo>>> groupEntry : registry.
        entrySet()) {
        Map<String, Lease<InstanceInfo>> leaseMap = groupEntry.getValue();
        if (leaseMap != null) {
            for (Entry<String, Lease<InstanceInfo>> leaseEntry : leaseMap.
                entrySet()) {
                Lease<InstanceInfo> lease = leaseEntry.getValue();
                // 1
                if (lease.isExpired(additionalLeaseMs) && lease.getHolder() != null) {
                    expiredLeases.add(lease);
                }
            }
        }
    }
    // 计算最大允许剔除的租约的数量，获取注册表租约总数
    int registrySize = (int) getLocalRegistrySize();
    // 计算注册表租约的阈值，与自我保护相关
    int registrySizeThreshold = (int) (registrySize * serverConfig.getRenewalPercentThreshold());
    int evictionLimit = registrySize - registrySizeThreshold;
    // 计算剔除租约的数量
```

```java
        int toEvict = Math.min(expiredLeases.size(), evictionLimit);
        if (toEvict > 0) {
            Random random = new Random(System.currentTimeMillis());
            // 逐个随机剔除
            for (int i = 0; i < toEvict; i++) {
                int next = i + random.nextInt(expiredLeases.size() - i);
                Collections.swap(expiredLeases, i, next);
                Lease<InstanceInfo> lease = expiredLeases.get(i);
                String appName = lease.getHolder().getAppName();
                String id = lease.getHolder().getId();
                EXPIRED.increment();
                // 逐个剔除
                internalCancel(appName, id, false);
            }
        }
    }
}
```

服务剔除将会遍历 registry 注册表，找出其中所有的过期租约，然后根据配置文件中续租百分比阀值和当前注册表的租约总数量计算出最大允许的剔除租约的数量（当前注册表中租约总数量减去当前注册表租约阀值），分批次剔除过期的服务实例租约。对过期的服务实例租约调用 AbstractInstanceRegistry#internalCancel 服务下线的方法将其从注册表中清除掉。

服务剔除 #evict 方法中有很多限制，都是为了保证 Eureka Server 的可用性：

- 自我保护时期不能进行服务剔除操作。
- 过期操作是分批进行。
- 服务剔除是随机逐个剔除，剔除均匀分布在所有应用中，防止在同一时间内同一服务集群中的服务全部过期被剔除，以致大量剔除发生时，在未进行自我保护前促使了程序的崩溃。

服务剔除是一个定时的任务，所以 AbstractInstanceRegistry 中定义了一个 EvictionTask 用于定时执行服务剔除，默认为 60 秒一次。服务剔除的定时任务一般在 AbstractInstanceRegistry 初始化结束后进行，按照执行频率 evictionIntervalTimerInMs 的设定，定时剔除过期的服务实例租约。

自我保护机制主要在 Eureka Client 和 Eureka Server 之间存在网络分区的情况下发挥保护作用，在服务器端和客户端都有对应实现。假设在某种特定的情况下（如网络故障），Eureka Client 和 Eureka Server 无法进行通信，此时 Eureka Client 无法向 Eureka Server 发起注册和续约请求，Eureka Server 中就可能因注册表中的服务实例租约出现大量过期而面临被剔除的危险，然而此时的 Eureka Client 可能是处于健康状态的（可接受服务访问），如果直接将注册表中大量过期的服务实例租约剔除显然是不合理的。

针对这种情况，Eureka 设计了"自我保护机制"。在 Eureka Server 处，如果出现大量的服务实例过期被剔除的现象，那么该 Server 节点将进入自我保护模式，保护注册表中的信息不再被剔除，在通信稳定后再退出该模式；在 Eureka Client 处，如果向 Eureka Server

注册失败，将快速超时并尝试与其他的 Eureka Server 进行通信。"自我保护机制"的设计大大提高了 Eureka 的可用性。

4.4.5 服务下线

Eureka Client 在应用销毁时，会向 Eureka Server 发送服务下线请求，清除注册表中关于本应用的租约，避免无效的服务调用。在服务剔除的过程中，也是通过服务下线的逻辑完成对单个服务实例过期租约的清除工作。

服务下线的主要实现代码位于 AbstractInstanceRegistry#internalCancel 方法中，仅需要服务实例的服务名和服务实例 id 即可完成服务下线。具体代码如下所示：

```java
// AbstractInstanceRegistry.java
@Override
public boolean cancel(String appName, String id, boolean isReplication) {
    return internalCancel(appName, id, isReplication);
}

protected boolean internalCancel(String appName, String id, boolean isReplication) {
    try {
        // 获取读锁，防止被其他线程进行修改
        read.lock();
        CANCEL.increment(isReplication);
        // 根据appName获取服务实例的集群
        Map<String, Lease<InstanceInfo>> gMap = registry.get(appName);
        Lease<InstanceInfo> leaseToCancel = null;
        // 移除服务实例的租约
        if (gMap != null) {
            leaseToCancel = gMap.remove(id);
        }
        // 将服务实例信息添加到最近下线服务实例统计队列
        synchronized (recentCanceledQueue) {
            recentCanceledQueue.add(new Pair<Long, String>(System.currentTimeMillis(),
                appName + "(" + id + ")"));
        }
        // 租约不存在，返回false
        if (leaseToCancel == null) {
            CANCEL_NOT_FOUND.increment(isReplication);
            return false;
        } else {
            // 设置租约的下线时间
            leaseToCancel.cancel();
            InstanceInfo instanceInfo = leaseToCancel.getHolder();
            ...
            if (instanceInfo != null) {
                instanceInfo.setActionType(ActionType.DELETED);
                // 添加最近租约变更记录队列，标识ActionType为DELETED
                // 这将用于Eureka Client增量式获取注册表信息
                recentlyChangedQueue.add(new RecentlyChangedItem(leaseToCancel));
                instanceInfo.setLastUpdatedTimestamp();
```

```
            }
            // 设置response缓存过期
            invalidateCache(appName, vip, svip);
            // 下线成功
            return true;
        }
    } finally {
        // 释放锁
        read.unlock();
    }
}
```

internalCancel 方法与 register 方法的行为过程很类似，首先通过 registry 根据服务名和服务实例 id 查询关于服务实例的租约 Lease 是否存在，统计最近请求下线的服务实例用于 Eureka Server 主页展示。如果租约不存在，返回下线失败；如果租约存在，从 registry 注册表中移除，设置租约的下线时间，同时在最近租约变更记录队列中添加新的下线记录，以用于 Eureka Client 的增量式获取注册表信息，最后设置 repsonse 缓存过期。

internalCancel 方法中同样通过读锁保证 registry 注册表中数据的一致性，避免脏读。

4.4.6 集群同步

如果 Eureka Server 是通过集群的方式进行部署，那么为了维护整个集群中 Eureka Server 注册表数据的一致性，势必需要一个机制同步 Eureka Server 集群中的注册表数据。

Eureka Server 集群同步包含两个部分，一部分是 Eureka Server 在启动过程中从它的 peer 节点中拉取注册表信息，并将这些服务实例的信息注册到本地注册表中；另一部分是 Eureka Server 每次对本地注册表进行操作时，同时会将操作同步到它的 peer 节点中，达到集群注册表数据统一的目的。

1. Eureka Server 初始化本地注册表信息

在 Eureka Server 启动的过程中，会从它的 peer 节点中拉取注册表来初始化本地注册表，这部分主要通过 PeerAwareInstanceRegistry#syncUp 方法完成，它将从可能存在的 peer 节点中，拉取 peer 节点中的注册表信息，并将其中的服务实例信息注册到本地注册表中，如下所示：

```
// PeerAwareInstanceRegistry.java
public int syncUp() {
    // 从临近的peer中复制整个注册表
    int count = 0;
    // 如果获取不到，线程等待
    for (int i = 0; ((i < serverConfig.getRegistrySyncRetries()) && (count == 0));
        i++) {
        if (i > 0) {
            try {
                Thread.sleep(serverConfig.getRegistrySyncRetryWaitMs());
            } catch (InterruptedException e) {
```

```java
                        break;
                }
            }
            // 获取所有的服务实例
            Applications apps = eurekaClient.getApplications();
            for (Application app : apps.getRegisteredApplications()) {
                for (InstanceInfo instance : app.getInstances()) {
                    try {
                        // 判断是否可注册,主要用于AWS环境下进行,若部署在其他的环境,直接返回true
                        if (isRegisterable(instance)) {
                            // 注册到自身的注册表中
                            register(instance, instance.getLeaseInfo().getDurationInSecs(),
                                true);
                            count++;
                        }
                    } catch (Throwable t) {
                        logger.error("During DS init copy", t);
                    }
                }
            }
        }
        return count;
    }
```

Eureka Server 也是一个 Eureka Client,在启动的时候也会进行 DiscoveryClient 的初始化,会从其对应的 Eureka Server 中拉取全量的注册表信息。在 Eureka Server 集群部署的情况下,Eureka Server 从它的 peer 节点中拉取到注册表信息后,将遍历这个 Applications,将所有的服务实例通过 AbstractRegistry#register 方法注册到自身注册表中。

在初始化本地注册表时,Eureka Server 并不会接受来自 Eureka Client 的通信请求(如注册、或者获取注册表信息等请求)。在同步注册表信息结束后会通过 PeerAwareInstanceRegistryImpl#openForTraffic 方法允许该 Server 接受流量。代码如下所示:

```java
// PeerAwareInstanceRegistryImpl.java
public void openForTraffic(ApplicationInfoManager applicationInfoManager, int count) {
    // 初始化自我保护机制统计参数
    this.expectedNumberOfRenewsPerMin = count * 2;
    this.numberOfRenewsPerMinThreshold = (int) (this.expectedNumberOfRenewsPerMin *
        serverConfig.getRenewalPercentThreshold());

    this.startupTime = System.currentTimeMillis();
    // 如果同步的应用实例数量为0,将在一段时间内拒绝Client获取注册信息
    if (count > 0) {
        this.peerInstancesTransferEmptyOnStartup = false;
    }
    DataCenterInfo.Name selfName = applicationInfoManager.getInfo().getDataCenterInfo().
        getName();
    boolean isAws = Name.Amazon == selfName;
    // 判断是否是AWS运行环境,此处忽略
    if (isAws && serverConfig.shouldPrimeAwsReplicaConnections()) {
        primeAwsReplicas(applicationInfoManager);
```

```
        }
        // 修改服务实例的状态为健康上线，可以接受流量
        applicationInfoManager.setInstanceStatus(InstanceStatus.UP);
        super.postInit();
    }
```

在 Eureka Server 中有一个 StatusFilter 过滤器，用于检查 Eureka Server 的状态，当 Server 的状态不为 UP 时，将拒绝所有的请求。在 Client 请求获取注册表信息时，Server 会判断此时是否允许获取注册表中的信息。上述做法是为了避免 Eureka Server 在 #syncUp 方法中没有获取到任何服务实例信息时（Eureka Server 集群部署的情况下），Eureka Server 注册表中的信息影响到 Eureka Client 缓存的注册表中的信息。如果 Eureka Server 在 #syncUp 方法中没有获得任何的服务实例信息，它将把 peerInstancesTransferEmptyOnStartup 设置为 true，这时该 Eureka Server 在 WaitTimeInMsWhenSyncEmpty（可以通过 eureka.server.wait-time-in-ms-when-sync-empty 设置，默认是 5 分钟）时间后才能被 Eureka Client 访问获取注册表信息。

2. Eureka Server 之间注册表信息的同步复制

为了保证 Eureka Server 集群运行时注册表信息的一致性，每个 Eureka Server 在对本地注册表进行管理操作时，会将相应的操作同步到所有 peer 节点中。

在 PeerAwareInstanceRegistryImpl 中，对 AbstractInstanceRegistry 中的 #register、#cancel 和 #renew 等方法都添加了同步到 peer 节点的操作，使 Server 集群中注册表信息保持最终一致性，如下所示：

```java
//PeerAwareInstanceRegistryImpl.java
@Override
public boolean cancel(final String appName, final String id, final boolean isReplication) {
    if (super.cancel(appName, id, isReplication)) {
            // 同步下线状态
            replicateToPeers(Action.Cancel, appName, id, null, null, isReplication);
            ...
    }
    ...
}
public void register(final InstanceInfo info, final boolean isReplication) {
    int leaseDuration = Lease.DEFAULT_DURATION_IN_SECS;
    if (info.getLeaseInfo() != null && info.getLeaseInfo().getDurationInSecs() > 0) {
            leaseDuration = info.getLeaseInfo().getDurationInSecs();
    }
    super.register(info, leaseDuration, isReplication);
    // 同步注册状态
    replicateToPeers(Action.Register, info.getAppName(), info.getId(), info,
        null, isReplication);
}

public boolean renew(final String appName, final String id, final boolean isReplication) {
    if (super.renew(appName, id, isReplication)) {
            // 同步续约状态
            replicateToPeers(Action.Heartbeat, appName, id, null, null, isReplication);
```

```
        return true;
    }
    return false;
}
```

同步的操作主要有：

```
public enum Action {
    Heartbeat, Register, Cancel, StatusUpdate, DeleteStatusOverride;
    ...
}
```

对此需要关注 replicateToPeers 方法，它将遍历 Eureka Server 中 peer 节点，向每个 peer 节点发送同步请求。代码如下所示：

```
//PeerAwareInstanceRegistryImpl.java
private void replicateToPeers(Action action, String appName, String id,
                              InstanceInfo info /* optional */,
                              InstanceStatus newStatus /* optional */, boolean
                                  isReplication) {
    Stopwatch tracer = action.getTimer().start();
    try {
        if (isReplication) {
            numberOfReplicationsLastMin.increment();
        }
        // 如果peer集群为空，或者这本来就是复制操作，那么就不再复制，防止造成循环复制
        if (peerEurekaNodes == Collections.EMPTY_LIST || isReplication) {
            return;
        }
        // 向peer集群中的每一个peer进行同步
        for (final PeerEurekaNode node : peerEurekaNodes.getPeerEurekaNodes()) {
            // 如果peer节点是自身的话，不进行同步复制
            if (peerEurekaNodes.isThisMyUrl(node.getServiceUrl())) {
                continue;
            }
            // 根据Action调用不同的同步请求
            replicateInstanceActionsToPeers(action, appName, id, info, newStatus,
                node);
        }
    } finally {
        tracer.stop();
    }
}
```

PeerEurekaNode 代表一个可同步共享数据的 Eureka Server。在 PeerEurekaNode 中，具有 register、cancel、heartbeat 和 statusUpdate 等诸多用于向 peer 节点同步注册表信息的操作。

在 replicateInstanceActionsToPeers 方法中将根据 action 的不同，调用 PeerEurekaNode 的不同方法进行同步复制，代码如下所示：

```
//PeerAwareInstanceRegistryImpl.java
```

```java
    private void replicateInstanceActionsToPeers(Action action, String appName,
        String id, InstanceInfo info, InstanceStatus newStatus, PeerEurekaNode node) {
        try {
            InstanceInfo infoFromRegistry = null;
            CurrentRequestVersion.set(Version.V2);
            switch (action) {
                case Cancel:
                    // 同步下线
                    node.cancel(appName, id);
                    break;
                case Heartbeat:
                    InstanceStatus overriddenStatus = overriddenInstanceStatusMap.get(id);
                    infoFromRegistry = getInstanceByAppAndId(appName, id, false);
                    // 同步心跳
                    node.heartbeat(appName, id, infoFromRegistry, overriddenStatus, false);
                    break;
                case Register:
                    // 同步注册
                    node.register(info);
                    break;
                case StatusUpdate:
                    infoFromRegistry = getInstanceByAppAndId(appName, id, false);
                    // 同步状态更新
                    node.statusUpdate(appName, id, newStatus, infoFromRegistry);
                    break;
                case DeleteStatusOverride:
                    infoFromRegistry = getInstanceByAppAndId(appName, id, false);
                    node.deleteStatusOverride(appName, id, infoFromRegistry);
                    break;
            }
        } catch (Throwable t) {
            logger.error("Cannot replicate information to {} for action {}", node.
                getServiceUrl(), action.name(), t);
        }
    }
```

PeerEurekaNode 中的每一个同步复制都是通过批任务流的方式进行操作，同一时间段内相同服务实例的相同操作将使用相同的任务编号，在进行同步复制的时候根据任务编号合并操作，减少同步操作的数量和网络消耗，但是同时也造成同步复制的延时性，不满足 CAP 中的 C（强一致性）。

通过 Eureka Server 在启动过程中初始化本地注册表信息和 Eureka Server 集群间的同步复制操作，最终达到了集群中 Eureka Server 注册表信息一致的目的。

4.4.7 获取注册表中服务实例信息

Eureka Server 中获取注册表的服务实例信息主要通过两个方法实现：AbstractInstanceRegistry#getApplicationsFromMultipleRegions 从多地区获取全量注册表数据，AbstractInstanceRegistry#getApplicationDeltasFromMultipleRegions 从多地区获取增量式注册表数据。

1. getApplicationsFromMultipleRegions

getApplicationsFromMultipleRegions 方法将会从多个地区中获取全量注册表信息，并封装成 Applications 返回，实现代码如下所示：

```
//AbstractInstanceRegistry.java
public Applications getApplicationsFromMultipleRegions(String[] remoteRegions) {
    boolean includeRemoteRegion = null != remoteRegions && remoteRegions.length != 0;
    Applications apps = new Applications();
    apps.setVersion(1L);
    // 从本地registry获取所有的服务实例信息InstanceInfo
    for (Entry<String, Map<String, Lease<InstanceInfo>>> entry : registry.entrySet()) {
        Application app = null;
        if (entry.getValue() != null) {
            for (Entry<String, Lease<InstanceInfo>> stringLeaseEntry : entry.
                getValue().entrySet()) {
                Lease<InstanceInfo> lease = stringLeaseEntry.getValue();
                if (app == null) {
                    app = new Application(lease.getHolder().getAppName());
                }
                app.addInstance(decorateInstanceInfo(lease));
            }
        }
        if (app != null) {
            apps.addApplication(app);
        }
    }
    if (includeRemoteRegion) {
        // 获取远程Region中的Eureka Server中的注册表信息
        ...
    }
    apps.setAppsHashCode(apps.getReconcileHashCode());
    return apps;
}
```

它首先会将本地注册表 registry 中的所有服务实例信息提取出来封装到 Applications 中，再根据是否需要拉取远程 Region 中的注册表信息，将远程 Region 的 Eureka Server 注册表中的服务实例信息添加到 Applications 中。最后将封装了全量注册表信息的 Applications 返回给 Client。

2. getApplicationDeltasFromMultipleRegions

getApplicationDeltasFromMultipleRegions 方法将会从多个地区中获取增量式注册表信息，并封装成 Applications 返回，实现代码如下所示：

```
//AbstractInstanceRegistry.java
public Applications getApplicationDeltasFromMultipleRegions(String[] remoteRegions) {
    if (null == remoteRegions) {
        remoteRegions = allKnownRemoteRegions; // null means all remote regions.
    }

    boolean includeRemoteRegion = remoteRegions.length != 0;
```

```
Applications apps = new Applications();
apps.setVersion(responseCache.getVersionDeltaWithRegions().get());
Map<String, Application> applicationInstancesMap = new HashMap<String, Application>();
try {
    write.lock();// 开启写锁
    // 遍历recentlyChangedQueue队列获取最近变化的服务实例信息InstanceInfo
    Iterator<RecentlyChangedItem> iter = this.recentlyChangedQueue.iterator();
    while (iter.hasNext()) {
        //...
    }
    if (includeRemoteRegion) {
        // 获取远程Region中的Eureka Server的增量式注册表信息
        ...
    } finally {
        write.unlock();
    }
}
// 计算应用集合一致性哈希码,用以在Eureka Client拉取时进行对比
apps.setAppsHashCode(apps.getReconcileHashCode());
return apps;
}
```

获取增量式注册表信息将会从 recentlyChangedQueue 中获取最近变化的服务实例信息。recentlyChangedQueue 中统计了近 3 分钟内进行注册、修改和剔除的服务实例信息,在服务注册 AbstractInstanceRegistry#registry、接受心跳请求 AbstractInstanceRegistry#renew 和服务下线 AbstractInstanceRegistry#internalCancel 等方法中均可见到 recentlyChangedQueue 对这些服务实例进行登记,用于记录增量式注册表信息。#getApplicationsFromMultipleRegions 方法同样提供了从远程 Region 的 Eureka Server 获取增量式注册表信息的能力。

4.5 进阶应用

4.5.1 Eureka Instance 和 Client 的元数据

在 EurekaInstanceConfigBean 中,相当大一部分内容是关于 Eureka Client 服务实例的信息,这部分信息称为元数据,它是用来描述自身服务实例的相关信息。Eureka 中的标准元数据有主机名、IP 地址、端口号、状态页 url 和健康检查 url 等用于服务注册与发现的重要信息。开发者可以自定义元数据,这部分额外的数据可以通过键值对(key-value)的形式放在 eureka.instance.metadataMap,如下所示:

```
eureka:
    instance:
        metadataMap:
            metadata-map
                mymetaData: mydata
```

这里定义了一个键为 mymetaData,值为 mydata 的自定义元数据,metadata-map 在 EurekaInstanceConfigBean 会被配置为以下的属性:

```
private Map<String, String> metadataMap = new HashMap<>();
```

这些自定义的元数据可以按照自身业务需要或者根据其他的特殊需要进行定制。

4.5.2 状态页和健康检查页端口设置

Eureka 服务实例状态页和健康检查页的默认 url 是 /actuator/info 和 /actuator/health，通常是使用 spring-boot-actuator 中相关的端点提供实现。一般情况下这些端点的配置都不需要修改，但是当 spring-boot 没有使用默认的应用上下文路径（context path）或者主分发器路径（Dispatch path）时，将会影响 Eureka Server 无法通过 /actuator/health 对 Eureka Client 进行健康检查，以及无法通过 /actuator/info 访问 Eureka Client 的信息接口。例如设置成应用上下文路径为：

```
server:
    servlet:
        context-path: /path
```

或者设置主分发器路径为：

```
server:
    servlet:
        path: /path
```

为此需要对这些端点的 URL 进行更改，如下所示：

```
server:
servlet:
        path: /path

eureka:
    instance:
        statusPageUrlPath: ${server.servlet.path}/actuator/info
        healthCheckUrlPath: ${server.servlet.path}/actuator/health
```

同样可以通过绝对路径的方式进行更改。

4.5.3 区域与可用区

在基础应用中，我们介绍了 AWS 的区域以及可用区的概念。

一般来说一个 Eureka Client 只属于一个区域，一个区域下有多个可用区，每个可用区下可能有多个 Eureka Server（一个 Eureka Server 可以属于多个可用区）。

```
eureka:
    client:
        region: us-east-1
        availability-zones:
            us-east-1: us-east-zone-1, us-east-zone-2
            us-west-2: us-west-zone-1, us-west-zone-2
        service-url:
```

```
                us-east-zone-1: http://xxx1,http://xxx2
                us-east-zone-2: http://xxx1,http://xxx2
                us-west-zone-1: http://xxx1,http://xxx2
                us-west-zone-2: http://xxx1,http://xxx2
```

以上是一份比较完整的关于区域与可用区的配置。获取 serverUrls 的过程是层层递进的，从区域到可用区，优先添加服务实例所处的可用区的 serverUrls，为了保证容错性，又把本区域中的其他的可用区也添加到了 serverUrls。通过这种多层次的设计，提供 Eureka 在区域内的容错性，保证了网络分区容忍性（Partition tolerance）。

当然也可以直接告诉 Eureka Client 服务实例所处的可用区，并希望使用同一个可用区的 Eureka Server 进行注册（在 Eureka Client 无法与 Eureke Server 进行通信时，它将轮询向配置中其他的 Eureka Server 注册直到成功为止），可以添加如下配置：

```
eureka:
    instance:
        metadataMap:
            zone: us-east-zone-2
    client:
        prefer-same-zone-eureka: true
```

这样配置的话，使用的 Eureka Server 的 url 将会是 us-east-zone-2: http://xxx1,http://xxx2。

4.5.4　高可用性服务注册中心

Eureka Server 可以变得更有弹性和具备高可用性，通过部署多个注册中心实例，并让它们之间互相注册。在 Standalone 模式中，只能依赖 Server 和 Client 之间的缓存，并需要弹性的机制保证 Server 实例一直存活，单例的注册中心崩溃了，Client 之间就很难互相发现和调用。

在配置文件中添加如下配置：

```
---
spring:
    profiles: peer1
    application:
        name: eureka-server-peer
server:
    port: 8762
eureka:
    instance:
        hostname: peer1
        instance-id: ${spring.application.name}:${vcap.application.instance_id:$
            {spring.application.instance_id:${random.value}}}
    client:
        service-url:
            defaultZone: http://localhost:8763/eureka/
---
spring:
    profiles: peer2
```

```yaml
    application:
        name: eureka-server-peer
server:
    port: 8763
eureka:
    instance:
        hostname: peer2
            instance-id: ${spring.application.name}:${vcap.application.instance_id:$
                {spring.application.instance_id:${random.value}}}
    client:
        service-url:
            defaultZone: http://localhost:8762/eureka/
---
spring:
    profiles:
        active: peer1
```

可以通过设置不同的 spring.profiles.active 启动不同配置的 Eureka Server，上述配置声明了两个 Eureka Server 的配置，它们之间是相互注册的。可以添加多个 peer，只要这些 Eureka Server 中存在一个连通点，那么这些注册中心的数据就能够进行同步，这就通过服务器的冗余增加了高可用性，即使其中一台 Eureka Server 宕机了，也不会导致系统崩溃。

同时对 Eureka Client 添加多个注册节点，使得其能够尝试向其他的注册中心发起请求（当注册节点前面的 Eureka Server 无法通信时），如下所示：

```yaml
eureka:
    client:
        service-url:
            defaultZone: http://localhost:8761/eureka/, http://localhost:8762/eureka/
```

4.6 本章小结

Eureka 为 Spring Cloud 提供了高可用的服务发现与注册组件，利用 Eureka，Spring Cloud 开发者能够更快地融入到微服务的开发中。Eureka Server 作为服务注册中心，为 Eureka Client 提供服务注册和服务发现的能力，它既可单机部署，也可以通过集群的方式进行部署，通过自我保护机制和集群同步复制机制保证 Eureka 的高可用性和网络分区容忍性，保证 Eureka Server 集群的注册表数据的最终一致性；Eureka Client 方便了与 Eureka Server 的交互，它与 Eureka Server 的一切交互，包括服务注册、发送心跳续租、服务下线和服务发现，都是在后台自主完成的，简化了开发者的开发工作。

当然 Eureka 也存在缺陷。由于集群间的同步复制是通过 HTTP 的方式进行，基于网络的不可靠性，集群中的 Eureka Server 间的注册表信息难免存在不同步的时间节点，不满足 CAP 中的 C（数据一致性）。

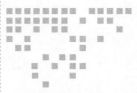

Chapter 5 第 5 章

声明式 RESTful 客户端：Spring Cloud OpenFeign

OpenFeign 是一个声明式 RESTful 网络请求客户端，使得编写 Web 服务客户端更加方便和快捷。只需要使用 OpenFeign 提供的注解修饰定义网络请求的接口类，就可以使用该接口的实例发送 RESTful 风格的网络请求。OpenFeign 还可以集成 Ribbon 和 Hytrix 来提供负载均衡和网络断路器的功能。

在本章中，第一小节主要讲解了微服务交互的常见方式以及 OpenFeign 的基础应用；第二小节对 OpenFeign 的源码进行了分析，分别讲述了动态注册 Spring 实例定义（BeanDefinition）、实例初始化和函数调用与网络请求三部分的代码实现逻辑；第三小节则介绍了 OpenFeign 相关的高级进阶用法。

5.1 基础应用

5.1.1 微服务之间的交互

微服务提倡将一个原本独立的系统分成众多小型服务系统，这些小型服务系统都在独立的进程中运行，通过各个小型服务系统之间的协作来实现原本独立系统的所有业务功能。小型服务系统使用多种跨进程的方式进行通信协作，而 RESTful 风格的网络请求是最为常见的交互方式之一。

RESTful 网络请求是指 RESTful 风格的网络请求，其中 REST 是 Resource Representational State Transfer 的缩写，直接翻译即"资源表现层状态转移"。

Resource 代表互联网资源。所谓"资源"是网络上的一个实体，或者说网上的一个具体信息。它可以是一段文本、一首歌曲、一种服务，可以使用一个 URI 指向它，每种"资源"对应一个 URI。

Representational 是"表现层"意思。"资源"是一种消息实体，它可以有多种外在的表现形式，我们把"资源"具体呈现出来的形式叫作它的"表现层"。比如说文本可以用 TXT 格式进行表现，也可以使用 XML 格式、JSON 格式和二进制格式；视频可以用 MP4 格式表现，也可以用 AVI 格式表现。URI 只代表资源的实体，不代表它的形式。它的具体表现形式，应该由 HTTP 请求的头信息 Accept 和 Content-Type 字段指定，这两个字段是对"表现层"的描述。

State Transfer 是指"状态转移"。客户端访问服务的过程中必然涉及数据和状态的转化。如果客户端想要操作服务端资源，必须通过某种手段，让服务器端资源发生"状态转移"。而这种转化是建立在表现层之上的，所以被称为"表现层状态转移"。客户端通过使用 HTTP 协议中的四个动词来实现上述操作，它们分别是：获取资源的 GET、新建或更新资源的 POST、更新资源的 PUT 和删除资源的 DELETE。

5.1.2 OpenFeign 简介

OpenFeign 是一个声明式 RESTful 网络请求客户端。OpenFeign 会根据带有注解的函数信息构建出网络请求的模板，在发送网络请求之前，OpenFeign 会将函数的参数值设置到这些请求模板中。虽然 OpenFeign 只能支持基于文本的网络请求，但是它可以极大简化网络请求的实现，方便编程人员快速构建自己的网络请求应用。

如图 5-1 所示，使用 OpenFeign 的 Spring 应用架构一般分为三个部分，分别为服务注册中心、服务提供者和服务消费者。服务提供者向服务注册中心注册自己，然后服务消费者通过 OpenFeign 发送请求时，OpenFeign 会向服务注册中心获取关于服务提供者的信息，然后再向服务提供者发送网络请求。

5.1.3 代码示例

1. 服务注册中心

OpenFeign 可以配合 Eureka 等服务注册中心同时使用。Eureka 作为服务注册中心，为 OpenFeign 提供服务端信息的获取，比如说服务的 IP 地址和端口。关于 Eureka 的具体使用可以参考 4.1 节。

2. 服务提供者

Spring Cloud OpenFeign 是声明式 RESTful 网络请求客户端，所以对服务提供者的实现方式没有任何影响。也就是说，服务提供者只需要提

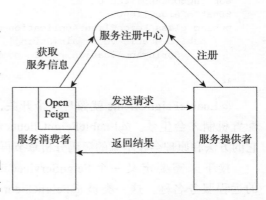

图 5-1 OpenFeign 调用架构示意图

供对外的网络请求接口就可，至于其具体实现既可以使用 Spring MVC，也可以使用 Jersey。只需要确保该服务提供者被注册到服务注册中心上即可。

```
@RestController
@RequestMapping("/feign-service")
public class FeignServiceController {
    @RequestMapping(value = "/instance/{serviceId}", method = RequestMethod.GET)
    public Instance getInstanceByServiceId(@PathVariable("serviceId") String serviceId){
        return new Instance(serviceId);
    }
}
```

如上述代码所示，通过 @RestController 和 @RequestMapping 声明了获取 Instance 资源的网络接口。除了实现网络 API 接口之外，还需要将该服务注册到 Eureka Server 上。需要在 application.yml 文件中设置服务注册中心的相关信息和该应用的名称，相关配置如下所示：

```
eureka:
  instance:
    instance-id: server:1
  client:
    service-url:
      default-zone: http://localhost:8761/eureka/
spring:
  application:
    name: feign-service
server:
  port: 9000
```

3. 服务消费者

OpenFeign 是声明式 RESTful 客户端，所以构建 OpenFeign 项目的关键在于构建服务消费者。通过下面的方法可以创建一个 Spring Cloud OpenFeign 的服务消费者项目。

首先需要在 pom 文件中添加 Eureka 和 OpenFeign 相关的依赖。然后在工程的入口类上添加 @EnableFeignClients 注解开启 Spring Cloud OpenFeign 的自动化配置功能，代码如下所示：

```
@SpringBootApplication
@EnableFeignClients
public class FeignClientApplication {
    public static void main(String[] args) {
        SpringApplication.run(ChapterFeignClientApplication.class, args);
    }
}
```

@EnableFeignClients 就像是一个开关，只有使用了该注解，OpenFeign 相关的组件和配置机制才会生效。@EnableFeignClients 还可以对 OpenFeign 相关组件进行自定义配置，它的方法和原理会在本章的源码分析章节再做具体的讲解。

接下来需要定义一个 FeignServiceClient 接口，通过 @FeignClient 注解来指定调用的远程服务名称。这一类被 @FeignClient 注解修饰的接口类一般被称为 FeignClient。在 FeignClient 接口类中，可以使用 @RequestMapping 定义网络请求相关的方法，如下所示：

```
@FeignClient("feign-service")
@RequestMapping("/feign-service")
public interface FeignServiceClient {
    @RequestMapping(value = "/instance/{serviceId}", method = RequestMethod.GET)
    public Instance getInstanceByServiceId(@PathVariable("serviceId") String serviceId);
}
```

如上面代码片段所显示的，如果你调用 FeignServiceClient 对象的 getInstanceByServiceId 方法，那么 OpenFeign 就会向 feign-service 服务的 /feign-service/instance/{serviceId} 接口发送网络请求。

最后，在服务消费端项目的 application.yml 文件中配置 Eureka 服务注册中心的相关属性，具体配置如下所示：

```
eureka:
    instance:
        instance-id: ${spring.application.name}:${vcap.application.instance_
           id:${spring.application.instance_id:${random.value}}}
    client:
        service-url:
            default-zone: http://localhost:8761/eureka/
spring:
    application:
        name: feign-client
server:
    port: 8770
```

相信读者通过搭建 OpenFeign 的项目，已经对 OpenFeign 的相关使用原理有了一定的了解，这个过程将对理解 OpenFeign 相关的工作原理大有裨益。

5.2 源码分析

5.2.1 核心组件与概念

读者在阅读 OpenFeign 源码时，可以沿着两条路线进行，一是 FeignServiceClient 这样的被 @FeignClient 注解修饰的接口类（后续简称为 FeignClient 接口类）如何创建，也就是其 Bean 实例是如何被创建的；二是调用 FeignServiceClient 对象的网络请求相关的函数时，OpenFeign 是如何发送网络请求的。而 OpenFeign 相关的类也可以以此来进行分类，一部分是用来初始化相应的 Bean 实例的，一部分是用来在调用方法时发送网络请求。

图 5-2 是 OpenFeign 相关的关键类图，其中比较重要的类为 FeignClientFactoryBean、FeignContext 和 SynchronousMethodHandler。FeignClientFactoryBean 是创建 @FeignClient 修饰的接口类 Bean 实例的工厂类；FeignContext 是配置组件的上下文环境，保存着相关组件的不同实例，这些实例由不同的 FeignConfiguration 配置类构造出来；SynchronousMethodHandler 是 MethodHandler 的子类，可以在 FeignClient 相应方法被调用时发送网络请求，然后再将请求响应转化为函数返回值进行输出。

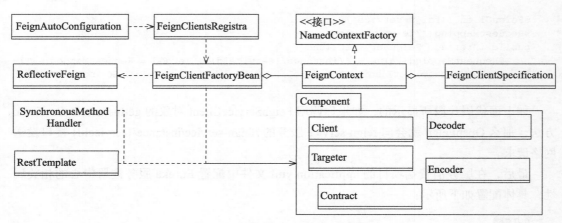

图 5-2　OpenFeign 关键类的类图

图 5-3 是后续源码讲解的流程图，OpenFeign 会首先进行相关 BeanDefinition 的动态注册，然后当 Spring 容器注入相关实例时会进行实例的初始化，最后当 FeignClient 接口类实例的函数被调用时会发送网络请求。

5.2.2　动态注册 BeanDefinition

OpenFeign 可以通过多种方式进行自定义配置，配置的变化会导致接口类初始化时使用不同的 Bean 实例，从而控制 OpenFeign 的相关行为，比如说网络请求的编解码、压缩和日志处理。可以说，了解 OpenFeign 配置和实例初始化的流程与原理对于我们学习和使用 OpenFeign 有着至关重要的作用，而且 Spring Cloud 的所有项目的配置和实例初始化过程的原理基本相同，了解了 OpenFeign 的原理，就可以触类旁通，一通百通了。

图 5-3　源码流程图

1. FeignClientsRegistrar

在 5.1 节已经介绍了 @EnableFeignClients 的基本作用，它就像是 OpenFeign 的开关一样，一切 OpenFeign 的相关操作都是从它开始的。@EnableFeignClients 有三个作用，一是引入 FeignClientsRegistrar；二是指定扫描 FeignClient 的包信息，就是指定 FeignClient 接口类所在的包名；三是指定 FeignClient 接口类的自定义配置类。@EnableFeignClients 注解的定义如下所示：

```
//EnableFeignClients.java
@Retention(RetentionPolicy.RUNTIME)
@Target(ElementType.TYPE)
@Documented
//ImportBeanDefinitionRegistrar的子类,用于处理@FeignClient注解
@Import(FeignClientsRegistrar.class)
public @interface EnableFeignClients {
    // 下面三个函数都是为了指定需要扫描的包
```

```
    String[] value() default {};
    String[] basePackages() default {};
    Class<?>[] basePackageClasses() default {};
    // 指定自定义feign client的自定义配置，可以配置Decoder、Encoder和Contract等组件，
        FeignClientsConfiguration是默认的配置类
    Class<?>[] defaultConfiguration() default {};
    // 指定被@FeignClient修饰的类，如果不为空，那么路径自动检测机制会被关闭
    Class<?>[] clients() default {};
}
```

上面的代码中，FeignClientsRegistrar 是 ImportBeanDefinitionRegistrar 的子类，Spring 用 ImportBeanDefinitionRegistrar 来动态注册 BeanDefinition。OpenFeign 通过 FeignClientsRegistrar 来处理 @FeignClient 修饰的 FeignClient 接口类，将这些接口类的 BeanDefinition 注册到 Spring 容器中，这样就可以使用 @Autowired 等方式来自动装载这些 FeignClient 接口类的 Bean 实例。FeignClientsRegistrar 的部分代码如下所示：

```
//FeignClientsRegistrar.java
class FeignClientsRegistrar implements ImportBeanDefinitionRegistrar,
        ResourceLoaderAware, BeanClassLoaderAware, EnvironmentAware {
    ...
    @Override
    public void registerBeanDefinitions(AnnotationMetadata metadata,
            BeanDefinitionRegistry registry) {
        //从EnableFeignClients的属性值来构建Feign的自定义Configuration进行注册
        registerDefaultConfiguration(metadata, registry);
        //扫描package，注册被@FeignClient修饰的接口类的Bean信息
        registerFeignClients(metadata, registry);
    }
    ...
}
```

如上述代码所示，FeignClientsRegistrar 的 registerBeanDefinitions 方法主要做了两个事情，一是注册 @EnableFeignClients 提供的自定义配置类中的相关 Bean 实例，二是根据 @EnableFeignClients 提供的包信息扫描 @FeignClient 注解修饰的 FeignCleint 接口类，然后进行 Bean 实例注册。

@EnableFeignClients 的自定义配置类是被 @Configuration 注解修饰的配置类，它会提供一系列组装 FeignClient 的各类组件实例。这些组件包括：Client、Targeter、Decoder、Encoder 和 Contract 等。接下来看看 registerDefaultConfiguration 的代码实现，如下所示：

```
//FeignClientsRegistrar.java
private void registerDefaultConfiguration(AnnotationMetadata metadata,
            BeanDefinitionRegistry registry) {
    //获取到metadata中关于EnableFeignClients的属性值键值对
    Map<String, Object> defaultAttrs = metadata
            .getAnnotationAttributes(EnableFeignClients.class.getName(), true);
    // 如果EnableFeignClients配置了defaultConfiguration类，那么才进行下一步操作，如果没有，
        会使用默认的FeignConfiguration
    if (defaultAttrs != null && defaultAttrs.containsKey("defaultConfiguration")) {
```

```
        String name;
        if (metadata.hasEnclosingClass()) {
            name = "default." + metadata.getEnclosingClassName();
        }
        else {
            name = "default." + metadata.getClassName();
        }
        registerClientConfiguration(registry, name,
                defaultAttrs.get("defaultConfiguration"));
    }
}
```

如上述代码所示，registerDefaultConfiguration 方法会判断 @EnableFeignClients 注解是否设置了 defaultConfiguration 属性。如果有，则将调用 registerClientConfiguration 方法，进行 BeanDefinitionRegistry 的注册。registerClientConfiguration 方法的代码如下所示。

```
// FeignClientsRegistrar.java
private void registerClientConfiguration(BeanDefinitionRegistry registry, Object name,
        Object configuration) {
    // 使用BeanDefinitionBuilder来生成BeanDefinition,并注册到registry上
    BeanDefinitionBuilder builder = BeanDefinitionBuilder
            .genericBeanDefinition(FeignClientSpecification.class);
    builder.addConstructorArgValue(name);
    builder.addConstructorArgValue(configuration);
    registry.registerBeanDefinition(
            name + "." + FeignClientSpecification.class.getSimpleName(),
            builder.getBeanDefinition());
}
```

BeanDefinitionRegistry 是 Spring 框架中用于动态注册 BeanDefinition 信息的接口，调用其 registerBeanDefinition 方法可以将 BeanDefinition 注册到 Spring 容器中，其中 name 属性就是注册 BeanDefinition 的名称。

FeignClientSpecification 类实现了 NamedContextFactory.Specification 接口，它是 OpenFeign 组件实例化的重要一环，它持有自定义配置类提供的组件实例，供 OpenFeign 使用。Spring Cloud 框架使用 NamedContextFactory 创建一系列的运行上下文（ApplicationContext），来让对应的 Specification 在这些上下文中创建实例对象。这样使得各个子上下文中的实例对象相互独立，互不影响，可以方便地通过子上下文管理一系列不同的实例对象。NamedContextFactory 有三个功能，一是创建 AnnotationConfigApplicationContext 子上下文；二是在子上下文中创建并获取 Bean 实例；三是当子上下文消亡时清除其中的 Bean 实例。在 OpenFeign 中，FeignContext 继承了 NamedContextFactory，用于存储各类 OpenFeign 的组件实例。图 5-4 就是 FeginContext 的相关类图。

FeignAutoConfiguration 是 OpenFeign 的自动配置类，它会提供 FeignContext 实例。并且将之前注册的 FeignClientSpecification 通过 setConfigurations 方法设置给 FeignContext 实例。这里处理了默认配置类 FeignClientsConfiguration 和自定义配置类的替换问题。如果 FeignClientsRegistrar 没有注册自定义配置类，那么 configurations 将不包含

FeignClientSpecification 对象，否则会在 setConfigurations 方法中进行默认配置类的替换。
FeignAutoConfiguration 的相关代码如下所示：

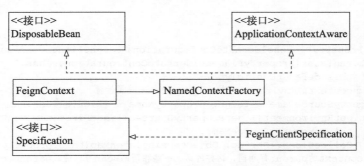

图 5-4　FeignContext 相关类图

```
//FeignAutoConfiguration.java
@Autowired(required = false)
private List<FeignClientSpecification> configurations = new ArrayList<>();
@Bean
public FeignContext feignContext() {
    FeignContext context = new FeignContext();
    context.setConfigurations(this.configurations);
    return context;
}
//FeignContext.java
public class FeignContext extends NamedContextFactory<FeignClientSpecification> {
    public FeignContext() {
        //将默认的FeignClientConfiguration作为参数传递给构造函数
        super(FeignClientsConfiguration.class, "feign", "feign.client.name");
    }
}
```

NamedContextFactory 是 FeignContext 的父类，其 createContext 方法会创建具有名称的 Spring 的 AnnotationConfigApplicationContext 实例作为当前上下文的子上下文。这些 AnnotationConfigApplicationContext 实例可以管理 OpenFeign 组件的不同实例。NamedContextFactory 的实现如下代码所示：

```
//NamedContextFactory.java
protected AnnotationConfigApplicationContext createContext(String name) {
    AnnotationConfigApplicationContext context = new AnnotationConfigApplicationContext();
    //获取该name所对应的configuration,如果有的话，就注册都子context中
    if (this.configurations.containsKey(name)) {
        for (Class<?> configuration : this.configurations.get(name)
                .getConfiguration()) {
            context.register(configuration);
        }
    }
    // 注册default的Configuration, 也就是FeignClientsRegistrar类的registerDefaultConfiguration
        方法中注册的Configuration
    for (Map.Entry<String, C> entry : this.configurations.entrySet()) {
```

```
        if (entry.getKey().startsWith("default.")) {
            for (Class<?> configuration : entry.getValue().getConfiguration()) {
                context.register(configuration);
            }
        }
    }
    // 注册PropertyPlaceholderAutoConfiguration和FeignClientsConfiguration配置类
    context.register(PropertyPlaceholderAutoConfiguration.class,
            this.defaultConfigType);
    // 设置子context的Environment的propertySource属性源
    // propertySourceName = feign; propertyName = feign.client.name
    context.getEnvironment().getPropertySources().addFirst(new MapPropertySource(
            this.propertySourceName,
            Collections.<String, Object> singletonMap(this.propertyName, name)));
    // 所有context的parent都相同,这样的话,一些相同的Bean可以通过parent context来获取
    if (this.parent != null) {
      context.setParent(this.parent);
    }
    context.setDisplayName(generateDisplayName(name));
    context.refresh();
    return context;
}
```

而由于 NamedContextFactory 实现了 DisposableBean 接口,当 NamedContextFactory 实例消亡时,Spring 框架会调用其 destroy 方法,清除掉自己创建的所有子上下文和自身包含的所有组件实例。NamedContextFactory 的 destroy 方法如下所示:

```
//NamedContextFactory.java
@Override
public void destroy() {
    Collection<AnnotationConfigApplicationContext> values = this.contexts.values();
    for(AnnotationConfigApplicationContext context : values) {
        context.close();
    }
    this.contexts.clear();
}
```

NamedContextFactory 会创建出 AnnotationConfigApplicationContext 实例,并以 name 作为唯一标识,然后每个 AnnotationConfigApplicationContext 实例都会注册部分配置类,从而可以给出一系列的基于配置类生成的组件实例,这样就可以基于 name 来管理一系列的组件实例,为不同的 FeignClient 准备不同配置组件实例,比如说 Decoder、Encoder 等。我们会在后续的讲解中详细介绍配置类 Bean 实例的获取。

2. 扫描类信息

FeignClientsRegistrar 做的第二件事情是扫描指定包下的类文件,注册 @FeignClient 注解修饰的接口类信息,如下所示:

```
//FeignClientsRegistrar.java
public void registerFeignClients(AnnotationMetadata metadata,
```

```java
        BeanDefinitionRegistry registry) {
    //生成自定义的ClassPathScanningProvider
    ClassPathScanningCandidateComponentProvider scanner = getScanner();
    scanner.setResourceLoader(this.resourceLoader);
    Set<String> basePackages;
    //获取EnableFeignClients所有属性的键值对
    Map<String, Object> attrs = metadata
            .getAnnotationAttributes(EnableFeignClients.class.getName());
    //依照Annotation来进行TypeFilter, 只会扫描出被FeignClient修饰的类
    AnnotationTypeFilter annotationTypeFilter = new AnnotationTypeFilter(
            FeignClient.class);
    final Class<?>[] clients = attrs == null ? null
            : (Class<?>[]) attrs.get("clients");
    //如果没有设置clients属性, 那么需要扫描basePackage, 所以设置了AnnotationTypeFilter,
    //  并且去获取basePackage
    if (clients == null || clients.length == 0) {
        scanner.addIncludeFilter(annotationTypeFilter);
        basePackages = getBasePackages(metadata);
    }
    //代码有删减, 遍历上述过程中获取的basePackages列表
    for (String basePackage : basePackages) {
        //获取basepackage下的所有BeanDefinition
        Set<BeanDefinition> candidateComponents = scanner
                .findCandidateComponents(basePackage);
        for (BeanDefinition candidateComponent : candidateComponents) {
            if (candidateComponent instanceof AnnotatedBeanDefinition) {
                AnnotatedBeanDefinition beanDefinition = (AnnotatedBeanDefinition)
                    candidateComponent;
                AnnotationMetadata annotationMetadata = beanDefinition.getMetadata();
                //从这些BeanDefinition中获取FeignClient的属性值
                Map<String, Object> attributes = annotationMetadata
                        .getAnnotationAttributes(
                                FeignClient.class.getCanonicalName());
                String name = getClientName(attributes);
                //对单独某个FeignClient的configuration进行配置
                registerClientConfiguration(registry, name,
                        attributes.get("configuration"));
                //注册FeignClient的BeanDefinition
                registerFeignClient(registry, annotationMetadata, attributes);
            }
        }
    }
}
```

如上述代码所示,FeignClientsRegistrar 的 registerFeignClients 方法依据 @EnableFeignClients 的属性获取要扫描的包路径信息,然后获取这些包下所有被 @FeignClient 注解修饰的接口类的 BeanDefinition,最后调用 registerFeignClient 动态注册 BeanDefinition。

registerFeignClients 方法中有一些细节值得认真学习,有利于加深了解 Spring 框架。首先是如何自定义 Spring 类扫描器,即如何使用 ClassPathScanningCandidateComponentProvider 和各类 TypeFilter。

OpenFeign 使用了 AnnotationTypeFilter，来过滤出被 @FeignClient 修饰的类，getScanner 方法的具体实现如下所示：

```java
//FeignClientsRegistrar.java
protected ClassPathScanningCandidateComponentProvider getScanner() {
    return new ClassPathScanningCandidateComponentProvider(false, this.environment) {
        @Override
        protected boolean isCandidateComponent(AnnotatedBeanDefinition beanDefinition) {
            boolean isCandidate = false;
            //判断beanDefinition是否为内部类，否则直接返回false
            if (beanDefinition.getMetadata().isIndependent()) {
                //判断是否为接口类，所实现的接口只有一个，并且该接口是Annotation。否则直接
                  返回true
                if (!beanDefinition.getMetadata().isAnnotation()) {
                    isCandidate = true;
                }
            }
            return isCandidate;
        }
    };
}
```

ClassPathScanningCandidateComponentProvider 的作用是遍历指定路径的包下的所有类。比如指定包路径为 com/test/openfeign，它会找出 com.test.openfeign 包下所有的类，将所有的类封装成 Resource 接口集合。Resource 接口是 Spring 对资源的封装，有 FileSystemResource、ClassPathResource、UrlResource 等多种实现。接着 ClassPathScanningCandidateComponentProvider 类会遍历 Resource 集合，通过 includeFilters 和 excludeFilters 两种过滤器进行过滤操作。includeFilters 和 excludeFilters 是 TypeFilter 接口类型实例的集合，TypeFilter 接口是一个用于判断类型是否满足要求的类型过滤器。excludeFilters 中只要有一个 TypeFilter 满足条件，这个 Resource 就会被过滤掉；而 includeFilters 中只要有一个 TypeFilter 满足条件，这个 Resource 就不会被过滤。如果一个 Resource 没有被过滤，它会被转换成 ScannedGenericBeanDefinition 添加到 BeanDefinition 集合中。

5.2.3 实例初始化

FeignClientFactoryBean 是工厂类，Spring 容器通过调用它的 getObject 方法来获取对应的 Bean 实例。被 @FeignClient 修饰的接口类都是通过 FeignClientFactoryBean 的 getObject 方法来进行实例化的，具体实现如下代码所示：

```java
//FeignClientFactoryBean.java
public Object getObject() throws Exception {
    FeignContext context = applicationContext.getBean(FeignContext.class);
    Feign.Builder builder = feign(context);
    if (StringUtils.hasText(this.url) && !this.url.startsWith("http")) {
        this.url = "http://" + this.url;
    }
```

```
    String url = this.url + cleanPath();
//调用FeignContext的getInstance方法获取Client对象
Client client = getOptional(context, Client.class);
//因为有具体的Url,所以就不需要负载均衡,所以除去LoadBalancerFeignClient实例
if (client != null) {
    if (client instanceof LoadBalancerFeignClient) {
        client = ((LoadBalancerFeignClient)client).getDelegate();
    }
    builder.client(client);
}
Targeter targeter = get(context, Targeter.class);
return targeter.target(this, builder, context, new HardCodedTarget<>(
        this.type, this.name, url));
}
```

这里就用到了 FeignContext 的 getInstance 方法,我们在前边已经讲解了 FeignContext 的作用,getOptional 方法调用了 FeignContext 的 getInstance 方法,从 FeignContext 的对应名称的子上下文中获取到 Client 类型的 Bean 实例,其具体实现如下所示:

```
//NamedContextFactory.java
public <T> T getInstance(String name, Class<T> type) {
    AnnotationConfigApplicationContext context = getContext(name);
    if (BeanFactoryUtils.beanNamesForTypeIncludingAncestors(context,
            type).length > 0) {
        //从对应的context中获取Bean实例,如果对应的子上下文没有则直接从父上下文中获取
        return context.getBean(type);
    }
    return null;
}
```

默认情况下,子上下文并没有这些类型的 BeanDefinition,只能从父上下文中获取,而父上下文 Client 类型的 BeanDefinition 是在 FeignAutoConfiguration 中进行注册的。但是当子上下文注册的配置类提供了 Client 实例时,子上下文会直接将自己配置类的 Client 实例进行返回,否则都是由父上下文返回默认 Client 实例。Client 在 FeignAutoConfiguration 中的配置如下所示。

```
//FeignAutoConfiguration.java
@Bean
@ConditionalOnMissingBean(Client.class)
public Client feignClient(HttpClient httpClient) {
    return new ApacheHttpClient(httpClient);
}
```

Targeter 是一个接口,它的 target 方法会生成对应的实例对象。它有两个实现类,分别为 DefaultTargeter 和 HystrixTargeter。OpenFeign 使用 HystrixTargeter 这一层抽象来封装关于 Hystrix 的实现。DefaultTargeter 的实现如下所示,只是调用了 Feign.Builder 的 target 方法:

```
//DefaultTargeter.java
```

```java
class DefaultTargeter implements Targeter {
@Override
    public <T> T target(FeignClientFactoryBean factory, Feign.Builder feign,
        FeignContext context,
                        Target.HardCodedTarget<T> target) {
        return feign.target(target);
    }
}
```

而 Feign.Builder 是由 FeignClientFactoryBean 对象的 feign 方法创建的。Feign.Builder 会设置 FeignLoggerFactory、EncoderDecoder 和 Contract 等组件，这些组件的 Bean 实例都是通过 FeignContext 获取的，也就是说这些实例都是可配置的，你可以通过 OpenFeign 的配置机制为不同的 FeignClient 配置不同的组件实例。feign 方法的实现如下所示：

```java
//FeignClientFactoryBean.java
protected Feign.Builder feign(FeignContext context) {
    FeignLoggerFactory loggerFactory = get(context, FeignLoggerFactory.class);
    Logger logger = loggerFactory.create(this.type);
    Feign.Builder builder = get(context, Feign.Builder.class)
        .logger(logger)
        .encoder(get(context, Encoder.class))
        .decoder(get(context, Decoder.class))
        .contract(get(context, Contract.class));
    configureFeign(context, builder);
    return builder;
}
```

Feign.Builder 负责生成被 @FeignClient 修饰的 FeignClient 接口类实例。它通过 Java 反射机制，构造 InvocationHandler 实例并将其注册到 FeignClient 上，当 FeignClient 的方法被调用时，InvocationHandler 的回调函数会被调用，OpenFeign 会在其回调函数中发送网络请求。build 方法如下所示：

```java
//Feign.Builder
public Feign build() {
    SynchronousMethodHandler.Factory synchronousMethodHandlerFactory =
        new SynchronousMethodHandler.Factory(client, retryer, requestInterceptors,
        logger, logLevel, decode404);
    ParseHandlersByName handlersByName = new ParseHandlersByName(contract, options,
        encoder, decoder, errorDecoder, synchronousMethodHandlerFactory);
    return new ReflectiveFeign(handlersByName, invocationHandlerFactory);
}
```

ReflectiveFeign 的 newInstance 方法是生成 FeignClient 实例的关键实现。它主要做了两件事情，一是扫描 FeignClient 接口类的所有函数，生成对应的 Handler；二是使用 Proxy 生成 FeignClient 的实例对象，代码如下所示：

```java
//ReflectiveFeign.java
public <T> T newInstance(Target<T> target) {
    Map<String, MethodHandler> nameToHandler = targetToHandlersByName.apply(target);
    Map<Method, MethodHandler> methodToHandler = new LinkedHashMap<Method,
```

```
        MethodHandler>();
    List<DefaultMethodHandler> defaultMethodHandlers = new LinkedList<DefaultMet
        hodHandler>();
    for (Method method : target.type().getMethods()) {
        if (method.getDeclaringClass() == Object.class) {
            continue;
        } else if(Util.isDefault(method)) {
            //为每个默认方法生成一个DefaultMethodHandler
            defaultMethodHandler handler = new DefaultMethodHandler(method);
            defaultMethodHandlers.add(handler);
            methodToHandler.put(method, handler);
        } else {
            methodToHandler.put(method, nameToHandler.get(Feign.configKey(target.
                type(), method)));
        }
    }
    //生成java reflective的InvocationHandler
    InvocationHandler handler = factory.create(target, methodToHandler);
    T proxy = (T) Proxy.newProxyInstance(target.type().getClassLoader(), new Class<?>[]
        {target.type()}, handler);
    //将defaultMethodHandler绑定到proxy中
    for(DefaultMethodHandler defaultMethodHandler : defaultMethodHandlers) {
        defaultMethodHandler.bindTo(proxy);
    }
    return proxy;
}
```

1. 扫描函数信息

在扫描 FeignClient 接口类所有函数生成对应 Handler 的过程中，OpenFeign 会生成调用该函数时发送网络请求的模板，也就是 RequestTemplate 实例。RequestTemplate 中包含了发送网络请求的 URL 和函数参数填充的信息。@RequestMapping、@PathVariable 等注解信息也会包含到 RequestTemplate 中，用于函数参数的填充。ParseHandlersByName 类的 apply 方法就是这一过程的具体实现。它首先会使用 Contract 来解析接口类中的函数信息，并检查函数的合法性，然后根据函数的不同类型来为每个函数生成一个 BuildTemplateByResolvingArgs 对象，最后使用 SynchronousMethodHandler.Factory 来创建 MethodHandler 实例。ParseHandlersByName 的 apply 实现如下代码所示：

```
//ParseHandlersByName.java
public Map<String, MethodHandler> apply(Target key) {
    // 获取type的所有方法的信息,会根据注解生成每个方法的RequestTemplate
    List<MethodMetadata> metadata = contract.parseAndValidatateMetadata(key.type());
    Map<String, MethodHandler> result = new LinkedHashMap<String, MethodHandler>();
    for (MethodMetadata md : metadata) {
    BuildTemplateByResolvingArgs buildTemplate;
    if (!md.formParams().isEmpty() && md.template().bodyTemplate() == null) {
        buildTemplate = new BuildFormEncodedTemplateFromArgs(md, encoder);
    } else if (md.bodyIndex() != null) {
        buildTemplate = new BuildEncodedTemplateFromArgs(md, encoder);
    } else {
```

```
            buildTemplate = new BuildTemplateByResolvingArgs(md);
        }
        result.put(md.configKey(),
            factory.create(key, md, buildTemplate, options, decoder, errorDecoder));
    }
    return result;
}
```

OpenFeign 默认的 Contract 实现是 SpringMvcContract。SpringMvcContract 的父类为 BaseContract，而 BaseContract 是 Contract 众多子类中的一员，其他还有 JAXRSContract 和 HystrixDelegatingContract 等。Contract 的 parseAndValidateMetadata 方法会解析与 HTTP 请求相关的所有函数的基本信息和注解信息，代码如下所示：

```
//SpringMvcContract.java
@Override
public MethodMetadata parseAndValidateMetadata(Class<?> targetType, Method method) {
    this.processedMethods.put(Feign.configKey(targetType, method), method);
    //调用父类BaseContract的函数
    MethodMetadata md = super.parseAndValidateMetadata(targetType, method);
    RequestMapping classAnnotation = findMergedAnnotation(targetType,
            RequestMapping.class);
    //处理RequestMapping注解
    if (classAnnotation != null) {
        if (!md.template().headers().containsKey(ACCEPT)) {
            parseProduces(md, method, classAnnotation);
        }
        if (!md.template().headers().containsKey(CONTENT_TYPE)) {
            parseConsumes(md, method, classAnnotation);
        }
        parseHeaders(md, method, classAnnotation);
    }
    return md;
}
```

BaseContract 的 parseAndValidateMetadata 方法会依次解析接口类的注解，函数注解和函数的参数注解，将这些注解包含的信息封装到 MethodMetadata 对象中，然后返回，代码如下所示：

```
//BaseContract.java
protected MethodMetadata parseAndValidateMetadata(Class<?> targetType, Method method) {
    MethodMetadata data = new MethodMetadata();
    //函数的返回值
    data.returnType(Types.resolve(targetType, targetType, method.getGenericReturnType()));
    //函数Feign相关的唯一配置键
    data.configKey(Feign.configKey(targetType, method));
    //获取并处理修饰class的注解信息
    if(targetType.getInterfaces().length == 1) {
        processAnnotationOnClass(data, targetType.getInterfaces()[0]);
    }
    //调用子类processAnnotationOnClass的实现
    processAnnotationOnClass(data, targetType);
```

```java
//处理修饰method的注解信息
for (Annotation methodAnnotation : method.getAnnotations()) {
    processAnnotationOnMethod(data, methodAnnotation, method);
}
//函数参数类型
Class<?>[] parameterTypes = method.getParameterTypes();
Type[] genericParameterTypes = method.getGenericParameterTypes();
//函数参数的注解类型
Annotation[][] parameterAnnotations = method.getParameterAnnotations();
int count = parameterAnnotations.length;
//依次处理各个函数参数注解
for (int i = 0; i < count; i++) {
    boolean isHttpAnnotation = false;
    if (parameterAnnotations[i] != null) {
        // 处理参数的注解,并且返回该参数来指明是否为将要发送请求的body。除了body之外,还
        //   可能是path, param等
        isHttpAnnotation = processAnnotationsOnParameter(data, parameterAnnotations[i], i);
    }
    if (parameterTypes[i] == URI.class) {
        data.urlIndex(i);
    } else if (!isHttpAnnotation) {
        //表明发送请求body的参数位置和参数类型
        data.bodyIndex(i);
        data.bodyType(Types.resolve(targetType, targetType, genericParameterTypes[i]));
    }
}
return data;
}
```

processAnnotationOnClass 方法用于处理接口类注解。该函数在 parseAndValidateMetadata 方法中可能会被调用两次,如果 targetType 只继承或者实现一种接口时,先处理该接口的注解,再处理 targetType 的注解;否则只会处理 targetType 的注解。@RequestMapping 在修饰 FeignClient 接口类时,其 value 所代表的值会被记录下来,它是该 FeignClient 下所有请求 URL 的前置路径,处理接口类注解的函数代码如下所示:

```java
//SpringMvcContract.java
protected void processAnnotationOnClass(MethodMetadata data, Class<?> clz) {
    if (clz.getInterfaces().length == 0) {
        //获取RequestMapping的注解信息,并设置MethodMetadata.template的数据
        RequestMapping classAnnotation = findMergedAnnotation(clz,
                RequestMapping.class);
        if (classAnnotation != null) {
            if (classAnnotation.value().length > 0) {
                String pathValue = emptyToNull(classAnnotation.value()[0]);
                pathValue = resolve(pathValue);
                if (!pathValue.startsWith("/")) {
                    pathValue = "/" + pathValue;
                }
                //处理@RequestMapping的value,一般都是发送请求的path
                data.template().insert(0, pathValue);
            }
```

```
            }
        }
    }
```

processAnnotationOnMethod 方法的主要作用是处理修饰函数的注解。它会首先校验该函数是否被 @RequestMapping 修饰，如果没有就会直接返回。然后获取该函数所对应的 HTTP 请求的方法，默认的方法是 GET。接着会处理 @RequestMapping 中的 value 属性，解析 value 属性中的 pathValue，比如说 value 属性值为 /instance/{instanceId}，那么 pathValue 的值就是 instanceId。最后处理消费（consumes）和生产（produces）相关的信息，记录媒体类型（media types），代码如下所示：

```
//SpringMvcContract.java
protected void processAnnotationOnMethod(MethodMetadata data,
        Annotation methodAnnotation, Method method) {
    if (!RequestMapping.class.isInstance(methodAnnotation) && !methodAnnotation
            .annotationType().isAnnotationPresent(RequestMapping.class)) {
        return;
    }
    RequestMapping methodMapping = findMergedAnnotation(method, RequestMapping.class);
    // 处理HTTP Method
    RequestMethod[] methods = methodMapping.method();
    //默认的method是GET
    if (methods.length == 0) {
        methods = new RequestMethod[] { RequestMethod.GET };
    }
    data.template().method(methods[0].name());
    // 处理请求的路径
    checkAtMostOne(method, methodMapping.value(), "value");
    if (methodMapping.value().length > 0) {
        String pathValue = emptyToNull(methodMapping.value()[0]);
        if (pathValue != null) {
            pathValue = resolve(pathValue);
            // Append path from @RequestMapping if value is present on method
            if (!pathValue.startsWith("/")
                    && !data.template().toString().endsWith("/")) {
                pathValue = "/" + pathValue;
            }
            data.template().append(pathValue);
        }
    }
    // 处理生产
    parseProduces(data, method, methodMapping);
    // 处理消费
    parseConsumes(data, method, methodMapping);
    // 处理头部
    parseHeaders(data, method, methodMapping);
    data.indexToExpander(new LinkedHashMap<Integer, Param.Expander>());
}
```

而 processAnnotationsOnParameter 方法则主要处理修饰函数参数的注解。它会根据注解类型来调用不同的 AnnotatedParameterProcessor 的实现类，解析注解的属性信息。函数参数的注解类型包括 @RequestParam、@RequestHeader 和 @PathVariable。processAnnotationsOnParameter 方法的具体实现如下代码所示：

```java
//SpringMvcContract.java
protected boolean processAnnotationsOnParameter(MethodMetadata data,
        Annotation[] annotations, int paramIndex) {
    boolean isHttpAnnotation = false;
    AnnotatedParameterProcessor.AnnotatedParameterContext context = new SimpleAn
        notatedParameterContext(
            data, paramIndex);
    Method method = this.processedMethods.get(data.configKey());
    //遍历所有的参数注解
    for (Annotation parameterAnnotation : annotations) {
        //不同的注解类型有不同的Processor
        AnnotatedParameterProcessor processor = this.annotatedArgumentProcessors
                .get(parameterAnnotation.annotationType());
        if (processor != null) {
            Annotation processParameterAnnotation;
            //如果没有缓存的Processor，则生成一个
            processParameterAnnotation = synthesizeWithMethodParameterNameAsFall
                backValue(
                    parameterAnnotation, method, paramIndex);
            isHttpAnnotation |= processor.processArgument(context,
                    processParameterAnnotation, method);
        }
    }
    return isHttpAnnotation;
}
```

AnnotatedParameterProcessor 是一个接口，有三个实现类：PathVariableParameterProcessor、RequestHeaderParameterProcessor 和 RequestParamParameterProcessor，三者分别用于处理 @RequestParam、@RequestHeader 和 @PathVariable 注解。三者的类图如图 5-5 所示，我们具体看一下 PathVariableParameterProcessor 的实现。

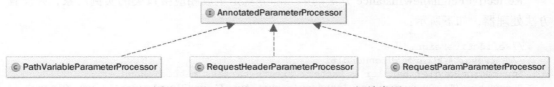

图 5-5　AnnotatedParameterProcessor 相关类图

```java
//PathVariableParameterProcessor.java
public boolean processArgument(AnnotatedParameterContext context, Annotation
    annotation, Method method) {
    //ANNOTATION就是@PathVariable,所以就获取它的值,也就是@RequestMapping value中{}内的值
    String name = ANNOTATION.cast(annotation).value();
    //将name设置为ParameterName
    context.setParameterName(name);
```

```
    MethodMetadata data = context.getMethodMetadata();
    //当varName在url、queries、headers中不存在时,将name添加到formParams中。因为无法找到
       对应的值
    String varName = '{' + name + '}';
    if (!data.template().url().contains(varName)
           && !searchMapValues(data.template().queries(), varName)
           && !searchMapValues(data.template().headers(), varName)) {
       data.formParams().add(name);
    }
    return true;
}
```

如上述代码所示,PathVariableParameterProcessor 的 processArgument 方法用于处理被 @PathVariable 注解修饰的参数。

ParseHandlersByName 的 apply 方法通过 Contract 的 parseAndValidatateMetadata 方法获得了接口类中所有方法的元数据,这些信息中包含了每个方法所对应的网络请求信息。比如说请求的路径(path)、参数(params)、头部(headers)和 body。接下来 apply 方法会为每个方法生成一个 MethodHandler。SynchronousMethodHandler.Factory 的 create 方法能直接创建 SynchronousMethodHandler 对象并返回,如下所示:

```
//SynchronousMethodHandler.Factory
public MethodHandler create(Target<?> target, MethodMetadata md,
    RequestTemplate.Factory buildTemplateFromArgs,
    Options options, Decoder decoder, ErrorDecoder errorDecoder) {
    return new SynchronousMethodHandler(target, client, retryer, requestInterceptors,
       logger, logLevel, md, buildTemplateFromArgs,
                    options, decoder, errorDecoder, decode404);
}
```

ParseHandlersByName 的 apply 方法作为 ReflectiveFeign 的 newInstance 方法的第一部分,其作用就是解析对应接口类的所有方法信息,并生成对应的 MethodHandler。

2. 生成 Proxy 接口类

ReflectiveFeign#newInstance 方法的第二部分就是生成相应接口类的实例对象,并设置方法处理器,如下所示:

```
//ReflectiveFeign.java
//生成Java反射的InvocationHandler
InvocationHandler handler = factory.create(target, methodToHandler);
T proxy = (T) Proxy.newProxyInstance(target.type().getClassLoader(), new Class<?>[]
    {target.type()}, handler);
//将defaultMethodHandler绑定到proxy中。
for(DefaultMethodHandler defaultMethodHandler : defaultMethodHandlers) {
    defaultMethodHandler.bindTo(proxy);
}
return proxy;
```

OpenFeign 使用 Proxy 的 newProxyInstance 方法来创建 FeignClient 接口类的实例,然

后将 InvocationHandler 绑定到接口类实例上，用于处理接口类函数调用，如下所示：

```
//Default.java
static final class Default implements InvocationHandlerFactory {
    @Override
    public InvocationHandler create(Target target, Map<Method, MethodHandler> dispatch) {
        return new ReflectiveFeign.FeignInvocationHandler(target, dispatch);
    }
}
```

Default 实现了 InvocationHandlerFactory 接口，其 create 方法返回 ReflectiveFeign.FeignInvocationHandler 实例。

ReflectiveFeign 的内部类 FeignInvocationHandler 是 InvocationHandler 的实现类，其主要作用是将接口类相关函数的调用分配给对应的 MethodToHandler 实例，即 SynchronousMethodHandler 来处理。当调用接口类实例的函数时，会直接调用到 FeignInvocationHandler 的 invoke 方法。invoke 方法会根据函数名称来调用不同的 MethodHandler 实例的 invoke 方法，如下所示：

```
//FeignInvocationHandler.java
public Object invoke(Object proxy, Method method, Object[] args) throws Throwable {
    if ("equals".equals(method.getName())) {
        try {
            Object
                otherHandler =
                args.length > 0 && args[0] != null ? Proxy.getInvocationHandler(args[0]):
                    null;
            return equals(otherHandler);
        } catch (IllegalArgumentException e) {
            return false;
        }
    } else if ("hashCode".equals(method.getName())) {
        return hashCode();
    } else if ("toString".equals(method.getName())) {
        return toString();
    }
    //dispatch就是Map<Method, MethodHandler>，所以就是将某个函数的调用交给对应的MethodHandler
      来处理
    return dispatch.get(method).invoke(args);
}
```

5.2.4 函数调用和网络请求

在配置和实例生成结束之后，就可以直接使用 FeignClient 接口类的实例，调用它的函数来发送网络请求。在调用其函数的过程中，由于设置了 MethodHandler，所以最终函数调用会执行 SynchronousMethodHandler 的 invoke 方法。在该方法中，OpenFeign 会将函数的实际参数值与之前生成的 RequestTemplate 进行结合，然后发送网络请求。

图 5-6 是 OpenFeign 发送网络请求时几个关键类的交互流程图，大概分为三个阶段：一是将函数实际参数值添加到 RequestTemplate 中；二是调用 Target 生成具体的 Request 对象；三是调用 Client 来发送网络请求，然后将 Response 转化为对象进行返回。

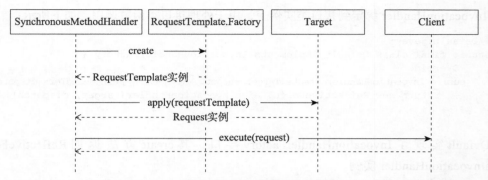

图 5-6　OpenFeign 的网络请求流程图

invoke 方法的代码如下所示:

```java
//SynchronousMethodHandler.java
final class SynchronousMethodHandler implements MethodHandler {
    public Object invoke(Object[] argv) throws Throwable {
        //根据函数参数创建RequestTemplate实例,buildTemplateFromArgs是RequestTemplate.
           Factory接口的实例,在当前状况下是
        //BuildTemplateByResolvingArgs类的实例
        RequestTemplate template = buildTemplateFromArgs.create(argv);
        Retryer retryer = this.retryer.clone();
        while (true) {
            try {
                return executeAndDecode(template);
            } catch (RetryableException e) {
                retryer.continueOrPropagate(e);
                if (logLevel != Logger.Level.NONE) {
                    logger.logRetry(metadata.configKey(), logLevel);
                }
                continue;
            }
        }
    }
}
```

如上代码所示,SynchronousMethodHandler 的 invoke 方法先创建了 RequestTemplate 对象。在该对象的创建过程中,使用到之前收集的函数信息 MethodMetadata。遍历 MethodMetadata 中参数相关的 indexToName,然后根据索引从 invoke 的参数数组中获得对应的值,将其填入对应的键值对中。然后依次处理查询和头部相关的参数值。invoke 方法调用 RequestTemplate.Factory 的 create 方法创建 RequestTemplate 对象,代码如下所示:

```java
//RequestTemplate.Factory
public RequestTemplate create(Object[] argv) {
    RequestTemplate mutable = new RequestTemplate(metadata.template());
    //设置URL
        if (metadata.urlIndex() != null) {
        int urlIndex = metadata.urlIndex();
        checkArgument(argv[urlIndex] != null, "URI parameter %s was null", urlIndex);
```

```
        mutable.insert(0, String.valueOf(argv[urlIndex]));
    }
    Map<String, Object> varBuilder = new LinkedHashMap<String, Object>();
    //遍历MethodMetadata中所有关于参数的索引及其对应名称的配置信息
    for (Entry<Integer, Collection<String>> entry : metadata.indexToName().
        entrySet()) {
        int i = entry.getKey();
        //entry.getKey就是参数的索引
        Object value = argv[entry.getKey()];
        if (value != null) { // Null values are skipped.
            //indexToExpander保存着将各种类型参数的值转换为string类型的Expander转换器
            if (indexToExpander.containsKey(i)) {
                //将value值转换为string
                value = expandElements(indexToExpander.get(i), value);
            }
            for (String name : entry.getValue()) {
                varBuilder.put(name, value);
            }
        }
    }
    RequestTemplate template = resolve(argv, mutable, varBuilder);
    //设置queryMap参数
    if (metadata.queryMapIndex() != null) {
        template = addQueryMapQueryParameters((Map<String, Object>) argv[metadata.
            queryMapIndex()], template);
    }
    //设置headersMap参数
    if (metadata.headerMapIndex() != null) {
        template = addHeaderMapHeaders((Map<String, Object>) argv[metadata.
            headerMapIndex()], template);
    }
    return template;
}
```

resolve 首先会替换 URL 中的 pathValues，然后对 URL 进行编码，接着将所有头部信息进行转化，最后处理请求的 Body 数据，如下所示：

```
//RequestTemplate.Factory
RequestTemplate resolve(Map<String, ?> unencoded, Map<String, Boolean> alreadyEncoded) {
    //替换query数值，将{queryVariable}替换成实际值
    replaceQueryValues(unencoded, alreadyEncoded);
    Map<String, String> encoded = new LinkedHashMap<String, String>();
    //把所有的参数都进行编码
    for (Entry<String, ?> entry : unencoded.entrySet()) {
        final String key = entry.getKey();
        final Object objectValue = entry.getValue();
        String encodedValue = encodeValueIfNotEncoded(key, objectValue, alreadyEncoded);
        encoded.put(key, encodedValue);
    }
    //编码url
    String resolvedUrl = expand(url.toString(), encoded).replace("+", "%20");
    if (decodeSlash) {
        resolvedUrl = resolvedUrl.replace("%2F", "/");
```

```java
        }
        url = new StringBuilder(resolvedUrl);
        Map<String, Collection<String>> resolvedHeaders = new LinkedHashMap<String,
            Collection<String>>();
        //将头部都进行串行化
        for (String field : headers.keySet()) {
            Collection<String> resolvedValues = new ArrayList<String>();
            for (String value : valuesOrEmpty(headers, field)) {
                String resolved = expand(value, unencoded);
                resolvedValues.add(resolved);
            }
            resolvedHeaders.put(field, resolvedValues);
        }
        headers.clear();
        headers.putAll(resolvedHeaders);
        //处理body
        if (bodyTemplate != null) {
            body(urlDecode(expand(bodyTemplate, encoded)));
        }
        return this;
    }
```

executeAndDecode 方法会根据 RequestTemplate 生成 Request 对象，然后交给 Client 实例发送网络请求，最后返回对应的函数返回类型的实例。executeAndDecode 方法的具体实现如下所示：

```java
//SynchronousMethodHandler.java
Object executeAndDecode(RequestTemplate template) throws Throwable {
    //根据RequestTemplate生成Request
    Request request = targetRequest(template);
    Response response;
    //client发送网络请求，client可能为okhttpclient和apacheClient
    try {
        response = client.execute(request, options);
        response.toBuilder().request(request).build();
    } catch (IOException e) {
        //...
    }
    try {
        //如果response的类型就是函数返回类型，那么可以直接返回
        if (Response.class == metadata.returnType()) {
            if (response.body() == null) {
                return response;
            }
            // 设置body
            byte[] bodyData = Util.toByteArray(response.body().asInputStream());
            return response.toBuilder().body(bodyData).build();
        }
    } catch (IOException e) {
        //...
    }
}
```

OpenFeign 也提供了 RequestInterceptor 机制，在由 RequestTemplate 生成 Request 的过程中，会调用所有 RequestInterceptor 对 RequestTemplate 进行处理。而 Target 是生成 JAXRS 2.0 网络请求 Request 的接口类。RequestInterceptor 处理的具体实现如下所示：

```
//SynchronousMethodHandler.java
//按照RequestTemplate来创建Request
Request targetRequest(RequestTemplate template) {
    //使用请求拦截器为每个请求添加固定的header信息。例如BasicAuthRequestInterceptor,
    //它是添加Authorization header字段的
    for (RequestInterceptor interceptor : requestInterceptors) {
        interceptor.apply(template);
    }
    return target.apply(new RequestTemplate(template));
}
```

Client 是用来发送网络请求的接口类，有 OkHttpClient 和 RibbonClient 两个子类。OkhttpClient 调用 OkHttp 的相关组件进行网络请求的发送。OkHttpClient 的具体实现如下所示：

```
//OkHttpClient.java
public feign.Response execute(feign.Request input, feign.Request.Options options)
        throws IOException {
    //将feign.Request转换为Oktthp的Request对象
    Request request = toOkHttpRequest(input);
    //使用Okhttp的同步操作发送网络请求
    Response response = requestOkHttpClient.newCall(request).execute();
    //将Okhttp的Response转换为feign.Response
    return toFeignResponse(response).toBuilder().request(input).build();
}
```

5.3 进阶应用

本小节主要讲解 OpenFeign 相关的进阶应用，包括 Client 编解码器的自定义和请求/响应的压缩。

5.3.1 Decoder 与 Encoder 的定制化

Encoder 用于将 Object 对象转化为 HTTP 的请求 Body，而 Decoder 用于将网络响应转化为对应的 Object 对象。对于二者，OpenFeign 都提供了默认的实现，但是使用者可以根据自己的业务来选择其他的编解码方式。只需要在自定义配置类中给出 Decoder 和 Encoder 的自定义 Bean 实例，那么 OpenFeign 就可以根据配置，自动使用我们提供的自定义实例进行编解码操作。如下代码所示，CustomFeignConfig 配置类将 ResponseEntityDecoder 和 SpringEncoder 配置为 Feign 的 Decoder 与 Encoder 实例。

```
public class CustomFeignConfig {
    @Bean
    public Decoder feignDecoder() {
        HttpMessageConverter jacksonConverter = new MappingJackson2HttpMessageConverter
```

```
            (customObjectMapper());
        ObjectFactory<HttpMessageConverters> objectFactory = () -> new HttpMessage
            Converters(jacksonConverter);
        return new ResponseEntityDecoder(new SpringDecoder(objectFactory));
    }
    @Bean
    public Encoder feignEncoder(){
        HttpMessageConverter jacksonConverter = new MappingJackson2HttpMessageConverter
            (customObjectMapper());
        ObjectFactory<HttpMessageConverters> objectFactory = () -> new HttpMessage
            Converters(jacksonConverter);
        return new SpringEncoder(objectFactory);
    }
    public ObjectMapper customObjectMapper(){
        ObjectMapper objectMapper = new ObjectMapper();
        objectMapper.configure(DeserializationFeature.ACCEPT_EMPTY_STRING_AS_
            NULL_OBJECT, true);
        return objectMapper;
    }
}
```

MappingJackson2HttpMessageConverter 是转换 JSON 的底层转换器，除了该转换器之外，还有如表 5-1 所示的转换器类型，基本上涵盖了大多数网络请求编解码场景。

表 5-1 Spring Cloud 编解码转换器信息表

类 型	解 释
ByteArrayHttpMessageConverter	二进制数据的转换器，支持所有的媒体类型
StringHttpMessageConverter	字符串类型数据的转换器
ResourceHttpMessageConverter	Resource 类型数据的转换器
SourceHttpMessageConverter	javax.xml.transform.Source 类型数据的转换器
FormHttpMessageConverter	普通 HTML 表单和 multipart data 的转换器

5.3.2 请求 / 响应压缩

可以通过下面的属性配置来让 OpenFeign 在发送请求时进行 GZIP 压缩：

```
feign.compression.request.enabled=true
feign.compression.response.enabled=true
```

OpenFeign 的压缩配置属性和一般的 Web Server 配置类似。这些属性允许选择性地压缩某种类型的请求并设置最小的请求阈值，配置如下所示：

```
feign.compression.request.enabled=true
feign.compression.request.mime-types=text/xml,application/xml,application/json
feign.compression.request.min-request-size=2048
```

你也可以使用 FeignContentGzipEncodingInterceptor 来实现请求的压缩，需要在自定义配置文件中初始化该类型的实例，供 OpenFeign 使用，具体实现如下所示：

```
public class FeignContentGzipEncodingAutoConfiguration {
    @Bean
    public FeignContentGzipEncodingInterceptor feignContentGzipEncodingInterceptor
        (FeignClientEncodingProperties properties) {
        return new FeignContentGzipEncodingInterceptor(properties);
    }
}
```

5.4 本章小结

在 Spring Cloud 中，各个微服务一般以 HTTP 接口的形式暴露自身服务，因此在调用远程服务时推荐使用 HTTP 客户端。OpenFeign 是一种声明式、模板化的 HTTP 客户端。在 Spring Cloud 中使用 OpenFeign，就可以在使用 HTTP 请求远程服务时获得与调用本地方法一样的编码体验，开发者完全感知不到这是远程方法，更感知不到这是个 HTTP 请求。Spring Cloud OpenFeign 可以与 Spring Cloud Netflix Ribbon 和 Spring Cloud Netflix Hystrix 一起使用，来实现负载均衡与断路器机制。

Chapter 6 第 6 章

断路器：Hystrix

在分布式系统下，微服务之间不可避免地会发生相互调用，但是没有一个系统能够保证自身运行的绝对正确。微服务在调用过程中，很可能会面临依赖服务失效的问题，这些问题的发生有很多原因，有可能是因为微服务之间的网络通信出现较大的延迟，或者是被调用的微服务发生了调用异常，还有可能是因为依赖的微服务负载过大无法及时响应请求等。因此希望有一个公共组件能够在服务通过网络请求访问其他微服务时，对延迟和失败提供强大的容错能力，为服务间调用提供保护和控制。

Hystrix 是 Netflix 的一个开源项目，它能够在依赖服务失效的情况下，通过隔离系统依赖服务的方式，防止服务级联失败；同时 Hystrix 提供失败回滚机制，使系统能够更快地从异常中恢复。

本章中，第一小节将会搭建用于演示 Hystrix 服务断路作用的简单例子；第二小节将会对 Hystrix 中相关术语和设计原理进行介绍；第三小节将从源码的角度分析 Hystrix 的实现机制和运行原理；第四小节将对 Hystrix 中的配置属性和高级特性进行介绍。

6.1 基础应用

spring-cloud-netflix-hystrix 对 Hystrix 进行封装和适配，使 Hystrix 能够更好地运行于 Spring Cloud 环境中，为微服务间的调用提供强有力的容错机制。

Hystrix 具有如下的功能：

- 在通过第三方客户端访问（通常是通过网络）依赖服务出现高延迟或者失败时，为系统提供保护和控制。
- 在复杂的分布式系统中防止级联失败（服务雪崩效应）。

- 快速失败（Fail fast）同时能快速恢复。
- 提供失败回滚（Fallback）和优雅的服务降级机制。
- 提供近实时的监控、报警和运维控制手段。

6.1.1 RestTemplate 与 Hystrix

可以搭建包含 Hystrix 依赖的 SpringBoot 项目。首先添加 eureka-client 和 hystrix 的相关依赖，如下所示：

```
<dependency> <!--eureka-client相关依赖-->
    <groupId>org.springframework.cloud</groupId>
    <artifactId>spring-cloud-starter-netflix-eureka-client</artifactId>
</dependency>
<dependency> <!--hystrix相关依赖-->
    <groupId>org.springframework.cloud</groupId>
    <artifactId>spring-cloud-starter-netflix-hystrix</artifactId>
</dependency>
```

在 application.yml 中为 hystrix-service 服务配置注册中心地址，代码如下所示：

```
# application.yml
eureka:
    instance:
        instance-id: ${spring.application.name}:${vcap.application.instance_id:${spring.application.instance_id:${random.value}}}
    client:
        service-url:
            default-zone: http://localhost:8761/eureka/
spring:
    application:
        name: hystrix-service
server:
    port: 8876
```

这里的 Eureka Server 将使用第 4 章中构建的 Eureka Server 项目。通过 @EnableCircuitBreaker 注解开启 Hystrix，同时注入一个可以进行负载均衡的 RestTemplate，代码如下所示：

```
@SpringBootApplication
@EnableCircuitBreaker // 开启Hystrix
public class Chapter6HystrixApplication {
    public static void main(String[] args) {
        SpringApplication.run(Chapter6HystrixApplication.class, args);
    }
    // 注入可以进行负载均衡的RestTemplate
    @Bean
    @LoadBalanced
    RestTemplate restTemplate(){
        return new RestTemplate();
    }
}
```

编写相关的服务，代码如下所示：

```java
@Service
public class InstanceService {
    private static String DEFAULT_SERVICE_ID = "application";
    private static String DEFAULT_HOST = "localhost";
    private static int DEFAULT_PORT = 8080;
    private static Logger logger = LoggerFactory.getLogger(InstanceService.class);
    @Autowired
    RestTemplate restTemplate;
    @HystrixCommand(fallbackMethod = "instanceInfoGetFail")
    public Instance getInstanceByServiceIdWithRestTemplate(String serviceId){

        Instance instance = restTemplate.getForEntity("http://FEIGN-SERVICE/
            feign-service/instance/{serviceId}", Instance.class, serviceId).getBody();
        return instance;
    }
    private Instance instanceInfoGetFail(String serviceId){
        logger.info("Can not get Instance by serviceId {}", serviceId);
        return new Instance("error", "error", 0);
    }
}
```

通过 @HystrixCommand 注解为 getInstanceByServiceIdWithRestTemplate 方法指定回滚的方法 instanceInfoGetFail，该方法返回了全是 error 的实体类信息。在 getInstanceByServiceIdWithRestTemplate 方法中通过 restTemplate 调用 feign-service 服务的相关的接口，期望返回结果，通过 @HystrixCommand 注解将该方法纳入到 Hystrix 的监控中。

编写相关的控制器类，调用 getInstanceByServiceIdWithRestTemplate 方法，获得结果并返回。代码如下所示：

```java
@RestController
@RequestMapping("/instance")
public class InstanceController {
    private static final Logger logger = LoggerFactory.getLogger(InstanceController.
        class);
    @Autowired
    InstanceService instanceService;
    @RequestMapping(value = "rest-template/{serviceId}", method = RequestMethod.
        GET)
    public Instance getInstanceByServiceIdWithRestTemplate(@PathVariable("serviceId")
        String serviceId){
        logger.info("Get Instance by serviceId {}", serviceId);
        return instanceService.getInstanceByServiceIdWithRestTemplate(serviceId);
    }
}
```

依次启动 eureka-server、feign-service（feign-service 为第 5 章的 feign-service 项目）以及本服务。访问 http://localhost:8876/instance/rest-template/my-application 接口。结果如下所示：

```
{"serviceId":"my-application","host":"localhost","port":8080}
```

访问成功执行，返回预期结果。关闭 feign-service，再次访问 http://localhost:8876/

instance/rest-template/my-application 接口。结果如下所示：

```
{"serviceId":"error","host":"error","port":0}
```

这说明在 feign-service 服务不可用时，系统执行失败回滚方法，返回"error"结果。

6.1.2 OpenFeign 与 Hystrix

使用 OpenFeign 需要添加相关依赖，在上一小节的基础上添加以下依赖：

```
<dependency> <!--openfegin的相关依赖-->
    <groupId>org.springframework.cloud</groupId>
    <artifactId>spring-cloud-starter-openfeign</artifactId>
</dependency>
```

OpenFeign 是自带 Hystrix，但是默认没有打开，在 application.yml 中添加以下配置开启 Hystrix：

```
feign:
  hystrix:
    enabled: true
```

在启动类中添加 @EnableFeignClients 注解启动 OpenFeign，如下所示：

```
@SpringBootApplication
@EnableCircuitBreaker
@EnableFeignClients
public class Chapter6HystrixApplication {
    public static void main(String[] args) {
        SpringApplication.run(Chapter6HystrixApplication.class, args);
    }
    @Bean
    @LoadBalanced
    RestTemplate restTemplate(){
        return new RestTemplate();
    }
}
```

添加 FeginClient 接口，调用 feign-service 服务，同时将 InstanceClientFallBack 指定为失败回滚类。具体代码如下所示：

```
@FeignClient(value = "feign-service", fallback = InstanceClientFallBack.class)
public interface InstanceClient {
    @RequestMapping(value = "/feign-service/instance/{serviceId}", method =
        RequestMethod.GET)
    public Instance getInstanceByServiceId(@PathVariable("serviceId") String
        serviceId);
}
```

InstanceClientFallBack 继承 InstanceClient 接口，提供相关回滚方法。具体代码如下所示：

```
@Component
public class InstanceClientFallBack implements InstanceClient {
```

```
private static Logger logger = LoggerFactory.getLogger(InstanceClientFallBack.
    class);

@Override
public Instance getInstanceByServiceId(String serviceId) {
    logger.info("Can not get Instance by serviceId {}", serviceId);
    return new Instance("error", "error", 0);
}
}
```

在 InstanceService 添加相关的服务，直接调用 instanceClient#getInstanceByServiceId 方法。代码如下所示：

```
@Autowired
InstanceClient instanceClient;

public Instance getInstanceByServiceIdWithFeign(String serviceId){
    Instance instance = instanceClient.getInstanceByServiceId(serviceId);
    return instance;
}
```

在 InstanceController 中添加相关的查看接口，代码如下所示：

```
@RequestMapping(value = "feign/{serviceId}", method = RequestMethod.GET)
public Instance getInstanceByServiceIdWithFeign(@PathVariable("serviceId") String
    serviceId){
    logger.info("Get Instance by serviceId {}", serviceId);
    return instanceService.getInstanceByServiceIdWithFeign(serviceId);
}
```

依次启动 eureka-server、feign-service 以及本服务。访问 http://localhost:8876/instance/feign/my-application 接口。结果如下所示：

```
{"serviceId":"my-application","host":"localhost","port":8080}
```

访问成功执行，返回预期的结果。关闭 feign-service，再次访问 http://localhost:8876/instance/feign/my-application 接口。结果如下所示：

```
{"serviceId":"error","host":"error","port":0}
```

这说明 OpenFeign 中的失败回滚发挥了作用。

6.2 Hystrix 原理

6.2.1 服务雪崩

服务雪崩效应是一种因服务提供者的不可用导致服务调用者的不可用，并将不可用逐渐放大的过程，如图 6-1 所示。

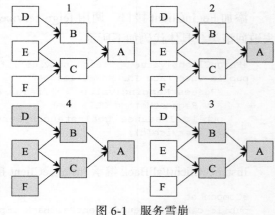

图 6-1 服务雪崩

其中，A 作为基础的服务提供者，为 B 和 C 提供服务，D、E、F 是 B 和 C 服务的调用者，当 A 不可用时，将引起 B 和 C 的不可用，并将这种不可用放大到 D、E、F，从而可能导致整个系统的不可用，服务雪崩的产生可能导致分布式系统的瘫痪。

服务雪崩效应的产生一般有三个流程，首先是服务提供者不可用，然后重试会导致网络流量加大，最后导致服务调用者不可用。

导致服务提供者不可用的原因有很多：可能是因为服务器的宕机或者网络故障；也可能是因为程序存在的缺陷；也有可能是大量的请求导致服务提供者的资源受限无法及时响应；还有可能是因为缓存击穿造成服务提供者超负荷运行等等，毕竟没有人能保证软件的完全正确。

在服务提供者不可用发生之后，用户可能无法忍受长时间的等待，不断地发送相同的请求，服务调用者重新调用服务提供者，同时服务提供者中可能存在对异常的重试机制，这些都会加大对服务提供者的请求流量。然而此时的服务提供者已经是一艘破船，它也无能为力，无法返回有效的结果。

最后是服务调用者因为服务提供者的不可用导致了自身的崩溃。当服务调用者使用同步调用的时候，大量的等待线程将会耗尽线程池中的资源，最终导致服务调用者的宕机，无法响应用户的请求，服务雪崩效应就此发生了。

6.2.2　断路器

在分布式系统中，不同服务之间的调用非常常见，当服务提供者不可用时就很有可能发生服务雪崩效应，导致整个系统的不可用。所以为了预防这种情况的发生，可以使用断路器模式进行预防（类比电路中的断路器，在电路过大的时候自动断开，防止电线过热损害整条电路）。

断路器将远程方法调用包装到一个断路器对象中，用于监控方法调用过程的失败。一旦该方法调用发生的失败次数在一段时间内达到一定的阀值，那么这个断路器将会跳闸，在接下来时间里再次调用该方法将会被断路器直接返回异常，而不再发生该方法的真实调用。这样就避免了服务调用者在服务提供者不可用时发送请求，从而减少线程池中资源的消耗，保护了服务调用者。图 6-2 为断路器时序图。

如图 6-2 所示，虽然断路器在打开的时候避免了被保护方法的无效调用，但是当情况恢复正常时，需要外部干预来重置断路器，使得方法调用可以重新发生。所以合理的断路器应该具备一定的开关转化逻辑，它需要一个机制来控制它的重新闭合，图 6-3 展示了一个通过重置时间来决定断路器的重新闭合的逻辑。

- **关闭状态**：断路器处于关闭状态，统计调用失败次数，在一段时间内达到一定的阀值后断路器打开。
- **打开状态**：断路器处于打开状态，对方法调用直接返回失败错误，不发生真正的方法调用。设置了一个重置时间，在重置时间结束后，断路器来到半开状态。

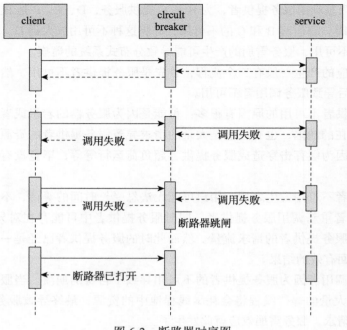

图 6-2　断路器时序图

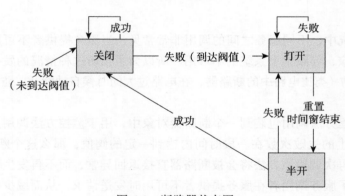

图 6-3　断路器状态图

- **半开状态**：断路器处于半开状态，此时允许进行方法调用，当调用都成功了（或者成功到达一定的比例），关闭断路器，否则认为服务没有恢复，重新打开断路器。

　　断路器的打开能保证服务调用者在调用异常服务时，快速返回结果，避免大量的同步等待，减少服务调用者的资源消耗。并且断路器能在打开一段时间后继续侦测请求执行结果，判断断路器是否能关闭，恢复服务的正常调用。

6.2.3　服务降级操作

　　断路器为隔断服务调用者和异常服务提供者防止服务雪崩的现象，提供了一种保护措施。而服务降级是为了在整体资源不够的时候，适当放弃部分服务，将主要的资源投放到核心服务中，待渡过难关之后，再重启已关闭的服务，保证了系统核心服务的稳定。

在 Hystrix 中，当服务间调用发生问题时，它将采用备用的 Fallback 方法代替主方法执行并返回结果，对失败服务进行了服务降级。当调用服务失败次数在一段时间内超过了断路器的阀值时，断路器将打开，不再进行真正的方法调用，而是快速失败，直接执行 Fallback 逻辑，服务降级，减少服务调用者的资源消耗，保护服务调用者中的线程资源，如图 6-4 所示。

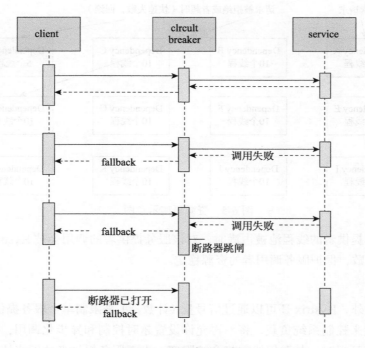

图 6-4　服务降级时序图

6.2.4　资源隔离

在货船中，为了防止漏水和火灾的扩散，一般会将货仓进行分割，避免了一个货仓出事导致整艘船沉没的悲剧。同样的，在 Hystrix 中，也采用了舱壁模式，将系统中的服务提供者隔离起来，一个服务提供者延迟升高或者失败，并不会导致整个系统的失败，同时也能够控制调用这些服务的并发度。

1. 线程与线程池

Hystrix 通过将调用服务线程与服务访问的执行线程分隔开来，调用线程能够空出来去做其他的工作而不至于因为服务调用的执行阻塞过长时间。

在 Hystrix 中，将使用独立的线程池对应每一个服务提供者，用于隔离和限制这些服务。

于是，某个服务提供者的高延迟或者饱和资源受限只会发生在该服务提供者对应的线程池中。

如图 6-5 所示，Dependency D 的调用失败或者高延迟仅会导致自身对应的线程池中的 5 个线程阻塞，并不会影响其他服务提供者的线程池。系统完全与服务提供者请求隔离开来，即使服务提供者对应的线程完全耗尽，并不会影响系统中的其他请求。

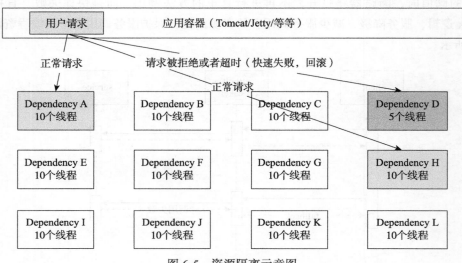

图 6-5　资源隔离示意图

注意在服务提供者的线程池被占满时，对该服务提供者的调用会被 Hystrix 直接进入回滚逻辑，快速失败，保护服务调用者的资源稳定。

2. 信号量

除了线程池外，Hystrix 还可以通过信号量（计数器）来限制单个服务提供者的并发量。如果通过信号量来控制系统负载，将不再允许设置超时控制和异步化调用，这就表示在服务提供者出现高延迟时，其调用线程将会被阻塞，直至服务提供者的网络请求超时。如果对服务提供者的稳定性有足够的信心，可以通过信号量来控制系统的负载。

6.2.5　Hystrix 实现思路

结合上面的介绍，我们可以简单理解一下 Hystrix 的实现思路：

- ❏ 它将所有的远程调用逻辑封装到 HystrixCommand 或者 HystrixObservableCommand 对象中，这些远程调用将会在独立的线程中执行（资源隔离），这里使用了设计模式中的命令模式。
- ❏ Hystrix 对访问耗时超过设置阀值的请求采用自动超时的策略。该策略对所有的命令都有效（如果资源隔离的方式为信号量，该特性将失效），超时的阀值可以通过命令配置进行自定义。
- ❏ 为每一个服务提供者维护一个线程池（或者信号量），当线程池被占满时，对于该服务提供者的请求将会被直接拒绝（快速失败）而不是排队等待，减少系统的资源等待。
- 针对请求服务提供者划分出成功、失效、超时和线程池被占满等四种可能出现的情况。
- 断路器机制将在请求服务提供者失败次数超过一定阀值后手动或者自动切断服务一段时间。
- 当请求服务提供者出现服务拒绝、超时和短路（多个服务提供者依次顺序请求，前面的服务提供者请求失败，后面的请求将不会发出）等情况时，执行其 Fallback 方法，服务降级。
- 提供接近实时的监控和配置变更服务。

6.3 源码解析

使用 Hystrix 后的远程调用流程如图 6-6 所示。

简单的流程如下：

1）构建 HystrixCommand 或者 HystrixObservableCommand 对象。
2）执行命令。
3）检查是否有相同命令执行的缓存。
4）检查断路器是否打开。
5）检查线程池或者信号量是否被消耗完。
6）调用 HystrixObservableCommand#construct 或 HystrixCommand#run 执行被封装的远程调用逻辑。
7）计算链路的健康情况。
8）在命令执行失败时获取 Fallback 逻辑。
9）返回成功的 Observable。

接着我们通过源码来逐步理解这些过程。

6.3.1 封装 HystrixCommand

1. @HystrixCommand 注解

在基础应用中我们使用 @HystrixCommand 注解来包装需要保护的远程调用方法。首先查看该注解的相关属性，代码如下所示：

```
//HystrixCommand.java
@Target({ElementType.METHOD})
@Retention(RetentionPolicy.RUNTIME)
@Inherited
@Documented
public @interface HystrixCommand {
    // 命令分组键用于报告、预警以及面板展示
```

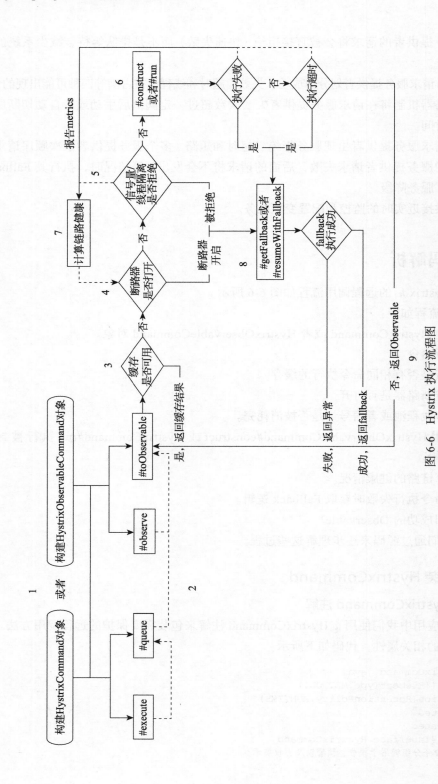

图 6-6 Hystrix 执行流程图

```java
// 默认为被注解方法的运行时类名
String groupKey() default "";
// Hystrix的命令键，用于区分不同的注解方法
// 默认为注解方法的名称
String commandKey() default "";
// 线程池键用来指定命令执行的HystrixThreadPool
String threadPoolKey() default "";
// 指定Fallback方法名，Fallback方法也可以被HystrixCommand注解
String fallbackMethod() default "";
// 自定义命令的相关配置
HystrixProperty[] commandProperties() default {};
// 自定义线程池的相关配置
HystrixProperty[] threadPoolProperties() default {};
// 定义忽略哪些异常
Class<? extends Throwable>[] ignoreExceptions() default {};
// 默认的fallback
String defaultFallback() default "";
...
}
```

一般来说，对于 HystrixCommand 的配置，仅需要关注 fallbackMethod 方法，当然如果对命令和线程池有特定需要，可以进行额外的配置。

除了 @HystrixCommand 还有一个 @HystrixCollapser 注解用于请求合并操作，但是需要与 @HystrixCommand 结合使用，批量操作的方法必须被 @HystrixCommand 注解。例子如下所示：

```java
//HystrixCollapser.java
@HystrixCollapser(batchMethod = "getInstanceBuServiceIds")
public Future<Instance> getInstanceByServiceIds(String serviceId) {
    return null;
}
@HystrixCommand
public List<Instance> getInstanceBuServiceIds(List<String> serviceIds){
    List<Instance> instances = new ArrayList<>();
    for(String s : serviceIds){
        instances.add(new Instance(s, DEFAULT_HOST, DEFAULT_PORT));
    }
    return instances;
}
```

2. HystrixCommandAspect 切面

被注解修饰的方法将会被 HystrixCommand 包装执行，在 Hystrix 中通过 Aspectj 切面的方式来将被注解修饰的方法进行封装调用。具体代码如下所示：

```java
//HystrixCommandAspect.java
//切面定义
@Around("hystrixCommandAnnotationPointcut() || hystrixCollapserAnnotationPointcut()")
public Object methodsAnnotatedWithHystrixCommand(final ProceedingJoinPoint
    joinPoint) throws Throwable {
    ...
    // 通过工厂的方式构建metaHolder
```

```java
        MetaHolderFactory metaHolderFactory = META_HOLDER_FACTORY_MAP.get(HystrixPointcutType.
            of(method));
        MetaHolder metaHolder = metaHolderFactory.create(joinPoint); //1
        HystrixInvokable invokable = HystrixCommandFactory.getInstance().create
            (metaHolder);
        ExecutionType executionType = metaHolder.isCollapserAnnotationPresent() ?
            metaHolder.getCollapserExecutionType() : metaHolder.getExecutionType();

        Object result;
        try {
            if (!metaHolder.isObservable()) {
                result = CommandExecutor.execute(invokable, executionType, metaHolder);
            } else {
                result = executeObservable(invokable, executionType, metaHolder);
            }
        } catch (HystrixBadRequestException e) {
            throw e.getCause() != null ? e.getCause() : e;
        } catch (HystrixRuntimeException e) {
            throw hystrixRuntimeExceptionToThrowable(metaHolder, e);
        }
        return result;
    }
```

上面代码主要执行步骤如下：

1）通过 MetaHolderFactory 构建出被注解修饰方法中用于构建 HystrixCommand 必要信息集合类 MetaHolder。

2）根据 MetaHolder 通过 HystrixCommandFactory 构建出合适的 HystrixCommand。

3）委托 CommandExecutor 执行 HystrixCommand，得到结果。

MetaHolder 持有用于构建 HystrixCommand 和与被包装方法相关的必要信息，如被注解的方法、失败回滚执行的方法和默认的命令键等属性。其属性代码如下所示：

```java
//MetaHolder.java
@Immutable
public final class MetaHolder {
    ...
    private final Method method; //被注解的方法
    private final Method cacheKeyMethod;
    private final Method ajcMethod;
    private final Method fallbackMethod; // 失败回滚执行的方法
    ...
    private final String defaultGroupKey; // 默认的group键
    private final String defaultCommandKey; // 默认的命令键
    private final String defaultCollapserKey; // 默认的合并请求键
    private final String defaultThreadPoolKey; // 默认的线程池键
    private final ExecutionType executionType; // 执行类型
    ...
}
```

在 HystrixCommandFactory 类中，用于创建 HystrixCommand 的方法如下所示：

```java
//HystrixCommandFactory.java
```

```java
public HystrixInvokable create(MetaHolder metaHolder) {
    HystrixInvokable executable;
    // 构建请求合并的命令
    if (metaHolder.isCollapserAnnotationPresent()) {
        executable = new CommandCollapser(metaHolder);
    } else if (metaHolder.isObservable()) {
        executable = new GenericObservableCommand(HystrixCommandBuilderFactory.
            getInstance().create(metaHolder));
    } else {
        executable = new GenericCommand(HystrixCommandBuilderFactory.getInstance().
            create(metaHolder));
    }
    return executable;
}
```

根据 MetaHolder#isObservable 方法返回属性的不同，将会构建不同的命令，比如 HystrixCommand 或者 HystrixObservableCommand，前者将同步或者异步执行命令，后者异步回调执行命令。Hystrix 根据被包装方法的返回值来决定命令的执行方式，判断代码如下：

```java
//CommandMetaHolderFactory.java
...
ExecutionType executionType = ExecutionType.getExecutionType(method.getReturnType());
...
public enum ExecutionType {
    // 异步执行命令
    ASYNCHRONOUS,
    // 同步执行命令
    SYNCHRONOUS,
    // 响应式执行命令(异步回调)
    OBSERVABLE;
    private static final Set<? extends Class> RX_TYPES = ImmutableSet.of(Observable.
        class, Single.class, Completable.class);
    // 根据方法的返回类型返回对应的ExecutionType
    public static ExecutionType getExecutionType(Class<?> type) {
        // Future为异步执行
        if (Future.class.isAssignableFrom(type)) {
            return ExecutionType.ASYNCHRONOUS;
        } else if (isRxType(type)) {
        // 属于 rxType为异步回调执行
            return ExecutionType.OBSERVABLE;
        } else {
        // 其他为同步执行
            return ExecutionType.SYNCHRONOUS;
        }
    }
}
```

根据被包装方法的返回值类型决定命令执行的 ExecutionType，从而决定构建 HystrixCommand 还是 HystrixObservableCommand。其中 Future 类型的返回值将会被异步执行，rx 类型的返回值将会被异步回调执行，其他的类型将会被同步执行。

CommandExecutor 根据 MetaHolder 中 ExecutionType 执行类型的不同，选择同步执行、

异步执行还是异步回调执行,返回不同的执行结果。同步执行,直接返回结果对象;异步执行,返回 Future,封装了异步操作的结果;异步回调执行将返回 Observable,封装响应式执行的结果,可以通过它对执行结果进行订阅,在执行结束后进行特定的操作。

图 6-7 为本节介绍的类的相关类图结构。

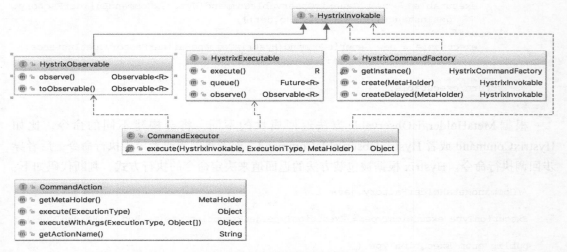

图 6-7　Hystrix 中的命令模式

通过代码和类图,会发现上述类结构中使用了设计模式中的命令模式进行设计。这其中 HystrixInvokable 是 HystrixCommand 的标记接口,继承了该接口的类都是可以被执行的 HystrixCommand。提供具体方法的接口为 HystrixExecutable,用于同步执行和异步执行命令,HystrixObservable 用于异步回调执行命令,它们对应命令模式中的 Command 和 ConcreteCommand。CommandExecutor 将调用 HystrixInvokable 执行命令,相当于命令模式中的 Invoker。HystrixCommandFactory 将生成命令,而 HystrixCommandAspect 相当于命令模式中的客户端情景类 Client。CommandAction 中持有 Fallback 方法或者被 @HystrixCommand 注解的远程调用方法,相当于命令模式中的 Receiver。图 6-8 为通用命令模式类图,可以将其与图 6-7 进行对比。

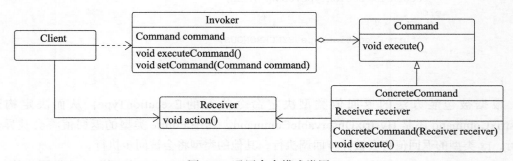

图 6-8　通用命令模式类图

6.3.2 HystrixCommand 类结构

下面将介绍 Hystrix 中整个命令的类结构体系，以及对其中的关键实现代码进行讲解。HystrixCommamd 核心类图如图 6-9 所示。

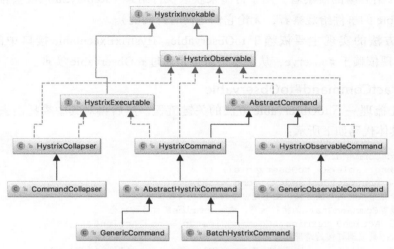

图 6-9　HystrixCommand 核心类图

虽然类图很复杂，但是最终实现类只有三个，分别是同步或异步执行命令的 GenericCommand；请求合并执行命令的 BatchHystrixCommand，以及异步回调执行命令的 GenericObservableCommand。以上三个类的关键实现都位于 AbstractCommand 抽象类中，所以我们会对 AbstractCommand 中源码进行重点讲解。

6.3.3 异步回调执行命令

本节我们将进入 AbstractCommand 抽象类中，了解 observe 和 toObservable 方法异步回调执行命令的具体实现。

1. AbstractCommand#observe

observe 实现如下代码所示：

```
//AbstractCommand.java
public Observable<R> observe() {
    ReplaySubject<R> subject = ReplaySubject.create();
    final Subscription sourceSubscription = toObservable().subscribe(subject);
    return subject.doOnUnsubscribe(new Action0() {
        @Override
        public void call() {
            sourceSubscription.unsubscribe();
        }
    });
}
```

在observe方法中，首先将创建一个方法ReplaySubject，rx中的Subject既是一个Observable也是一个Observer。接着调用toObservable方法获取到懒执行的Observable，通过创建的ReplaySubject订阅该Observable，启动Observable中相关命令，同时返回ReplaySubject给后续的观察者，用于订阅来获取执行结果（ReplaySubject会推送所有来自原始Observable的事件给观察者，无论它们是何时订阅的）。

observe方法的实现主要依赖于toObservable。HystrixExecutable接口中的execute和queue方法实现依赖于#observe，从根本上讲也是通过toObservable实现。

2. AbstractCommand#toObservable

我们首先梳理一下toObservable方法的关键流程，然后再对其中委托出去的具体实现进行分解，具体代码如下所示：

```java
//AbstractCommand.java
public Observable<R> toObservable() {
    final AbstractCommand<R> _cmd = this;
    // 命令结束时的回调方法，主要是命令调用后的清理工作
    // 根据CommandState的执行状态，通过Metrics统计各种状态
    final Action0 terminateCommandCleanup = new Action0() {...};
    // 命令被取消订阅的清理回调方法
    final Action0 unsubscribeCommandCleanup = new Action0() {};
    // 构建执行命令,封装断路器、资源隔离逻辑
    final Func0<Observable<R>> applyHystrixSemantics = new Func0<Observable<R>>(){
        @Override
        public Observable<R> call() {
            // 如果没有订阅返回既不会开始也不会结束的Observable
            if (commandState.get().equals(CommandState.UNSUBSCRIBED)) {
                return Observable.never();
            }
            // 通过applyHystrixSemantics声明Observable
            return applyHystrixSemantics(_cmd);
        }
    };
    ...
    return Observable.defer(new Func0<Observable<R>>() {
        @Override
        public Observable<R> call() {
            // 执行状态转化有误，抛出异常
            if (!commandState.compareAndSet(CommandState.NOT_STARTED, CommandState.
                OBSERVABLE_CHAIN_CREATED)) {
                // 命令被多次执行，抛出异常
                throw new HystrixRuntimeException(...);
            }
            // 记录命令开始时间
            commandStartTimestamp = System.currentTimeMillis();
            ...
            final boolean requestCacheEnabled = isRequestCachingEnabled();
            final String cacheKey = getCacheKey();
            // 尝试从缓存中获取结果
            if (requestCacheEnabled) {
```

```
        ...
    // 如果缓存不为空,直接返回缓存结果
    }
    // 构建执行命令的Observable
    Observable<R> hystrixObservable =
    Observable.defer(applyHystrixSemantics)
        .map(wrapWithAllOnNextHooks);
    Observable<R> afterCache;
    // 将Observable封装成HystrixCachedObservable放到缓存中
    if (requestCacheEnabled && cacheKey != null) {
        ....
    } else {
        afterCache = hystrixObservable;
    }
     return afterCache
    .doOnTerminate(terminateCommandCleanup)
    .doOnUnsubscribe(unsubscribeCommandCleanup)
    .doOnCompleted(fireOnCompletedHook);
    }
});
}
```

这段代码非常长,我们一步步进行分解:

1)首先通过 Observable#defer 方法来构建返回的 Observable。以 Observable#defer 方式声明的 Observable 只有当有观察者订阅才会真正开始创建,并且是为每一个观察者创建一个新的 Observable,这就保证了 toObservable 方法返回的 Observable 是纯净的,并没有开始执行命令。

2)在构建 Observable 过程中,先通过 commandState 查看当前的命令执行状态,保证命令未开始执行并且每条命令只能执行一次。

3)如果允许请求缓存并且缓存存在的话,将尝试从缓存中获取对应的执行结果,并直接返回结果。

4)如果无法获取缓存,通过 applyHystrixSemantics 方法构建用于返回的 Observable。

5)如果允许请求缓存,将 Observable 放置到缓存中用于下一次调用。

6)最后为返回 Observable 添加提前定义好的回调方法。

在上述的流程中,需要重点关注两个地方,一个是 HystrixRequestCache,其内封装了缓存 Observable 的逻辑;另一个是 applyHystrixSemantics 回调方法,其内封装了断路、资源隔离等核心断路器逻辑。

3. HystrixRequestCache 请求缓存

HystrixRequestCache 对 Observable 进行缓存操作,使用每个命令特有的 cacheKey 对 Observable 进行缓存,通过 ConcurrentHashMap 保存缓存结果以保证线程安全。

HystrixRequestCache 中缓存的并不是直接的 Observable,而是被封装好的 HystrixCachedObservable。在 HystrixCachedObservable 中,通过 ReplaySubject 订阅需要缓

存的 Observable，保证了缓存的 Observable 能够多次执行，代码如下所示：

```java
//HystrixCachedObservable.java
public class HystrixCachedObservable<R> {
protected final Subscription originalSubscription;
protected final Observable<R> cachedObservable;
private volatile int outstandingSubscriptions = 0;
    protected HystrixCachedObservable(final Observable<R> originalObservable) {
    // 使用ReplaySubject订阅原始的Observable，并返回ReplaySubject，
    // 保证其从缓存取出后订阅者依然能够接受对应的事件，即命令依然能够执行
    ReplaySubject<R> replaySubject = ReplaySubject.create();
    this.originalSubscription = originalObservable
            .subscribe(replaySubject);
    this.cachedObservable = replaySubject
            .doOnUnsubscribe(new Action0() {...})
            .doOnSubscribe(new Action0() {...});
    }
}
```

4. applyHystrixSemantics 断路器判断与获取信号量

在 applyHystrixSemantics 回调方法中，通过 AbstractCommand#applyHystrixSemantics 方法声明 Observable。它主要工作是判断断路器是否打开，以及尝试获取信号量用于执行命令（仅在信号量隔离模式下生效），具体代码如下所示：

```java
//AbstractCommand.java
private Observable<R> applyHystrixSemantics(final AbstractCommand<R> _cmd) {
    // 标记在ExecutionHook中执行
    executionHook.onStart(cmd);
    // 判断HystrixCircuitBreaker判断命令是否可以执行
    if (circuitBreaker.attemptExecution()) {
        // 获得信号量
        final TryableSemaphore executionSemaphore = getExecutionSemaphore();
        final AtomicBoolean semaphoreHasBeenReleased = new AtomicBoolean(false);
        // 释放信号量的回调方法
        final Action0 singleSemaphoreRelease = new Action0() {...};
        // 标记异常的回调方法，对异常进行推送
        final Action1<Throwable> markExceptionThrown = new Action1<Throwable>() {
            @Override
            public void call(Throwable t) {
                eventNotifier.markEvent(HystrixEventType.EXCEPTION_THROWN, commandKey);
            }
        };
        // 尝试获取信号量
        if (executionSemaphore.tryAcquire()) {
            try {
                // 标记executionResult开始时间
                executionResult = executionResult.setInvocationStartTime(System.
                    currentTimeMillis());
                // 获取执行命令的Observable
                return executeCommandAndObserve(_cmd).doOnError(markExceptionThrown)
                    .doOnTerminate(singleSemaphoreRelease)
                    .doOnUnsubscribe(singleSemaphoreRelease);
```

```
                } catch (RuntimeException e) {
                    return Observable.error(e);
                }
            } else {
                return handleSemaphoreRejectionViaFallback();
            }
        } else {
            return handleShortCircuitViaFallback();
        }
    }
```

在 AbstractCommand#applyHystrixSemantics 中，首先通过断路器 HystrixCircuitBreaker 检查链路中的断路器是否开启，如果开启的话，执行断路失败逻辑 handleShortCircuitViaFallback 方法。如果通过断路器的检查，将会尝试获取信号量。如果不能获取信号量，那么执行信号量获取失败逻辑 handleSemaphoreRejectionViaFallback 方法。当上述检查都通过了，才执行 executeCommandAndObserve 方法获取执行命令的 Observable，并为该 Observable 配置回调操作，该回调操作在命令执行结束后以及取消订阅时用于释放信号量。

在介绍 executeCommandAndObserve 方法之前，我们先了解 ExecutionResult，它是一个用来记录命令执行中各种状态的类，主要记录以下属性：

```
//ExecutionResult.java
public class ExecutionResult {
    private final EventCounts eventCounts;
    // 执行失败时抛出的异常
    private final Exception failedExecutionException;
    // 执行异常
    private final Exception executionException;
    // 开始执行命令的时间
    private final long startTimestamp;
    // run()方法执行时间,即被HystrixCommand包装的方法执行的时间
    private final int executionLatency;
    ...
}
```

通过 ExecutionResult，Hystrix 可以记录 HystrixCommand 在不同执行阶段的状态和相关执行记录，用于统计和分析。

applyHystrixSemantics 方法最后将委托 executeCommandAndObserve 方法为命令配置执行异常回调方法从而为命令的执行保驾护航。

5. executeCommandAndObserve 配置执行异常回调方法

executeCommandAndObserve 方法主要用于为执行命令 Observable 配置执行失败的回调方法，对执行失败的结果进行记录和处理。具体代码如下所示：

```
//AbstractCommand.java
private Observable<R> executeCommandAndObserve(final AbstractCommand<R> _cmd) {
    final HystrixRequestContext currentRequestContext = HystrixRequestContext.
        getContextForCurrentThread();
```

```java
// 标记命令开始执行的回调方法
final Action1<R> markEmits = new Action1<R>() {...};
// 标记命令执行结束的回调方法
final Action0 markOnCompleted = new Action0() {...};
// 失败回滚逻辑
final Func1<Throwable, Observable<R>> handleFallback = new Func1<Throwable,
    Observable<R>>() {
    @Override
    public Observable<R> call(Throwable t) {
        // 断路器标记命令执行失败
        circuitBreaker.markNonSuccess();
        Exception e = getExceptionFromThrowable(t);
        executionResult = executionResult.setExecutionException(e);
        if (e instanceof RejectedExecutionException) {
        return handleThreadPoolRejectionViaFallback(e);
        } else if (t instanceof HystrixTimeoutException) {
            return handleTimeoutViaFallback();
        } else if (t instanceof HystrixBadRequestException) {
            return handleBadRequestByEmittingError(e);
        } else {
            if (e instanceof HystrixBadRequestException) {
                eventNotifier.markEvent(HystrixEventType.BAD_REQUEST, commandKey);
                return Observable.error(e);
            }
            return handleFailureViaFallback(e);
        }
    }
};
...
Observable<R> execution;
if (properties.executionTimeoutEnabled().get()) {
    execution = executeCommandWithSpecifiedIsolation(_cmd)
            .lift(new HystrixObservableTimeoutOperator<R>(_cmd));
} else {
    execution = executeCommandWithSpecifiedIsolation(_cmd);
}

return execution.doOnNext(markEmits)
    .doOnCompleted(markOnCompleted)
    .onErrorResumeNext(handleFallback)
    .doOnEach(setRequestContext);
}
```

上述代码中，handleFallback 失败回滚回调方法将根据执行过程中抛出的异常调用不同的方法进行处理，对线程获取失败处理用 handleThreadPoolRejectionViaFallback 方法，超时处理用 handleTimeoutViaFallback 方法，远程调用请求失败处理用 handleBadRequestByEmittingError 方法，Hystrix 自身执行异常处理用 handleFailureViaFallback 方法。另外，我们还知道在 applyHystrixSemantics 方法中的断路失败处理 handleShortCircuitViaFallback 方法和获取信号量失败处理 handleSemaphoreRejectionViaFallback 方法。至此我们已经知道了不同的执行失败结果将调用不同的方式进行处理，在后面的章节再对这些方法进行讲解。

在 executeCommandAndObserve 方法的最后，调用 executeCommandWithSpecifiedIsolation 方法为命令的执行配置资源隔离和添加超时控制。

6. executeCommandWithSpecifiedIsolation 配置线程隔离和超时控制

executeCommandWithSpecifiedIsolation 方法为命令构造了隔离的执行环境，提供两种资源隔离的方式，线程隔离和信号量隔离；如果 Hystrix 配置中开启了超时控制，还会通过 Observable#lift 方法将现有的 Observable 转化为添加了超时检查的 Observable。

executeCommandWithSpecifiedIsolation 方法根据配置中的隔离策略对命令执行采用了不同的资源隔离方式：ExecutionIsolationStrategy.THREAD 将使用线程隔离的方式，ExecutionIsolationStrategy.SEMAPHORE 将使用信号量隔离的方式。具体代码如下：

```java
//AbstractCommand.java
private Observable<R> executeCommandWithSpecifiedIsolation(final AbstractCommand<R> _cmd) {
    if (properties.executionIsolationStrategy().get() == ExecutionIsolationStrategy.
        THREAD) {
        return Observable.defer(new Func0<Observable<R>>() {
            @Override
            public Observable<R> call() {
                executionResult = executionResult.setExecutionOccurred();
                ...
                // 标记命令是通过线程隔离资源执行
                metrics.markCommandStart(commandKey, threadPoolKey, ExecutionIsolationStrategy.
                    THREAD);
                ...
                // 标记线程已经执行
                HystrixCounters.incrementGlobalConcurrentThreads();
                threadPool.markThreadExecution();
                endCurrentThreadExecutingCommand = Hystrix.startCurrentThreadExe
                    cutingCommand(getCommandKey());
                // 记录是使用线程隔离执行的
                executionResult = executionResult.setExecutedInThread();
                return getUserExecutionObservable(_cmd);
                ...
            }
        })
        ...
        // 指定命令在哪个线程执行
        .subscribeOn(threadPool.getScheduler(new Func0<Boolean>() {
            @Override
            public Boolean call() {
                return properties.executionIsolationThreadInterruptOnTimeout().get()
                    && _cmd.isCommandTimedOut.get() == TimedOutStatus.TIMED_OUT;
            }
        }));
    } else {
        return Observable.defer(new Func0<Observable<R>>() {
            @Override
            public Observable<R> call() {
                ...
```

```
            // 标记命令是通过信号量隔离执行
            metrics.markCommandStart(commandKey, threadPoolKey, ExecutionIsolationStrategy.
            SEMAPHORE);
            endCurrentThreadExecutingCommand = Hystrix.startCurrentThreadExe
                cutingCommand(getCommandKey());
            ...
            return getUserExecutionObservable(_cmd);
            ...
        }
    });
}
```

当以线程的方式隔离资源时,需要指定命令在哪一个线程执行,主要通过 HystrixThreadPool#getScheduler 方法获取相应的线程调度。信号量的获取在 AbstractCommand #applyHystrixSemantics 方法中执行。

最后,executeCommandWithSpecifiedIsolation 通过 getUserExecutionObservable 方法拿到了被封装的远程调用方法,在 Hystrix 的重重保护下执行远程方法以获取结果。

7. getExecutionObservable 配置被封装的远程调用方法

getUserExecutionObservable 方法将为命令获取在声明 HystrixCommand 时被包装的具体远程调用方法。在 getUserExecutionObservable 方法中,通过 getExecutionObservable 抽象方法将具体实现延迟到子类中。getExecutionObservable 方法在 HystrixCommand 中的相关实现如下所示:

```
//HystrixCommand.java
@Override
final protected Observable<R> getExecutionObservable() {
    return Observable.defer(new Func0<Observable<R>>() {
        @Override
        public Observable<R> call() {
            try {
                // 执行被包装到HystrixCommand的远程调用逻辑
                return Observable.just(run());
            } catch (Throwable ex) {
                return Observable.error(ex);
            }
        }
    }).doOnSubscribe(new Action0() {
        @Override
        public void call() {
         executionThread.set(Thread.currentThread());
        }
    });
}
```

在上述代码中,run 方法也是延迟到子类中实现,在高级应用中,我们将尝试直接继承 HystrixCommand 和 HystrixObservableCommand 构建对应的 HystrixCommand,在

HystrixCommand 的默认实现 GenericCommand 中，run 方法是通过创建 HystrixCommand 时传递的 CommandActions 提供具体实现。CommandActions 持有 commandAction 和 fallbackAction，分别对应 HystrixCommand 中远程调用方法和失败回滚方法。

HystrixObservableCommand 中相关逻辑也类似，这里就不再描述和展示代码实现。

6.3.4 异步执行命令和同步执行命令

了解了 HystrixObservable 中的两个关键接口在 AbstractCommand 中的实现后，接下来我们需要到 HystrixCommand 中了解 execute 同步执行命令和 queue 异步执行命令的相关实现。

1. HystrixCommand#queue

代码如下所示：

```java
//HystrixCommand.java
public Future<R> queue() {
    final Future<R> delegate = toObservable().toBlocking().toFuture();
    final Future<R> f = new Future<R>() {
        ...
    }
    return f;
}
```

queue 方法中将 AbstractCommand#toObservable 获取到的 Observable 通过 toBlocking 转化为具备阻塞功能的 BlockingObservable，再通过 toFuture 方法获取到能够执行 run 抽象方法的 Future，最后通过 Future 得到正在异步执行的命令的执行结果。

2. HystrixCommand#execute

代码如下所示：

```java
//HystrixCommand.java
public R execute() {
    try {
        return queue().get();
    } catch (Exception e) {
        throw Exceptions.sneakyThrow(decomposeException(e));
    }
}
```

exeute 方法通过 queue 获取到 Future，使用 Future#get 方法获取到命令的执行结果，它将一直阻塞线程直到有执行结果返回。

6.3.5 断路器逻辑

HystrixCircuitBreaker 是 Hystrix 提供断路器逻辑的核心接口，它通过 HystrixCommandKey（由 @HystrixCommand 的 commandKey 构造而成）与每一个 HystrixCommand 绑定。在 HystrixCircuitBreaker.Factory 中使用 ConcurrentHashMap 维持了基于 HystrixCommandKey

的 HystrixCircuitBreaker 的单例映射表，保证具备相同 CommandKey 的 HystrixCommand 对应同一个断路器。

HystrixCircuitBreaker 提供的接口如下代码所示：

```java
//HystrixCircuitBreaker.java
public interface HystrixCircuitBreaker {
    // 是否允许命令执行
    boolean allowRequest();
    // 断路器是否打开(断路)
    boolean isOpen();
    // 在半开状态时作为命令执行成功反馈
    void markSuccess();
    // 在半开状态时作为命令执行失败反馈
    void markNonSuccess();
    // 尝试执行命令，该接口可能会修改断路器的状态
    boolean attemptExecution();
}
```

attemptExecution 方法在命令构建执行的过程中（AbstractCommand#applyHystrixSemantics 方法中）用于判断断路器是否打开，它的功能基本与 allowRequest 方法类似，但是它可能会修改断路器的状态，如将其从打开状态修改到半开状态。

HystrixCircuitBreaker 有两个默认实现，一个是 NoOpCircuitBreaker，顾名思义即空实现，不会发挥任何断路器的功能，另一个实现为 HystrixCircuitBreakerImpl，为断路器的真正实现。

1. HystrixCircuitBreakerImpl 断路器具体实现

在 HystrixCircuitBreakerImpl 中定义了三种状态：关闭、开启、半开，与在 Hystrix 原理中介绍的断路器的三种状态相对应，如下所示：

```java
enum Status {
    CLOSED, OPEN, HALF_OPEN;
}
```

allowRequest 方法代码如下所示：

```java
//HystrixCircuitBreakerImpl.java
@Override
    public boolean allowRequest() {
        // 断路器强制打开
        if (properties.circuitBreakerForceOpen().get()) {
            return false;
        }
        // 断路器强制关闭
        if (properties.circuitBreakerForceClosed().get()) {
            return true;
        }
        // 打开时间为空，即未打开
        if (circuitOpened.get() == -1) {
            return true;
        } else {
```

```java
        // 半开状态
        if (status.get().equals(Status.HALF_OPEN)) {
            return false;
        } else {
            // 重置时间是否结束，这将允许尝试执行
            return isAfterSleepWindow();
        }
    }
}

private boolean isAfterSleepWindow() {
    final long circuitOpenTime = circuitOpened.get();
    final long currentTime = System.currentTimeMillis();
    final long sleepWindowTime = properties.circuitBreakerSleepWindowInMilli
        seconds().get();
    return currentTime > circuitOpenTime + sleepWindowTime;
}
```

断路器强制开始和关闭的相关配置可以通过配置中心的方式动态修改，这样就可以人为干预断路器的状态，方便调试。断路器打开的时候将会记录一个打开时间，用于判断断路器是否打开，通过它与配置中的 circuitBreakerSleepWindowInMilliseconds 重置时间结合判断：在断路器打开一段时间后（重置时间结束），允许尝试执行命令，检查远程调用是否恢复到可使用的状态。

attemptExecution 方法代码如下所示：

```java
//HystrixCircuitBreakerImpl.java
@Override
public boolean attemptExecution() {
    if (properties.circuitBreakerForceOpen().get()) {
        return false;
    }
    if (properties.circuitBreakerForceClosed().get()) {
        return true;
    }
    if (circuitOpened.get() == -1) {
        return true;
    } else {
        if (isAfterSleepWindow()) {
            // 只在第一次重置时间窗结束时被执行
            if (status.compareAndSet(Status.OPEN, Status.HALF_OPEN)) {
                return true;
            } else {
                return false;
            }
        } else {
            return false;
        }
    }
}
```

attemptExecution 方法与 #allowRequest 方法基本一致，但是在第一次发现重置时间结

束时，会尝试将断路器的状态从打开修改为半开，方便在命令执行正常或者失败后关闭断路器或者重新打开断路器。

markSuccess 与 #markNonSuccess 方法代码如下所示：

```java
//HystrixCircuitBreakerImpl.java
@Override
public void markSuccess() {
    if (status.compareAndSet(Status.HALF_OPEN, Status.CLOSED)) {
        metrics.resetStream();
        Subscription previousSubscription = activeSubscription.get();
        if (previousSubscription != null) {
            previousSubscription.unsubscribe();
        }
        Subscription newSubscription = subscribeToStream();
        activeSubscription.set(newSubscription);
        circuitOpened.set(-1L);
    }
}
@Override
public void markNonSuccess() {
    if (status.compareAndSet(Status.HALF_OPEN, Status.OPEN)) {
        circuitOpened.set(System.currentTimeMillis());
    }
}
```

markSuccess 方法在命令执行成功后进行调用，将断路器从半开状态转换为关闭状态，同时重置断路器在 HystrixCommandMetrics 的统计记录和设置断路器打开时间为 –1（即关闭断路器）。markNonSuccess 方法在命令执行失败后将断路器从半开状态转换为打开状态，同时重置断路器的打开时间，用于下一次的 attemptExecution 方法的执行。

2. HystrixCommandMetrics 统计命令执行情况

断路器通过向 HystrixCommandMetrics 中的请求执行统计 Observable 发起订阅来完成断路器自动打开的相关逻辑。

HystrixCommandMetrics 统计了同一 HystrixCommand 请求的指标数据，包括链路健康统计流 HealthCountsStream。HealthCountsStream 中使用滑动窗口的方式对各项数据（HealthCounts）进行统计，在一个滑动窗口时间中又划分了若干个 bucket（滑动窗口时间与 bucket 成整数倍关系），滑动窗口的移动是以 bucket 为单位，每个 bucket 仅统计该时间间隔内的请求数据。最后按照滑动窗口的大小对每个 bucket 中的统计数据进行聚合，得到周期时间内的统计数据 HealthCounts。以下是 HealthCounts 中统计的数据项：

```java
//HealthCounts.java
public static class HealthCounts {
    private final long totalCount; //执行总次数
    private final long errorCount; //失败次数
    private final int errorPercentage;  // 失败百分比
```

```
    ...
}
```

Hystrix 使用 rx 中的 Observable#window 实现滑动窗口，通过 rx 中单线程的无锁特性保证计数变更时的线程安全，后台线程创建新 bucket，避免并发情况。

下面是 HealthCountsStream 的父类创建滑动窗口的相关代码：

```
//BucketedRollingCounterStream.java
...
Func1<Observable<Bucket>, Observable<Output>> reduceWindowToSummary = new Func1
    <Observable<Bucket>, Observable<Output>>() {
    @Override
    public Observable<Output> call(Observable<Bucket> window) {
        // 合并20个bucket数据项操作
        return window.scan(getEmptyOutputValue(), reduceBucket).skip(numBuckets);
    }
};
this.sourceStream = bucketedStream
    .window(numBuckets, 1)               // 每次发送数据，都合并20个bucket中数据项
    .flatMap(reduceWindowToSummary)      // 将20个bucket中的数据合并为一个HealthCounts进
                                         //   行发送
    .doOnSubscribe(new Action0() {...})
    .doOnUnsubscribe(new Action0() {...})
    .share()                             // 订阅者获得相同的数据
    .onBackpressureDrop();
...
```

上述代码中，#window 定义了每发射一次数据（此时一个数据项将会被从滑动窗口中移除，以及创建一个新的 bucket 用于统计）都会聚合 numBuckets 个数据项，即整个滑动窗口的数据统计集合，numBuckets 的计算方式如下所示，一般为 20 个：

```
final int numHealthCountBuckets = properties.metricsRollingStatisticalWindowInMi
    lliseconds().get() / healthCountBucketSizeInMs;
```

metricsRollingStatisticalWindowInMilliseconds 是整个时间滑动窗口的时间，默认为 10 秒，healthCountBucketSizeInMs 是链路监控的时间间隔，默认是 0.5 秒，在 healthCountBucketSizeInMs 时间后会进行一次数据的发射，监控链路健康。

```
return window.scan(getEmptyOutputValue(), reduceBucket).skip(numBuckets);
    // 合并20个bucket数据项操作
```

合并数据项的代码位于 HealthCounts#plus，用于将 bucket 中的统计数据合并为 HealthCounts 发出。具体代码如下所示：

```
// HealthCounts.java
public HealthCounts plus(long[] eventTypeCounts) {
    long updatedTotalCount = totalCount;
    long updatedErrorCount = errorCount;
    long successCount = eventTypeCounts[HystrixEventType.SUCCESS.ordinal()];
                                                            // 成功次数
    long failureCount = eventTypeCounts[HystrixEventType.FAILURE.ordinal()];
```

```
                                                            // 失败次数
    long timeoutCount = eventTypeCounts[HystrixEventType.TIMEOUT.ordinal()];
                                                            // 超时次数
    long threadPoolRejectedCount = eventTypeCounts[HystrixEventType.THREAD_POOL_
        REJECTED.ordinal()]; // 请求线程失败次数
    long semaphoreRejectedCount = eventTypeCounts[HystrixEventType.SEMAPHORE_
        REJECTED.ordinal()]; // 请求信号量失败次数
    updatedTotalCount += (successCount + failureCount + timeoutCount +
        threadPoolRejectedCount + semaphoreRejectedCount); // 执行总次数
    updatedErrorCount += (failureCount + timeoutCount + threadPoolRejectedCount +
        semaphoreRejectedCount); // 执行错误次数
    return new HealthCounts(updatedTotalCount, updatedErrorCount);
}
```

上述代码将每个 bucket 数据统计合并成一个完整的 HealthCounts 用于发射，统计近 10 秒内的命令执行情况。

在 BucketedRollingCounterStream 的父类 BucketedCounterStream 中，有如下规定：healthCountBucketSizeInMs 间隔后发射一次数据，同时初始化 numBuckets 个 bucket 用于统计，代码如下所示：

```
//BucketedCounterStream.java
final List<Bucket> emptyEventCountsToStart = new ArrayList<Bucket>();
for (int i = 0; i < numBuckets; i++) {
emptyEventCountsToStart.add(getEmptyBucketSummary());
}

this.bucketedStream = Observable.defer(new Func0<Observable<Bucket>>() {
    @Override
    public Observable<Bucket> call() {
        return inputEventStream
        .observe()
            .window(bucketSizeInMs, TimeUnit.MILLISECONDS)   // 500ms发送一次
                .flatMap(reduceBucketToSummary)
            .startWith(emptyEventCountsToStart);    // 初始化20个bucket
    }
});
```

HystrixCircuitBreak 通过对 HealthCountsStream 进行订阅，监控链路健康，在一定条件下打开断路器，代码如下所示：

```
//HystrixCircuitBreak.java
private Subscription subscribeToStream() {
    return metrics.getHealthCountsStream()
        .observe()
        .subscribe(new Subscriber<HealthCounts>() {
            ....
            @Override
            public void onNext(HealthCounts hc) {
                if (hc.getTotalRequests() < properties.circuitBreakerRequestVolume
                    Threshold().get()) {
                    // 请求总次数未达到断路器响应的阀值
```

```
            } else {
                if (hc.getErrorPercentage() < properties.circuitBreakerErrorThreshold
                    Percentage().get()) {
                // 请求执行错误率低于配置文件中要求，无需操作
                } else {
                    // 请求执行错误率高于配置文件中要求，尝试打开断路器
                    if (status.compareAndSet(Status.CLOSED, Status.OPEN)) {
                        circuitOpened.set(System.currentTimeMillis());
                    }
                }
            }
        }
    });
}
```

上述代码中，在固定间隔时间（metricsHealthSnapshotIntervalInMilliseconds，默认为 500ms，与 healthCountBucketSizeInMs 代表相同意义）内，#onNext 方法会被定时调用，滑动窗口中聚合而成的链路健康统计数据的 HealthCounts 将会被用来检查是否打开断路器。只有在周期内（10 秒）请求的总数超过一定的阀值（circuitBreakerRequestVolumeThreshold），且执行失败的比例超过 circuitBreakerErrorThresholdPercentage 时，断路器将会被打开。

触发 HystrixCommandMetrics 的统计命令执行结果主要发生在 AbstractCommand 中，如下所示：

```
//AbstractCommand.java
private void handleCommandEnd(boolean commandExecutionStarted) {
    ...
    // 统计命令执行结果
    if (executionResultAtTimeOfCancellation == null) {
        metrics.markCommandDone(executionResult, commandKey, threadPoolKey,
            commandExecutionStarted);
    } else {
        metrics.markCommandDone(executionResultAtTimeOfCancellation, commandKey,
            threadPoolKey, commandExecutionStarted);
    }
    ...
}
```

AbstractCommand 在命令执行结束后的回调方法中，通过 HystrixCommandMetrics 统计相关命令的执行结果，这其中主要通过 HystrixCommandCompletion 数据类对命令执行结束后的事件流进行统计，其中的事件类型由 HystrixEventType 定义。

HystrixCommandCompletion 由 HystrixCommandCompletionStream 进行管理，最终在 HealthCountsStream 中用于统计一段时间内的链路健康情况。

6.3.6　资源隔离

在 AbstractCommand#applyHystrixSemantics 方法中，如果发现断路器关闭，将会尝试获取信号量。在 Hystrix 中，主要有两种策略进行资源隔离，一种是信号量隔离的策略，另

一种是线程隔离的策略,下面将对这两种资源隔离策略进行介绍。

1. 信号量隔离策略

信号量隔离主要由 TryableSemaphore 接口提供,如下所示:

```java
//TryableSemaphore.java
interface TryableSemaphore {
    // 尝试获取信号量
    public abstract boolean tryAcquire();
    // 释放信号量
    public abstract void release();
    // 获取已被使用信号量数量
    public abstract int getNumberOfPermitsUsed();
}
```

它有两个实现类,其中一个是 TryableSemaphoreNoOp,顾名思义即不进行信号量隔离,当采取线程隔离策略的时候将会注入该实现到 HystrixCommand 中,此时信号量隔离形同虚设;另一个具体的实现为 TryableSemaphoreActual,如果采用信号量的隔离策略时,将会注入 TryableSemaphoreActual 的实现,但此时命令的执行将无法进行超时控制和异步化执行,因为信号量资源隔离策略无法指定命令在特定的线程执行,命令执行的线程将由 rx 控制,Hystrix 无法在命令执行超时后获取到对应的线程进行强制中断。

TryableSemaphoreActual 的实现相当简单,通过 AtomicInteger 记录当前请求信号量的线程数,与初始化设置的允许最大信号量数 numberOfPermits(可以动态调整)进行比较,判断是否允许获取信号量。这种轻量级的实现,保证 TryableSemaphoreActual 无阻塞的操作方式。实现代码如下所示:

```java
//TryableSemaphoreActual.java
static class TryableSemaphoreActual implements TryableSemaphore {
    protected final HystrixProperty<Integer> numberOfPermits;
    private final AtomicInteger count = new AtomicInteger(0);
    public TryableSemaphoreActual(HystrixProperty<Integer> numberOfPermits) {
        this.numberOfPermits = numberOfPermits;
    }
    @Override
    public boolean tryAcquire() {
        // 获取信号量
        int currentCount = count.incrementAndGet();
        if (currentCount > numberOfPermits.get()) {
            // 信号量已满,无法获取
            count.decrementAndGet();
            return false;
        } else {
            // 信号量未满,可以获取
            return true;
        }
    }
    @Override
    public void release() {
```

```
            // 释放信号量
            count.decrementAndGet();
        }
        @Override
        public int getNumberOfPermitsUsed() {
            // 获取已使用的信号量数量
            return count.get();
        }
    }
```

这其中每一个 TryableSemaphore 通过 CommandKey 与 HystrixCommand 一一绑定，这在 AbstractCommand#getExecutionSemaphore 方法中有所体现。如果采用信号量隔离的策略，将尝试从缓存中获取该命令的 CommandKey 对应的 TryableSemaphoreActual（缓存中不存在则创建一个新的，并与 CommandKey 绑定放置到缓存中），否则返回 TryableSemaphoreNoOp，不进行信号量隔离相关操作。

2. 线程隔离策略

在 AbstractCommand#executeCommandWithSpecifiedIsolation 的方法中，线程隔离策略与信号量隔离策略的主要区别是，线程隔离策略将 Observable 的执行线程通过 HystrixThreadPool#getScheduler 方法进行了指定。

HystrixThreadPool 的作用是将 HystrixCommand#run 方法指定到隔离的线程中执行。HystrixThreadPool 是由 HystrixThreadPool.Factory 生成和管理的，通过 ThreadPoolKey（由 @HystrixCommand 中 threadPoolKey 指定）与 HystrixCommand 进行绑定，它的默认实现为 HystrixThreadPoolDefault。HystrixThreadPoolDefault 中的线程池 ThreadPoolExecutor 通过 HystrixConcurrencyStrategy 策略生成，生成方法代码如下所示：

```
//HystrixConcurrencyStrategy.java
public ThreadPoolExecutor getThreadPool(final HystrixThreadPoolKey threadPoolKey,
    HystrixThreadPoolProperties threadPoolProperties) {
    ...
    if (allowMaximumSizeToDivergeFromCoreSize) {
        // 如果允许配置的maximumSize生效
        final int dynamicMaximumSize = threadPoolProperties.maximumSize().get();
        // 比较dynamicCoreSize和dynamicMaximumSize的大小，决定线程池的最大线程数
        if (dynamicCoreSize > dynamicMaximumSize) {
            return new ThreadPoolExecutor(dynamicCoreSize, dynamicCoreSize,
                keepAliveTime, TimeUnit.MINUTES, workQueue, threadFactory);
        } else {
            return new ThreadPoolExecutor(dynamicCoreSize, dynamicMaximumSize,
                keepAliveTime, TimeUnit.MINUTES, workQueue, threadFactory);
        }
    } else {
        return new ThreadPoolExecutor(dynamicCoreSize, dynamicCoreSize,
            keepAliveTime, TimeUnit.MINUTES, workQueue, threadFactory);
    }
}
```

如果允许配置的 maximumSize 生效的话（即配置中 allowMaximumSizeToDivergeFromCoreSize 为 true），在 coreSize 小于 maximumSize 时，会创建一个线程数最大值为 maximumSize 的线程池，但会在非活跃期返回多余的线程到系统。否则就只应用 coreSize 来定义线程池中线程的数量。dynamic** 前缀说明这些配置都可以在运行时动态修改，如通过配置中心的方式进行运行时修改。

HystrixThreadPoolDefault#getScheduler 方法为 rx 提供线程调度器，为 Observable 指定执行线程，实现代码如下所示：

```
//HystrixThreadPoolDefault.java
@Override
public Scheduler getScheduler() {
    //默认在超时可中断线程
    return getScheduler(new Func0<Boolean>() {
        @Override
        public Boolean call() {
        return true;
        }
    });
}
@Override
public Scheduler getScheduler(Func0<Boolean> shouldInterruptThread) {
    touchConfig();
    return new HystrixContextScheduler(HystrixPlugins.getInstance().getConcurrencyStrategy(),
        this, shouldInterruptThread);
}
```

touchConfig 方法通过刷新配置的方式，动态调整线程池线程大小、线程存活时间等线程池的关键配置，以便在应用程序的相关配置发生改变时动态改变线程池配置。

HystrixContextScheduler 是 Hystrix 对 rx 中 Scheduler 调度器的重写，主要为了实现在 Observable 被退订时，不从线程池中获取线程执行命令，以及提供在命令执行过程中中断命令执行的能力（如在命令执行超时时中断命令执行）。Scheduler 相关类图如图 6-10 所示。

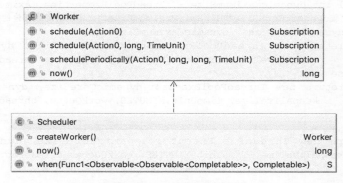

图 6-10　Scheduler 类图

在 Scheduler 中将生成对应的 Worker 给 Observable 用于执行命令，由 Worker 具体负责相关执行线程的调度。在 Hystrix 实现的 ThreadPoolWorker 中，线程调度的核心方法如下所示：

```java
//ThreadPoolWorker.java
@Override
public Subscription schedule(final Action0 action) {
    // 如果Observable被退订，取消执行，不分配线程
    if (subscription.isUnsubscribed()) {
        return Subscriptions.unsubscribed();
    }
    ScheduledAction sa = new ScheduledAction(action);
    subscription.add(sa);
    sa.addParent(subscription);
    // 分配线程提交任务
    ThreadPoolExecutor executor = (ThreadPoolExecutor) threadPool.getExecutor();
    FutureTask<?> f = (FutureTask<?>) executor.submit(sa);
    // 添加一个订阅者用于在取消任务时释放线程
    sa.add(new FutureCompleterWithConfigurableInterrupt(f, shouldInterruptThread,
        executor));
    return sa;
}
```

如果 Observable 被退订，ThreadPoolWorker 将取消任务的执行，返回被退订的 Subscription；如果 Observable 没被退订，ThreadPoolWorker 将为命令分配线程提交任务。注意在提交任务的过程中有可能会出现线程池中的线程已被占满，导致抛出 RejectedExecutionException 异常，拒绝任务提交。添加订阅者 FutureCompleterWithConfigurableInterrupt 是为了在取消任务的时候释放任务执行的线程。FutureCompleterWithConfigurableInterrupt 代码如下所示：

```java
//FutureCompleterWithConfigurableInterrupt.java
@Override
public void unsubscribe() {
    // 释放线程
    executor.remove(f);
    // 是否强制中断任务执行
    if (shouldInterruptThread.call()) {
        f.cancel(true);
    } else {
        f.cancel(false);
    }
}
```

取消任务的时候将从线程池中移除任务，释放线程，同时根据配置决定是否强制中断任务的执行。

通过线程隔离的方式，可以将调用线程与执行命令的线程分隔开来，避免了调用线程被阻塞。同时通过线程池的方式对每种命令的并发线程数量进行控制，避免了一种命令的阻塞影响系统的其他请求的执行，很好地保护了服务调用者的线程资源。

6.3.7 请求超时监控

在 AbstractCommand#executeCommandAndObserve 方法中，如果命令开启了执行超时控制的相关配置，Hystrix 将会为 Observable 配置超时监控，主要通过 lift(new HystrixObservableTimeoutOperator<R>(_cmd)) 方法将现有的 Observable 转化为添加了超时检查的 Observable。超时控制的主要实现逻辑位于 HystrixObservableTimeoutOperator 中。

HystrixObservableTimeoutOperator 中定义了一个超时监听器 TimerListener，代码如下所示：

```java
//HystrixObservableTimeoutOperator.java
TimerListener listener = new TimerListener() {
    @Override
    public void tick() {
    // 命令执行超时
        if (originalCommand.isCommandTimedOut.compareAndSet(TimedOutStatus.NOT_
            EXECUTED, TimedOutStatus.TIMED_OUT)) {
            originalCommand.eventNotifier.markEvent(HystrixEventType.TIMEOUT,
                originalCommand.commandKey);
            // 取消命令的执行
            s.unsubscribe();
            final HystrixContextRunnable timeoutRunnable = new HystrixContextRunnable
                (originalCommand.concurrencyStrategy, hystrixRequestContext, new
                Runnable() {
                @Override
                public void run() {
                    child.onError(new HystrixTimeoutException());
                }
            });
            timeoutRunnable.run();
        }
    }
    @Override
    public int getIntervalTimeInMilliseconds() {
        return originalCommand.properties.executionTimeoutInMilliseconds().get();
    }
};
```

TimerListener 将会在命令执行时间超过 executionTimeoutInMilliseconds 时被执行。如果此时命令的 TimedOutStatus 状态能够从 NOT_EXECUTED 设置为 TIMED_OUT，那么 Hystrix 断定命令执行超时（命令执行结束 Hystrix 会将 TimedOutStatus 设置为 COMPLETED）。此时将执行取消命令执行的操作，并且在原来的 Observable 中抛出 HystrixTimeoutException 异常，使 Observable 执行 onError 方法的逻辑。

TimedOutStatus 具备三种状态，分别是未执行、执行结束以及执行超时，如下所示：

```java
//TimedOutStatus.java
protected enum TimedOutStatus {
    NOT_EXECUTED, COMPLETED, TIMED_OUT
}
```

执行 TimerListener 的代码位于 HystrixTimer 中，如下所示：

```java
//HystrixTimer.java
public Reference<TimerListener> addTimerListener(final TimerListener listener) {
    startThreadIfNeeded();
    Runnable r = new Runnable() {

        @Override
        public void run() {
            try {
                // 检查是否超时以及进行超时处理
                listener.tick();
            } catch (Exception e) {
            }
        }
    };
    // 启动超时监控的定时任务
    ScheduledFuture<?> f = executor.get().getThreadPool().scheduleAtFixedRate(r,
        listener.getIntervalTimeInMilliseconds(), listener.getIntervalTimeInMill
iseconds(), TimeUnit.MILLISECONDS);
    return new TimerReference(listener, f);
}
```

通过 ScheduledThreadPoolExecutor#scheduleAtFixedRate 的方式启动定时任务，使 tick 方法能够在命令执行超时时执行，取消超时命令的执行并抛出超时异常。

在 HystrixObservableTimeoutOperator 中还对原来的 Observable 添加了一个 Subscriber 订阅者，监听 Observable 执行结果，在命令执行结束后清理 TimerListener。代码如下所示：

```java
//HystrixObservableTimeoutOperator.java
Subscriber<R> parent = new Subscriber<R>() {
    @Override
    public void onCompleted() {
        // 执行结束且未超时
        if (isNotTimedOut()) {
            // 清理 TimerListener
            tl.clear();
            child.onCompleted();
        }
    }

    @Override
    public void onError(Throwable e) {
        // 执行错误且未超时
        if (isNotTimedOut()) {
            // 清理 TimerListener
            tl.clear();
            child.onError(e);
        }
    }

    @Override
    public void onNext(R v) {
        // 传递给订阅者且未超时
        if (isNotTimedOut()) {
```

```
                // 清理TimerListener
                child.onNext(v);
            }
        }
        // 检查命令执行是否超时
        private boolean isNotTimedOut() {
            return originalCommand.isCommandTimedOut.get() == TimedOutStatus.COMPLETED ||
                originalCommand.isCommandTimedOut.compareAndSet(TimedOutStatus.NOT_
                    EXECUTED, TimedOutStatus.COMPLETED);
        }
    };
```

因为 parent 与 TimerListener 存在互相引用的关系，所以这里加多了一层 CompositeSubscription 的订阅者，child 才是最原始的 Observable。

isNotTimedOut 方法会尝试将 TimedOutStatus 从 NOT_EXECUTED 状态转换为 COMPLETED，从而保证在任务执行结束后，防止 TimerListener#tick 方法错误地认为命令执行超时而执行取消命令的相关逻辑。

6.3.8 失败回滚逻辑

Hystrix 中执行失败回滚的逻辑主要封装在 AbstractCommand#executeCommandAndObserve #handleFallback 的异常回调方法中，根据执行过程中抛出的异常调用不同的方法对其进行处理，返回带有失败回滚逻辑的 Observable。

在 handleFallback 方法中对不同的执行错误调用不同的处理方法，主要有：

- 对线程获取失败处理的 handleThreadPoolRejectionViaFallback 方法。
- 执行超时处理的 handleTimeoutViaFallback 方法。
- 远程调用请求失败处理的 handleBadRequestByEmittingError 方法。
- Hystrix 自身执行异常处理的 handleFailureViaFallback 方法。

除此之外，在 applyHystrixSemantics 中包括了断路失败处理方法和获取信号量失败处理方法：

- 断路失败处理的 handleShortCircuitViaFallback 方法。
- 获取信号量失败处理的 #handleSemaphoreRejectionViaFallback 方法。

这些方法的处理过程大同小异，最终都是通过 AbstractCommand#getFallbackOrThrow Exception 获取到包含失败逻辑的 Observable 或者异常 Observable（没有配置失败回滚逻辑的情况下），handleThreadPoolRejectionViaFallback 方法的代码如下所示：

```
//AbstractCommand.java
private Observable<R> handleThreadPoolRejectionViaFallback(Exception underlying) {
    // 推送线程获取失败事件
    eventNotifier.markEvent(HystrixEventType.THREAD_POOL_REJECTED, commandKey);
    // 标记线程请求失败,将由HystrixThreadPoolMetrics统计对应的数据
    threadPool.markThreadRejection();
    // 获取失败回滚逻辑,如果不存在的话抛出异常
```

```
            return getFallbackOrThrowException(this, HystrixEventType.THREAD_POOL_REJECTED,
                FailureType.REJECTED_THREAD_EXECUTION, "could not be queued for execution",
                underlying);
}
```

其他的失败处理函数也是类似。getFallbackOrThrowException 中主要检查失败回滚是否允许执行以及相应的失败回滚逻辑是否已配置。具体代码如下所示：

```
//AbstractCommand.java
private Observable<R> getFallbackOrThrowException(final AbstractCommand<R> _cmd,
    final HystrixEventType eventType, final FailureType failureType, final String
    message, final Exception originalException) {
    ...
    // 如果无法处理该类型异常，直接返回执行异常Observable
    if (isUnrecoverable(originalException)) {
        ...
    } else {
        // 是否允许失败回滚
        if (properties.fallbackEnabled().get()) {
            // 配置失败回滚Observable对应的回调方法，包括执行结束、执行错误等
            final TryableSemaphore fallbackSemaphore = getFallbackSemaphore();
            ...
            Observable<R> fallbackExecutionChain;
            // 获取失败回滚执行信号量
            if (fallbackSemaphore.tryAcquire()) {
                try {
                    // 如果用户定义了相关失败回滚方法
                    if (isFallbackUserDefined()) {
                        executionHook.onFallbackStart(this);
                        // 获取包含失败回滚逻辑的Observable
                        fallbackExecutionChain = getFallbackObservable();
                    } else {
                        //same logic as above without the hook invocation
                        fallbackExecutionChain = getFallbackObservable();
                    }
                } catch (Throwable ex) {
                    fallbackExecutionChain = Observable.error(ex);
                }
                return fallbackExecutionChain
                    ...
            } else {
                // 获取信号量失败，直接返回异常
                return handleFallbackRejectionByEmittingError();
            }
        } else {
            // 不允许失败回滚，直接返回异常
            return handleFallbackDisabledByEmittingError(originalException,
                failureType, message);
        }
    }
}
```

在获取 HystrixCommand 的失败回滚方法之前，会判断正常逻辑执行时抛出的异常是否

在 Hystrix 可以处理的范围内，接着判断配置中是否允许失败回滚以及是否为命令配置了失败回滚逻辑。失败回滚逻辑通过信号量方式隔离执行，所以默认是不存在超时控制和异步化执行，但是可以将失败回滚逻辑用 @HystrixCommand 注解进行封装，使得失败回滚逻辑的执行也可以采用线程隔离策略。

上述代码中，最终通过 AbstractCommand#getFallbackObservable 获取到封装有失败回滚逻辑的 Observable，该方法将获取相应的失败回滚逻辑延迟到子类实现。HystrixCommand 的实现如下所示：

```java
//HystrixCommand.java
@Override
final protected Observable<R> getFallbackObservable() {
    return Observable.defer(new Func0<Observable<R>>() {
        @Override
        public Observable<R> call() {
            try {
                return Observable.just(getFallback());
            } catch (Throwable ex) {
                return Observable.error(ex);
            }
        }
    });
}
```

上述方法中，getFallback 被延迟到子类中实现，在 HystrixCommand 的子类 GenericCommand 中，getFallback 方法通过创建 HystrixCommand 时传递的 CommandActions 完成对失败回滚逻辑的调用。CommandActions 内持有 commandAction 和 fallbackAction，分别是 HystrixCommand 中具体的远程调用方法和失败回滚方法。getFallback 方法使用 CommandActions 中的 fallbackAction 完成了预先定义的失败回滚逻辑的调用。

6.4 进阶应用

6.4.1 异步与异步回调执行命令

Hystrix 除了同步执行命令，还可以异步以及异步回调执行命令。异步执行命令需要定义函数的返回方式为 Future，如下面的例子所示：

```java
@HystrixCommand(fallbackMethod = "instanceInfoGetFailAsync")
public Future<Instance> getInstanceByServiceIdAsync(String serviceId){
    logger.info("Can not get Instance by serviceId {}", serviceId);
    return new AsyncResult<Instance>() {
        @Override
        public Instance invoke() {
            return restTemplate.getForEntity("http://FEIGN-SERVICE/feign-service/
                instance/{serviceId}", Instance.class, serviceId).getBody();
        }
```

            };
        }

在这种情况下,命令的失败回滚方法也需要通过 @HystrixCommand 进行注解,如下所示:

```
@HystrixCommand
public Future<Instance> instanceInfoGetFailAsync(String serviceId){
    logger.info("Can not get Instance by serviceId {}", serviceId);
    return new AsyncResult<Instance>() {
        @Override
        public Instance invoke() {
            return new Instance("error", "error", 0);
        }
    };
}
```

同样,想要异步回调执行命令,只需要将包装函数的返回值设定为 Observable 即可,如下所示:

```
@HystrixCommand(fallbackMethod = "instanceInfoGetFailObservable",
            observableExecutionMode = ObservableExecutionMode.LAZY)
public Observable<Instance> getInstanceByServiceIdObservable(String serviceId){
    return Observable.create(
        subscriber -> {
            if(!subscriber.isUnsubscribed()){

                subscriber.onNext(restTemplate.getForEntity("http://FEIGN-SERVICE/
                    feign-service/instance/{serviceId}", Instance.class, serviceId).
                    getBody());
                subscriber.onCompleted();
            }
        }
    );
}
```

异步回调执行命令并不需要 Fallback 方法被 @HystrixCommand 注解。可以通过设置 observableExecutionMode 来决定返回的 Observable 是否已经开始执行命令:EAGER 将会返回通过 AbstractCommand#observable 方法生成的 Observable,此时的命令已经开始执行了;而 LAZY 则会返回通过 AbstractCommand#toObservable 方法生成的 Observable,返回的 Observale 必须订阅后才会真正开始执行命令。上面的例子中设定的 observableExecutionMode 为 ObservableExecutionMode.LAZY。

6.4.2 继承 HystrixCommand

除了通过注解的方式声明 Hystrix 包装函数,还可以通过继承 HystrixCommand 以及 HystrixObservableCommand 抽象类接口来包装需要保护的远程调用函数。

1. 继承 HystrixCommand

下面将通过继承 HystrixCommand 抽象类来实现特定的 HystrixCommand 子类,代码如下所示：

```java
public class CustomHystrixCommand extends HystrixCommand<Instance>{
    private static Logger logger = LoggerFactory.getLogger(CustomHystrixCommand.class);
    private RestTemplate restTemplate;
    private String serviceId;
    protected CustomHystrixCommand(RestTemplate restTemplate, String serviceId) {
        super(HystrixCommandGroupKey.Factory.asKey("CustomServiceGroup"));
        this.restTemplate = restTemplate;
        this.serviceId = serviceId;
    }
    // 被保护的包装函数
    @Override
    protected Instance run() throws Exception {
        Instance instance = restTemplate.getForEntity("http://FEIGN-SERVICE/feign-service/instance/{serviceId}", Instance.class, serviceId).getBody();
        return instance;
    }
     // 失败回滚函数
    @Override
    protected Instance getFallback() {
        logger.info("Can not get Instance by serviceId {}", serviceId);
        return new Instance("error", "error", 0);
    }
}
```

run 方法中是需要进行包装的远程调用函数，是必须要实现的抽象方法，getFallback 方法是该命令执行失败后的失败回滚方法，属于可选实现。值得注意的是，在构造 HystrixCommand 时至少要为它指定一个 HystrixCommandGroupKey，在通过注解的方式生成 HystrixCommand 时，该值一般是注解方法所在类的运行时类名。

在使用 CustomHystrixCommand 时，会发现无法在 #run 方法中传递参数，所以需要在构造器中携带 #run 方法的相关参数，如上述例子中的构造函数，传递被保护方法执行所需要的参数：

```java
protected CustomHystrixCommand(RestTemplate restTemplate, String serviceId){...}
```

调用命令的方式很简单，如下所示：

```java
CustomHystrixCommand customHystrixCommand = new CustomHystrixCommand(restTemplate, serviceId);
Instance instance = customHystrixCommand.execute();
```

创建一个 CustomHystrixCommand，并调用它的 execute 方法，即可按照 Hystrix 的逻辑执行命令。如果想要以异步方式执行命令，可以调用它的 queue 方法，如下所示：

```java
CustomHystrixCommand customHystrixCommand = new CustomHystrixCommand(restTemplate, serviceId);
Future<Instance> future = customHystrixCommand.queue();
```

还需要注意的是，一个 CustomHystrixCommand 只能执行一次（execute 方法或者 queue

方法），所以每次使用都要创建一个新的 Command。为了满足自定义 HystrixCommand 配置的需求，HystrixCommand 抽象类中定义了多种构造器，用于对构建的 Command 配置进行设置。

下面将通过 HystrixCommand#Setter 的方式在构造函数中对 CustomHystrixCommand 的默认配置进行修改：

```
protected CustomHystrixCommand(RestTemplate restTemplate, String serviceId) {
    super(Setter.withGroupKey(HystrixCommandGroupKey.Factory.asKey("CustomServiceGroup"))
        .andCommandPropertiesDefaults(
            HystrixCommandProperties.Setter()
                .withExecutionIsolationStrategy(HystrixCommandProperties.ExecutionIsolationStrategy.
                    SEMAPHORE)
                .withExecutionIsolationSemaphoreMaxConcurrentRequests(20)
                .withCircuitBreakerErrorThresholdPercentage(80)
        )
        .andThreadPoolPropertiesDefaults(HystrixThreadPoolProperties.Setter()
            .withCoreSize(20))
    );

    this.restTemplate = restTemplate;
    this.serviceId = serviceId;
}
```

HystrixCommandProperties.Setter 可以对 HystrixCommand 中的诸多默认配置进行修改，包括 commandKey、threadPoolKey、HystrixCommand 配置以及线程池配置等，如下所示：

```
final public static class Setter {
    protected final HystrixCommandGroupKey groupKey;
    protected HystrixCommandKey commandKey; // 默认为getClass().getSimpleName();
    protected HystrixThreadPoolKey threadPoolKey;
    protected HystrixCommandProperties.Setter commandPropertiesDefaults;
    protected HystrixThreadPoolProperties.Setter threadPoolPropertiesDefaults;
    ...
}
```

在上面代码中，HystrixCommandProperties#Setter 和 HystrixThreadPoolProperties#Setter 为 HystrixCommand 和 Hystrix 线程池配置的 Setter。

上面的例子设置了信号量资源隔离策略，信号量隔离策略的最大并发数为 20，断路器打开的错误阀值为 80%，线程池线程数为 20。

2. 继承 HystrixObservableCommand

除了 HystrixCommand，还可以继承 HystrixObservableCommand 来构建以异步回调执行命令的 Command。如下例子所示：

```
public class CustomHystrixObservableCommand extends HystrixObservableCommand<Instance>{
    private static Logger logger = LoggerFactory.getLogger(CustomHystrixObservab
        leCommand.class);
    private RestTemplate restTemplate;
```

```
    private String serviceId;
    protected CustomHystrixObservableCommand(RestTemplate restTemplate, String serviceId) {
        super(HystrixCommandGroupKey.Factory.asKey("CustomServiceGroup"));
        this.restTemplate = restTemplate;
        this.serviceId = serviceId;
    }
    @Override
    protected Observable<Instance> construct() {
        return Observable.create(
                subscriber -> {
                    if(!subscriber.isUnsubscribed()){
                        subscriber.onNext(restTemplate.getForEntity("http://FEIGN-
                            SERVICE/feign-service/instance/{serviceId}", Instance.
                            class, serviceId).getBody());
                        subscriber.onCompleted();
                    }
                }
        );
    }
    @Override
    protected Observable<Instance> resumeWithFallback() {
        logger.info("Can not get Instance by serviceId {}", serviceId);
        return Observable.create(
                subscriber -> {
                    if(!subscriber.isUnsubscribed()){
                        subscriber.onNext(new Instance("error", "error", 0));
                        subscriber.onCompleted();
                    }
                }
        );
    }
}
```

调用方式如下所示：

```
CustomHystrixObservableCommand command1 = new CustomHystrixObservableCommand(res
    tTemplate, serviceId);
Observable<Instance> observable1 = command1.observe();
CustomHystrixObservableCommand command2 = new CustomHystrixObservableCommand(res
    tTemplate, serviceId);
Observable<Instance> observable2 = command2.toObservable();
```

和 CustomHystrixCommand 一样，每个 CustomHystrixObservableCommand 只能执行一次（observe 方法或者 toObservable 方法），所以每次使用都要创建一个新的命令。

对于返回的 Observable，在知道返回值是单个值的情况下，可以通过以下的方式直接同步获取结果，如下所示：

```
Instance instance = observable1.toBlocking().single();
```

也可以通过订阅的方式定义回调函数获取执行结果，如下所示：

```
observable2.subscribe(new Action1<Instance>() {
    @Override
```

```
    public void call(Instance instance) {
        System.out.println(instance);
    }
});
```

HystrixObservableCommand 也提供 Setter 用于修改默认配置,如下所示:

```
protected CustomHystrixObservableCommand(RestTemplate restTemplate, String serviceId) {
    super(Setter.withGroupKey(HystrixCommandGroupKey.Factory.asKey("CustomServiceGroup")));
    this.restTemplate = restTemplate;
    this.serviceId = serviceId;
}
```

6.4.3 请求合并

Hystrix 还提供了请求合并的功能。多个请求被合并为一个请求进行一次性处理,可以有效减少网络通信和线程池资源。请求合并之后,一个请求原本可能在 6 毫秒之内能够结束,现在必须等待请求合并周期后(10 毫秒)才能发送请求,增加了请求的时间(16 毫秒)。但是请求合并在处理高并发和高延迟命令上效果极佳。

它提供两种方式进行请求合并:request-scoped 收集一个 HystrixRequestContext 中的请求集合成一个批次;而 globally-scoped 将多个 HystrixRequestContext 中的请求集合成一个批次,这需要应用的下游依赖能够支持在一个命令调用中处理多个 HystrixRequestContext。

HystrixRequestContext 中包含和管理着 HystrixRequestVariableDefault,HystrixRequestVariableDefault 中提供了请求范围内的相关变量,所以在同一请求中的多个线程可以分享状态,HystrixRequestContext 也可以通过 HystrixRequestVariableDefault 收集到请求范围内相同的 HystrixCommand 进行合并。

1. 通过注解方式进行请求合并

单个请求需要使用 @HystrixCollapser 注解修饰,并指明 batchMethod 方法,这里我们设置请求合并的周期为 100 秒。由于请求合并中不能同步等待结果,所以单个请求返回的结果为 Future,即需要异步等待结果。

batchMethod 方法需要被 @HystrixCommand 注解,说明这是一个被 HystrixCommand 封装的方法,其内是一个批量的请求接口,为了方便展示,例中就直接虚假地构建了本地数据,同时有日志打印批量方法被执行。具体代码如下所示:

```
@HystrixCollapser(batchMethod = "getInstanceByServiceIds",
    collapserProperties = {@HystrixProperty(name ="timerDelayInMilliseconds",value = "100")})
public Future<Instance> getInstanceByServiceId(String serviceId) {
    return null;
}
@HystrixCommand
public List<Instance> getInstanceByServiceIds(List<String> serviceIds){
    List<Instance> instances = new ArrayList<>();
    logger.info("start batch!");
    for(String s : serviceIds){
```

```
        instances.add(new Instance(s, DEFAULT_HOST, DEFAULT_PORT));
    }
    return instances;
}
```

请求开始前需要为请求初始化 HystrixRequestContext,用于在同一请求中的线程间共享数据,封装批量请求,请求结束前需要关闭 HystrixRequestContext,如下所示:

```
@RequestMapping(value = "batch/test", method = RequestMethod.GET)
public Instance getInstancesBatch() throws ExecutionException, InterruptedException {
    HystrixRequestContext context = HystrixRequestContext.initializeContext();
    Future<Instance> future1 = instanceService.getInstanceByServiceId("test1");
    Future<Instance> future2 = instanceService.getInstanceByServiceId("test2");
    Future<Instance> future3 = instanceService.getInstanceByServiceId("test3");
    future1.get();
    future2.get();
    future3.get();
    TimeUnit.MILLISECONDS.sleep(1000); // 中断线程1s
    Future<Instance> future4 = instanceService.getInstanceByServiceId("test4");
    Instance instance = future4.get();
    context.close();
    return instance;
}
```

中间线程中断了 1 秒,是为了保证将两次请求合并分离开以展示效果。执行结果如下所示:

```
2018-03-19 21:37:42.926  INFO 73976 --- [stanceService-1] c.x.c.hystrix.service.
    InstanceService      : start batch!
2018-03-19 21:37:44.010  INFO 73976 --- [stanceService-2] c.x.c.hystrix.service.
    InstanceService      : start batch!
```

结果显示只执行了两次批量操作,前三个为一次,最后一个单独进行。如果在请求合并周期内直接调用 Future#get 方法阻塞等待同步结果,将会强行合并请求,增加远程请求次数,如下所示:

```
HystrixRequestContext context = HystrixRequestContext.initializeContext();
Future<Instance> future1 = instanceService.getInstanceByServiceId("test1");
instanceService.getInstanceByServiceId("test2").get();
Future<Instance> future3 = instanceService.getInstanceByServiceId("test3");
future1.get();
future3.get();
TimeUnit.MILLISECONDS.sleep(1000);
Future<Instance> future4 = instanceService.getInstanceByServiceId("test4");
Instance instance = future4.get();
context.close();
return instance;
```

主线程将会因为同步等待而阻塞,后面命令没有提交合并,请求合并只进行到 test2 中,所以上面代码中将会执行三次批量请求,前两个为一次,第三个和第四个分别为一次。

2. 继承 HystrixCollapser

请求合并命令同样也可以通过自定义的方式实现，只需继承 HystrixCollapser 抽象类，如下所示：

```java
public class CustomCollapseCommand extends HystrixCollapser<List<Instance>, Instance,
    String>{
    public String serviceId;
    public CustomCollapseCommand(String serviceId){
        super(Setter.withCollapserKey(HystrixCollapserKey.Factory.asKey("customC
            ollapseCommand")));
        this.serviceId = serviceId;
    }
    @Override
    public String getRequestArgument() {
        return serviceId;
    }
    @Override
    protected HystrixCommand<List<Instance>> createCommand(Collection<CollapsedR
        equest<Instance, String>> collapsedRequests) {
        List<String> ids = collapsedRequests.stream().map(CollapsedRequest::getA
            rgument).collect(Collectors.toList());
        return new InstanceBatchCommand(ids);
    }
    @Override
    protected void mapResponseToRequests(List<Instance> batchResponse, Collectio
        n<CollapsedRequest<Instance, String>> collapsedRequests) {
        int count = 0;
        for( CollapsedRequest<Instance, String> request : collapsedRequests){
            request.setResponse(batchResponse.get(count++));
        }
    }
    private static final class InstanceBatchCommand extends HystrixCommand<List<Instance>>{
        private List<String> serviceIds;
        private static String DEFAULT_SERVICE_ID = "application";
        private static String DEFAULT_HOST = "localhost";
        private static int DEFAULT_PORT = 8080;
        private static Logger logger = LoggerFactory.getLogger(InstanceBatchCommand.
            class);
        protected InstanceBatchCommand(List<String> serviceIds) {

            super(HystrixCommandGroupKey.Factory.asKey("instanceBatchGroup"));
            this.serviceIds = serviceIds;
        }
        @Override
        protected List<Instance> run() throws Exception {
            List<Instance> instances = new ArrayList<>();
            logger.info("start batch!");
            for(String s : serviceIds){
                instances.add(new Instance(s, DEFAULT_HOST, DEFAULT_PORT));
            }
            return instances;
        }
```

```
        }
    }
```

继承 HystrixCollapser 需要指定三个泛型，如下所示：

```
public abstract class HystrixCollapser<BatchReturnType, ResponseType, RequestArgumentType>
    implements HystrixExecutable<ResponseType>, HystrixObservable<ResponseType> {...}
```

BatchReturnType 是批量操作的返回值类型，例子中为 List<Instance>；ResponseType 是单个操作的返回值类型，例子中是 Instance；RequestArgumentType 是单个操作的请求参数类型，例子中是 String 类型。

在构造函数中需要指定 CollapserKey，用来标记被合并请求的键值。CustomCollapseCommand 同样可以通过 Setter 的方式修改默认配置。

同时还需要实现三个方法，getRequestArgument 方法获取被合并的单个请求的参数，createCommand 方法用来生成进行批量请求的命令 Command，mapResponseToRequests 方法是将批量请求的结果与合并的请求进行匹配以返回对应的请求结果。

为了在 createCommand 方法中能够返回合适的批量请求命令，还需要定义一个继承 HystrixCommand 的 InstanceBatchCommand。测试代码如下所示：

```
@RequestMapping(value = "batch/test2", method = RequestMethod.GET)
public Instance getInstancesBatch2() throws ExecutionException, InterruptedException
    HystrixRequestContext context = HystrixRequestContext.initializeContext();
    CustomCollapseCommand c1 = new CustomCollapseCommand("test1");
    CustomCollapseCommand c2 = new CustomCollapseCommand("test2");
    CustomCollapseCommand c3 = new CustomCollapseCommand("test3");
    CustomCollapseCommand c4 = new CustomCollapseCommand("test4");
    Future<Instance> future1 = c1.queue();
    Future<Instance> future2 = c2.queue();
    Future<Instance> future3 = c3.queue();
    future1.get();
    future2.get();
    future3.get();
    TimeUnit.MILLISECONDS.sleep(1000);
    Future<Instance> future4 = c4.queue();
    Instance instance = future4.get();
    context.close();
    return instance;
}
```

检查日志发现，只发生了两次批量操作，如下所示：

```
2018-03-19 23:40:26.234  INFO 13070 --- [nceBatchGroup-1] stomCollapseCommand$In
    stanceBatchCommand  : start batch!
2018-03-19 23:40:27.263  INFO 13070 --- [nceBatchGroup-2] stomCollapseCommand$In
    stanceBatchCommand  : start batch!
```

需要注意的是，因为 HystrixCollapser 继承了 HystrixInvokable，它同样提供 execute 同步获取结果，但是调用 execute 获取结果时，会因为同步等待的原因阻塞主线程，导致后面的请求无法提交合并，增加远程调用的次数。

6.5 本章小结

微服务架构下,每个微服务独立部署运行,业务之间的耦合使微服务间不可避免地会发生相互调用。由于网络故障或者程序故障,服务提供者很容易无法响应服务调用者的请求,引起服务雪崩,最坏情况下可能导致整个分布式系统的瘫痪。

Hystrix 为 Spring Cloud 中微服务间的相互调用提供了强大的容错保护。它通过将服务调用者和服务提供者隔离的方式,在服务提供者失效的情况下,保护服务调用者的线程资源,保证系统的整体稳定性,防止服务雪崩效应的发生。

Hystrix 提供了很多服务容错的机制和手段,如断路器、资源隔离、失败回滚等,同时也提供接近实时的执行监控,为服务的健康运行保驾护航。

Chapter 7 第 7 章

客户端负载均衡器：Spring Cloud Netflix Ribbon

Ribbon 是管理 HTTP 和 TCP 服务客户端的负载均衡器。Ribbon 具有一系列带有名称的客户端（Named Client），也就是带有名称的 Ribbon 客户端（Ribbon Client）。每个客户端由可配置的组件构成，负责一类服务的调用请求。Spring Cloud 通过 RibbonClientConfiguration 为每个 Ribbon 客户端创建一个 ApplicationContext 上下文来进行组件装配。Ribbon 作为 Spring Cloud 的负载均衡机制的实现，可以与 OpenFeign 和 RestTemplate 进行无缝对接，让二者具有负载均衡的能力。

本章的第一小节主要讲解了负载均衡的相关概念和实现；第二小节则展示了使用 Ribbon 的一些代码示例，讲解了 Ribbon 的基本使用方法；第三小节主要是 Ribbon 的实现原理和代码详解；第四小节讲解了有关 Ribbon 的进阶应用。

7.1 负载均衡

当系统面临大量的用户访问，负载过高的时候，通常会增加服务器数量来进行横向扩展，多个服务器的负载需要均衡，以免出现服务器负载不均衡，部分服务器负载较大，部分服务器负载较小的情况。通过负载均衡，使得集群中服务器的负载保持在稳定高效的状态，从而提高整个系统的处理能力。

系统的负载均衡分为软件负载均衡和硬件负载均衡。软件负载均衡使用独立的负载均衡程序或系统自带的负载均衡模块完成对请求的分配派发。硬件负载均衡通过特殊的硬件设备进行负载均衡的调配。

而软负载均衡一般分为两种类型，基于 DNS 的负载均衡和基于 IP 的负载均衡。利用 DNS 实现负载均衡，就是在 DNS 服务器配置多个 A 记录，不同的 DNS 请求会解析到不同的 IP 地址。大型网站一般使用 DNS 作为第一级负载均衡。基于 IP 的负载均衡根据请求的 IP 进行负载均衡。LVS 就是具有代表性的基于 IP 的负载均衡实现。

但是目前来说，大家最为熟悉的，最为主流的负载均衡技术还是反向代理负载均衡。所有主流的 Web 服务器都热衷于支持基于反向代理的负载均衡。它的核心工作是代理根据一定规则，将 HTTP 请求转发到服务器集群的单一服务器上。

Ribbon 使用的是客户端负载均衡。客户端负载均衡和服务端负载均衡最大的区别在于服务端地址列表的存储位置，在客户端负载均衡中，所有的客户端节点都有一份自己要访问的服务端地址列表，这些列表统统都是从服务注册中心获取的；而在服务端负载均衡中，客户端节点只知道单一服务代理的地址，服务代理则知道所有服务端的地址。在 Spring Cloud 中我们如果想要使用客户端负载均衡，方法很简单，使用 @LoadBalanced 注解即可，这样客户端在发起请求的时候会根据负载均衡策略从服务端列表中选择一个服务端，向该服务端发起网络请求，从而实现负载均衡。

7.2 基础应用

Ribbon 可以和 RestTemplate 一起使用，也可以集成到 Feign 中使用。本节将会分别给出 Ribbon 在这两种使用方式下的代码示例。

RestTemplate 是 Spring 提供的同步 HTTP 网络客户端接口，它可以简化客户端与 HTTP 服务器之间的交互，并且它强制使用 RESTful 风格。它会处理 HTTP 连接和关闭，只需要使用者提供服务器的地址（URL）和模板参数。

Spring Cloud 为客户端负载均衡创建了特定的注解 @LoadBalanced，我们只需要使用该注解修饰创建 RestTemplate 实例的 @Bean 函数，Spring Cloud 就会让 RestTemplate 使用相关的负载均衡策略，默认情况下是使用 Ribbon。

除了 @LoadBalanced 之外，Ribbon 还提供 @RibbonClient 注解。该注解可以为 Ribbon 客户端声明名称和自定义配置。name 属性可以设置客户端的名称，configuration 属性则会设置 Ribbon 相关的自定义配置类，如下所示：

```
@SpringBootApplication
@RestController
@RibbonClient(name = "say-hello", configuration = RibbonConfiguration.class)
public class RibbonApplication {
    @LoadBalanced
    @Bean
    RestTemplate restTemplate(){
        return new RestTemplate();
    }
    @Autowired
```

```
    RestTemplate restTemplate;
    @RequestMapping("/hi")
    public String hi(@RequestParam(value="name", defaultValue="Artaban") String name) {
        String greeting = this.restTemplate.getForObject("http://say-hello/greeting",
            String.class);
        return String.format("%s, %s!", greeting, name);
    }
}
```

如上面代码所示，由于 RestTemplate 的 Bean 实例化方法 restTemplate 被 @LoadBalanced 修饰，所以当调用 restTemplate 的 getForObject 方法发送 HTTP 请求时，会使用 Ribbon 进行负载均衡。@RibbonClient 修饰了代码中的 RibbonApplication 类，声明了一个名为 say-hello 的 Ribbon 客户端，并且设置它的配置类为 RibbonConfiguration，其代码如下所示：

```
public class RibbonConfiguration {
    @Autowired
    IClientConfig ribbonClientConfig;
    @Bean
    public IPing ribbonPing(IClientConfig config) {
        return new PingUrl();
    }
    @Bean
    public IRule ribbonRule(IClientConfig config) {
        return new AvailabilityFilteringRule();
    }
}
```

使用者可以通过配置类创建组件实例来覆盖 Ribbon 提供的默认组件实例。如上代码所示，RibbonConfiguration 配置类重载了 IPing 和 IRule 两个组件的实例，通过 @Bean 函数创建了 PingUrl 实例和 AvailabilityFilteringRule 实例，来替换 Ribbon 默认提供的 NoOpPing 实例和 ZoneAvoidanceRule 实例。通过这种方式，使用者可以依据自己的需求，更改 Ribbon 的组件实例，这也是 Ribbon 高可扩展性和高可修改性的体现。

使用者可以在 application.yml 文件中对 Ribbon 进行配置，比如设置服务端列表或者使用 Eureka 来获取服务端列表，如下所示：

```
say-hello:
    ribbon:
        eureka:
    # 将Eureka关闭，则Ribbon无法从Eureka中获取服务端列表信息
    enabled: false
    # listOfServers可以设置服务端列表
    listOfServers:localhost:8090,localhost:9092,localhost:9999
    serverListRefreshInterval: 15000
```

如上配置所示，Ribbon 可以和服务注册中心 Eureka 一起工作，从服务注册中心获取服务端的地址信息，也可以在配置文件中使用 listOfServers 字段来设置服务端地址。由于 listOfServers 字段可以随意指定服务端地址，所以使用者往往在项目开发和测试阶段使用该字段。

除了和 RestTemplate 进行配套使用之外，Ribbon 还默认被集成到了 OpenFeign 中，当使用 @FeignClient 时，OpenFeign 默认使用 Ribbon 来进行网络请求的负载均衡。

7.3 源码分析

本节首先会介绍 Ribbon 相关的配置和实例的初始化过程，然后讲解 Ribbon 是如何与 OpenFeign 集成的，接着讲解负载均衡器 LoadBalancerClient，最后依次讲解 ILoadBalancer 的实现和负载均衡策略 IRule 的实现，如图 7-1 所示。

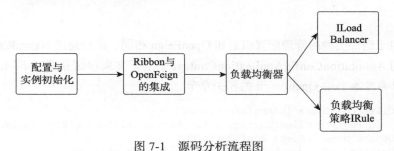

图 7-1　源码分析流程图

7.3.1 配置和实例初始化

@RibbonClient 注解可以声明 Ribbon 客户端，设置 Ribbon 客户端的名称和配置类，configuration 属性可以指定 @Configuration 的配置类，进行 Ribbon 相关的配置。@RibbonClient 还会导入（import）RibbonClientConfigurationRegistrar 类来动态注册 Ribbon 相关的 BeanDefinition。@RibbonClient 注解的具体实现如下所示：

```
@Import(RibbonClientConfigurationRegistrar.class)
public @interface RibbonClient {
    String value() default "";
    /**
     * 配置Ribbon客户端名称
     */
    String name() default "";
    /**
     * Ribbon客户端的自定义配置，可以配置生成客户端的各个组件，比如说ILoadBalancer、ServerListFilter
     *     和IRule。默认的配置为RibbonClientConfiguration.java
     */
    Class<?>[] configuration() default {};
}
```

RibbonClientConfigurationRegistrar 是 ImportBeanDefinitionRegistrar 的实现类，ImportBeanDefinitionRegistrar 是 Spring 动态注册 BeanDefinition 的接口，可以用来注册 Ribbon 所需的 BeanDefinition，比如说 Ribbon 客户端实例（Ribbon Client）。ImportBeanDefinitionRegistrar 的 registerBeanDefinitions 方法可以注册 Ribbon 客户端的配置类，也就是 @RibbonClient 的

configuration 属性值。registerBeanDefinitions 方法的具体实现如下所示：

```java
//RibbonClientConfigurationRegistrar.java
public void registerBeanDefinitions(AnnotationMetadata metadata,
        BeanDefinitionRegistry registry) {
    ...
    //获取@RibbonClient的参数数值，获取clientName后进行configuration的注册
    Map<String, Object> client = metadata.getAnnotationAttributes(
            RibbonClient.class.getName(), true);
    String name = getClientName(client);//获取RibbonClient的value或者name数值。
    if (name != null) {
        registerClientConfiguration(registry, name, client.get("configuration"));
    }
}
```

Ribbon 对于组件实例的管理配置机制和 OpenFeign 相同，都是通过 NamedContextFactory 创建带名称的 AnnotationConfigApplicationContext 子上下文来存储并管理不同的组件实例，具体细节可以参考本书中 OpenFeign 的相关章节。

```java
//RibbonClientConfigurationRegistrar.java
private void registerClientConfiguration(BeanDefinitionRegistry registry,
        Object name, Object configuration) {
    BeanDefinitionBuilder builder = BeanDefinitionBuilder
            .genericBeanDefinition(RibbonClientSpecification.class);
    builder.addConstructorArgValue(name);
    builder.addConstructorArgValue(configuration);
    registry.registerBeanDefinition(name + ".RibbonClientSpecification",
            builder.getBeanDefinition());
}
```

如上代码所示，registerClientConfiguration 方法会向 BeanDefinitionRegistry 注册一个 RibbonClientSpecification 的 BeanDefinition，其名称为 RibbonClient 的名称加上 .RibbonClientSpecification。RibbonClientSpecification 是 NamedContextFactory.Specification 的实现类，是供 SpringClientFactory 使用的。在 RibbonAutoConfiguration 里会进行 SpringClientFactory 实例的初始化，并将所有的 RibbonClientSpecification 实例都设置给 SpringClientFactory，供其在初始化 Ribbon 相关组件实例时使用。

```java
@Configuration
@RibbonClients
@AutoConfigureAfter(name = "org.springframework.cloud.netflix.eureka.EurekaClientAutoConfiguration")
@AutoConfigureBefore({LoadBalancerAutoConfiguration.class, AsyncLoadBalancerAuto
        Configuration.class})
//@AutoConfigureBefore表明该Configuration会在LoadBalancerAutoConfiguration配置类之前
    进行执行，因为后者会依赖前者
@EnableConfigurationProperties({RibbonEagerLoadProperties.class, ServerIntrospectorProperties.
        class})
public class RibbonAutoConfiguration {
    //在RibbonClientConfigurationRegistrar中注册的RibbonClientSpecification实例都会
        被注入到这里
    @Autowired(required = false)
```

```java
    private List<RibbonClientSpecification> configurations = new ArrayList<>();
    @Autowired
    private RibbonEagerLoadProperties ribbonEagerLoadProperties;
    @Bean
    public SpringClientFactory springClientFactory() {
        SpringClientFactory factory = new SpringClientFactory();
        factory.setConfigurations(this.configurations);
        return factory;
    }
    //LoadBalanceClient是核心的类
    @Bean
    @ConditionalOnMissingBean(LoadBalancerClient.class)
    public LoadBalancerClient loadBalancerClient() {
        return new RibbonLoadBalancerClient(springClientFactory());
    }
}
```

如上代码所示，RibbonAutoConfiguration 配置类也会进行 LoadBalancerClient 接口的默认实例的初始化。loadBalancerClient 方法被 @ConditionalOnMissingBean 注解修饰，意味着只有当 Spring 容器中没有 LoadBalancerClient 实例时，该方法才会初始化 RibbonLoadBalancerClient 对象，将其作为 LoadBalancerClient 的实例。

7.3.2 与 OpenFeign 的集成

Ribbon 是 RESTful HTTP 客户端 OpenFeign 负载均衡的默认实现。本书在 OpenFeign 的章节中讲解了相关实例的初始化过程。FeignClientFactoryBean 是创造 FeignClient 的工厂类，在其 getObject 方法中有一个分支判断，当请求 URL 不为空时，就会生成一个具有负载均衡的 FeignClient。在这个过程中，OpenFeign 就默认引入了 Ribbon 的负载均衡实现，OpenFegin 引入 Ribbon 的部分代码如下所示：

```java
//FeignClientFactoryBean.java
public Object getObject() throws Exception {
    FeignContext context = applicationContext.getBean(FeignContext.class);
    Feign.Builder builder = feign(context);
    //如果url不为空，则需要负载均衡
    if (!StringUtils.hasText(this.url)) {
        String url;
        if (!this.name.startsWith("http")) {
            url = "http://" + this.name;
        }
        else {
            url = this.name;
        }
        url += cleanPath();
        return loadBalance(builder, context, new HardCodedTarget<>(this.type,
                this.name, url));
    }
    //....生成普通的FeignClient
}
```

如 OpenFeign 的源码所示，loadBalance 方法会生成 LoadBalancerFeignClient 实例进行返回。LoadBalancerFeignClient 实现了 OpenFeign 的 Client 接口，负责 OpenFeign 网络请求的发送和响应的接收，并带有客户端负载均衡机制。loadBalance 方法实现如下所示：

```java
//FeignClientFactoryBean.java
protected <T> T loadBalance(Feign.Builder builder, FeignContext context,
        HardCodedTarget<T> target) {
    //会得到'LoadBalancerFeignClient'
    Client client = getOptional(context, Client.class);
    if (client != null) {
        builder.client(client);
        Targeter targeter = get(context, Targeter.class);
        return targeter.target(this, builder, context, target);
    }
}
```

LoadBalancerFeignClient#execute 方法会将普通的 Request 对象转化为 RibbonRequest，并使用 FeignLoadBalancer 实例来发送 RibbonRequest。execute 方法会首先将 Request 的 URL 转化为对应的服务名称，然后构造出 RibbonRequest 对象，接着调用 lbClient 方法来生成 FeignLoadBalancer 实例，最后调用 FeignLoadBalancer 实例的 executeWithLoadBalancer 方法来处理网络请求。LoadBalancerFeignClient#execute 方法的具体实现如下所示：

```java
//LoadBalancerFeignClient.java
public Response execute(Request request, Request.Options options) throws IOException {
    try {
        //负载均衡时，host就是需要调用的服务的名称
        URI asUri = URI.create(request.url());
        String clientName = asUri.getHost();
        URI uriWithoutHost = cleanUrl(request.url(), clientName);
        //构造RibbonRequest,delegate一般就是真正发送网络请求的客户端，比如说OkHttpClient
            和ApacheClient
        FeignLoadBalancer.RibbonRequest ribbonRequest = new FeignLoadBalancer.
            RibbonRequest(
                this.delegate, request, uriWithoutHost);
        IClientConfig requestConfig = getClientConfig(options, clientName);
        //executeWithLoadBalancer是进行负载均衡的关键
        return lbClient(clientName).executeWithLoadBalancer(ribbonRequest,
                requestConfig).toResponse();
    }
    catch (ClientException e) {
        IOException io = findIOException(e);
        if (io != null) {
            throw io;
        }
        throw new RuntimeException(e);
    }
}
private FeignLoadBalancer lbClient(String clientName) {
    //调用CachingSpringLoadBalancerFactory类的create方法。
    return this.lbClientFactory.create(clientName);
```

}

lbClientFactory 的参数是 CachingSpringLoadBalancerFactory 的实例,它是带有缓存机制的 FeignLoadBalancer 的工厂类。

create 方法的 clientName 参数是指 HTTP 请求对应的服务端名称,它会首先使用这个名称去缓存中查找是否已经存在对应的实例。如果没有,再根据系统是否支持请求重试来创建出不同的 FeignLoadBalancer 实例,最后将该实例存储到缓存中,create 方法的代码如下所示:

```java
//CachingSpringLoadBalancerFactory.java
public FeignLoadBalancer create(String clientName) {
    if (this.cache.containsKey(clientName)) {
        return this.cache.get(clientName);
    }
    IClientConfig config = this.factory.getClientConfig(clientName);
    ILoadBalancer lb = this.factory.getLoadBalancer(clientName);
    ServerIntrospector serverIntrospector = this.factory.getInstance(clientName,
        ServerIntrospector.class);
    //如果需要重试就是RetryableFeignLoadBalancer,否则是FeignLoadBalancer
    FeignLoadBalancer client = enableRetry ? new RetryableFeignLoadBalancer(lb,
        config, serverIntrospector,
        loadBalancedRetryPolicyFactory, loadBalancedBackOffPolicyFactory, load
            BalancedRetryListenerFactory) : new FeignLoadBalancer(lb, config,
            serverIntrospector);
    this.cache.put(clientName, client);
    return client;
}
```

FeignLoadBalancer 是 OpenFeign 在不需要重试机制的情况下默认的负载均衡实现。它的 execute 方法的实现很简单,使用 RibbonRequest 对象的客户端来发送网络请求,然后将 Response 包装成 RibbonResponse 进行返回。RibbonRequest 的 request 方法返回的对象就是构造 RibbonRequest 对象时传入的 delegate 参数。该参数是 Client 接口的实例,Client 接口是 OpenFeign 真正发送网络请求的客户端,比如说 OkHttpClient 和 ApacheClient。FeignLoadBalancer 的 execute 方法如下所示:

```java
//FeignLoadBalancer.java
public RibbonResponse execute(RibbonRequest request, IClientConfig configOverride)
        throws IOException {
    Request.Options options;
    if (configOverride != null) {
        RibbonProperties override = RibbonProperties.from(configOverride);
        options = new Request.Options(
                override.connectTimeout(this.connectTimeout),
                override.readTimeout(this.readTimeout));
    }
    else {
        options = new Request.Options(this.connectTimeout, this.readTimeout);
    }
    Response response = request.client().execute(request.toRequest(), options);
    return new RibbonResponse(request.getUri(), response);
```

}

FeignLoadBalancer 是 AbstractLoadBalancerAwareClient 的子类，其 executeWithLoadBalancer 方法会首先创建一个 LoadBalancerCommand 实例，然后在该实例的 submit 方法的回调中调用子类的 execute 方法，executeWithLoadBalancer 方法如下代码所示：

```java
//AbstractLoadBalancerAwareClient.java
public T executeWithLoadBalancer(final S request, final IClientConfig requestConfig)
    throws ClientException {
    //创建LoadBalancerCommand
    LoadBalancerCommand<T> command = buildLoadBalancerCommand(request, requestConfig);
    return command.submit(
        new ServerOperation<T>() {
            @Override
            public Observable<T> call(Server server) {
                URI finalUri = reconstructURIWithServer(server, request.getUri());
                S requestForServer = (S) request.replaceUri(finalUri);
                try {
                    //调用子类的execute方法，进行HTTP请求的处理
                    return Observable.just(AbstractLoadBalancerAwareClient.this.
                        execute(requestForServer, requestConfig));
                }
                catch (Exception e) {
                    return Observable.error(e);
                }
            }
        })
        .toBlocking()
        .single();
}
```

其中，buildLoadBalancerCommand 方法使用了 LoadBalancerCommand.Builder 来创建 LoadBalancerCommand 实例，并将 AbstractLoadBalancerAwareClient 作为 LoadBalancerContext 接口的实例设置给 LoadBalancerCommand 实例，如下所示：

```java
//LoadBalancerCommand.Builder
protected LoadBalancerCommand<T> buildLoadBalancerCommand(final S request, final
    IClientConfig config) {
    RequestSpecificRetryHandler handler = getRequestSpecificRetryHandler(req
        uest, config);
    LoadBalancerCommand.Builder<T> builder = LoadBalancerCommand.<T>builder()
        .withLoadBalancerContext(this)
        .withRetryHandler(handler)
        .withLoadBalancerURI(request.getUri());
    customizeLoadBalancerCommandBuilder(request, config, builder);
    return builder.build();
}
```

LoadBalancerCommand 的 submit 方法使用了响应式编程的原理，创建一个 Observable 实例来订阅，使用通过负载均衡器选出的服务器来进行异步的网络请求。LoadBalancerCommand 的 submit 方法的具体实现基于 Observable 机制，较为复杂，笔者将 submit 源码进行了简化，可

以让读者在不了解 Observable 的情况下了解源码的基本原理。submit 方法调用了 selectServer 方法来选择一个 server，这里就是 OpenFeign 进行负载均衡的地方，代码如下所示：

```java
//LoadBalancerCommand.java
public Observable<T> submit(final ServerOperation<T> operation) {
    final ExecutionInfoContext context = new ExecutionInfoContext();
    // 使用loadBalancerContext通过负载均衡来选择Server
    Server server = loadBalancerContext.getServerFromLoadBalancer(loadBalancerURI,
        loadBalancerKey);
    context.setServer(server);
    //调用operation对象，就是调用AbstractLoadBalancerAwareClient中executeWithLoadBalancer
      方法中创建的匿名Operation对象
    return operation.call(server);
}
```

上面代码中，LoadBalancerCommand 的 selectServer 方法调用了 LoadBalancerContext 的 getServerFromLoadBalancer 方法，代码如下所示：

```java
//LoadBalancerCommand.java
private Observable<Server> selectServer() {
    return Observable.create(new OnSubscribe<Server>() {
        @Override
        public void call(Subscriber<? super Server> next) {
            try {
                Server server = loadBalancerContext.getServerFromLoadBalancer(lo
                    adBalancerURI, loadBalancerKey);
                next.onNext(server);
                next.onCompleted();
            } catch (Exception e) {
                next.onError(e);
            }
        }
    });
}
```

LoadBalancerContext 的 getServerFromLoadBalancer 方法调用了 ILoadBalancer 的 chooseServer 方法，从而完成了负载均衡中服务器的选择。这部分的具体实现，在下一节会进行详细描述。

7.3.3　负载均衡器 LoadBalancerClient

本小节主要讲解 LoadBalancerClient 进行负载均衡的具体原理和实现。LoadBalancerClient 是 Ribbon 项目的核心类之一，可以在 RestTemplate 发送网络请求时替代 RestTemplate 进行网络调用。LoadBalancerClient 的定义如下所示：

```java
// LoadBalancerClient.java
public interface LoadBalancerClient extends ServiceInstanceChooser {
    // 从serviceId所代表的服务列表中选择一个服务器来发送网络请求
    <T> T execute(String serviceId, LoadBalancerRequest<T> request) throws IOException;
    <T> T execute(String serviceId, ServiceInstance serviceInstance, LoadBalancerRequest<T>
```

```
        request) throws IOException;
    // 构建网络请求URI
    URI reconstructURI(ServiceInstance instance, URI original);
}
```

LoadBalancerClient 接口继承了 ServiceInstanceChooser 接口，其 choose 方法可以从服务器列表中依据负载均衡策略选出一个服务器实例。ServiceInstanceChooser 的定义如下所示：

```
//实现该类来选择一个服务器用于发送请求
public interface ServiceInstanceChooser {
    /**
     * 根据serviceId从服务器列表中选择一个ServiceInstance
     **/
    ServiceInstance choose(String serviceId);
}
```

RibbonLoadBalancerClient 是 LoadBalancerClient 的实现类之一，它的 execute 方法会首先使用 ILoadBalancer 来选择服务器实例（Server），然后将该服务器实例封装成 RibbonServer 对象，最后再调用 LoadBalancerRequest 的 apply 方法进行网络请求的处理。excute 方法的具体实现如下所示：

```
//RibbonLoadBalancerClient.java
public <T> T execute(String serviceId, LoadBalancerRequest<T> request) throws 
    IOException {
    //每次发送请求都会获取一个ILoadBalancer,会涉及负载均衡规则(IRule)、服务器列表集群(ServerList)
        和检验服务是否存在(IPing)等细节实现
    ILoadBalancer loadBalancer = getLoadBalancer(serviceId);
    Server server = getServer(loadBalancer);
    if (server == null) {
        throw new IllegalStateException("No instances available for " + serviceId);
    }
    RibbonServer ribbonServer = new RibbonServer(serviceId, server, isSecure(server,
            serviceId), serverIntrospector(serviceId).getMetadata(server));

    return execute(serviceId, ribbonServer, request);
}
```

getLoadBalancer 方法直接调用了 SpringClientFactory 的 getLoadBalancer 方法。SpringClientFactory 是 NamedContextFactory 的实现类，关于 NamedContextFactory 的机制在第 5 章中已经详细讲解过了，通过它可以实现多套组件实例的管理，代码如下所示：

```
//RibbonLoadBalancerClient.java
protected ILoadBalancer getLoadBalancer(String serviceId) {
    return this.clientFactory.getLoadBalancer(serviceId);
}
```

getServer 方法则是直接调用了 ILoadBalancer 的 chooseServer 方法来使用负载均衡策略——从已知的服务器列表中选出一个服务器实例，其具体实现如下所示：

```
//RibbonLoadBalancerClient.java
protected Server getServer(ILoadBalancer loadBalancer) {
```

```
    if (loadBalancer == null) {
        return null;
    }
    return loadBalancer.chooseServer("default");
}
```

execute 方法调用 LoadBalancerRequest 实例的 apply 方法，将之前根据负载均衡策略选择出来的服务器作为参数传递进去，进行真正的 HTTP 请求发送，代码如下所示：

```
//RibbonLoadBalancerClient.java
public <T> T execute(String serviceId, ServiceInstance serviceInstance, LoadBalancerRequest<T>
    request) throws IOException {
    Server server = null;
    if(serviceInstance instanceof RibbonServer) {
        server = ((RibbonServer)serviceInstance).getServer();
    }
    RibbonLoadBalancerContext context = this.clientFactory
            .getLoadBalancerContext(serviceId);
    RibbonStatsRecorder statsRecorder = new RibbonStatsRecorder(context, server);
    try {
        T returnVal = request.apply(serviceInstance);
        statsRecorder.recordStats(returnVal);
        return returnVal;
    }
catch (IOException ex) {
    ...
    }
    return null;
}
```

LoadBalancerRequest 的 apply 方法的具体实现本书不再详细讲解，因为 Ribbon 最为重要的部分就是使用负载均衡策略来选择服务器，也就是 ILoadBalancer 的 chooseServer 方法的实现，本书会在接下来的小节里对其进行详细讲解。

7.3.4　ILoadBalancer

ILoadBalancer 是 Ribbon 的关键类之一，它是定义负载均衡操作过程的接口。Ribbon 通过 SpringClientFactory 工厂类的 getLoadBalancer 方法可以获取 ILoadBalancer 实例。根据 Ribbon 的组件实例化机制，ILoadBalnacer 实例是在 RibbonAutoConfiguration 中被创建生成的。

SpringClientFactory 中的实例都是 RibbonClientConfiguration 或者自定义 Configuration 配置类创建的 Bean 实例。RibbonClientConfiguration 还创建了 IRule、IPing 和 ServerList 等相关组件的实例。使用者可以通过自定义配置类给出上述几个组件的不同实例。

如图 7-2 所示，ZoneAwareLoadBalancer 是 ILoadBalancer 接口的实现类之一，它是 Ribbon 默认的 ILoadBalancer 接口的实例。

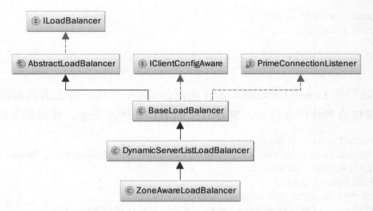

图 7-2　ZoneAwareLoadBalancer 类图

RibbonClientConfiguration 中有关 ZoneAwareLoadBalancer 的配置如下所示：

```
//RibbonClientConfiguration.java
@Bean
@ConditionalOnMissingBean
public ILoadBalancer ribbonLoadBalancer(IClientConfig config,
    ServerList<Server> serverList, ServerListFilter<Server> serverListFilter,
        IRule rule, IPing ping, ServerListUpdater serverListUpdater) {
    if (this.propertiesFactory.isSet(ILoadBalancer.class, name)) {
        return this.propertiesFactory.get(ILoadBalancer.class, config, name);
    }
    return new ZoneAwareLoadBalancer<>(config, rule, ping, serverList,
            serverListFilter, serverListUpdater);
}
```

图 7-3 是 IBalancer 相关的类图，其中的类都是 ZoneAwareLoadBalancer 构造方法所需参数实例的类型。

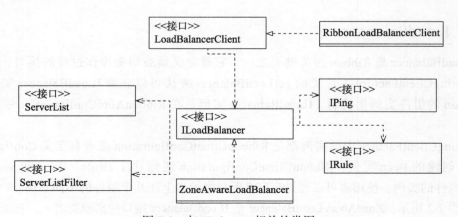

图 7-3　与 IBalancer 相关的类图

接下来按照 ZoneAwareLoadBalancer 构造函数的参数顺序来看一下与 ILoadBalancer 相

关的重要的类，它们分别是 IClientConfig、IRule、IPing、ServerList 和 ServerListFilter，默认配置如表 7-1 所示。

表 7-1 默认 Bean 信息表

Bean 类型	Bean 名称	类 名	解 释
ILoadBalancer	ribbonLoadBalancer	ZoneAwareLoadBalancer	负载均衡器的抽象，由 butong 的组件构成
IClientConfig	ribbonClientConfig	ribbonClientConfig	Client 的配置类
IRule	ribbonRule	RoundRobinRule	负载均衡策略
IPing	ribbonPing	DummyPing	服务可用性检测
ServerList	ribbonServerList	ConfigurationBasedServerList	服务列表的获取
ServerListFilter	ribbonServerListFilter	ZonePreferenceServerListFilter	服务器列表的过滤

ZoneAwareLoadBalancer 的 chooseServer 方法会首先使用 DynamicPropertyFactory 来获取平均负载（triggeringLoadPerServerThreshold）和实例故障率（avoidZoneWithBlackoutPercetage）两个阈值，然后调用 ZoneAvoidanceRule 的 getAvailableZones 方法使用这两个阈值来获取所有可用的服务区（Zone）列表，每个服务区实例中包含了一定数量的服务器实例。然后调用 ZoneAvoidanceRule 的 randomChooseZone 方法从上述的服务区列表中随机选出一个服务区，最后调用该服务区对应 BaseLoadBalancer 实例的 chooseServer 方法获取到最终的服务器实例。ZoneAwareLoadBalancer 会为不同的服务区调用不同的 BaseLoadBalancer 的 chooseServer 方法，这正体现了它类名的含义。chooseServer 方法是其中最重要的方法，具体实现如下所示：

```
//ZoneAwareLoadBalancer.java
public Server chooseServer(Object key) {
    if (!ENABLED.get() || getLoadBalancerStats().getAvailableZones().size() <= 1) {
            logger.debug("Zone aware logic disabled or there is only one zone");
            return super.chooseServer(key);
    }
    Server server = null;
    try {
        //获取当前有关负载均衡的服务器状态集合
        LoadBalancerStats lbStats = getLoadBalancerStats();
        Map<String, ZoneSnapshot> zoneSnapshot = ZoneAvoidanceRule.createSnapshot
            (lbStats);
        logger.debug("Zone snapshots: {}", zoneSnapshot);
        //使用'DynamicPropertyFactory'获取平均负载的阈值
        if (triggeringLoad == null) {
            triggeringLoad = DynamicPropertyFactory.getInstance().getDoubleProperty(
                "ZoneAwareNIWSDiscoveryLoadBalancer." + this.getName() + ".trigg
                    eringLoadPerServerThreshold", 0.2d);
        }
        //使用'DynamicPropertyFactory'获取平均实例故障率的阈值
        if (triggeringBlackoutPercentage == null) {
            triggeringBlackoutPercentage = DynamicPropertyFactory.getInstance().
```

```java
                getDoubleProperty(
                    "ZoneAwareNIWSDiscoveryLoadBalancer." + this.getName() + ".av
                        oidZoneWithBlackoutPercetage", 0.99999d);
        }
        //根据两个阈值来获取所有可用的服务区列表
        Set<String> availableZones = ZoneAvoidanceRule.getAvailableZones(zoneSnapshot,
            triggeringLoad.get(), triggeringBlackoutPercentage.get());
        logger.debug("Available zones: {}", availableZones);
        if (availableZones != null &&  availableZones.size() < zoneSnapshot.
            keySet().size()) {
            //随机从可用的服务区列表中选择一个服务区
            String zone = ZoneAvoidanceRule.randomChooseZone(zoneSnapshot,
                availableZones);
            logger.debug("Zone chosen: {}", zone);
            if (zone != null) {
                //得到zone对应的BaseLoadBalancer
                BaseLoadBalancer zoneLoadBalancer = getLoadBalancer(zone);
                server = zoneLoadBalancer.chooseServer(key);
            }
        }
    } catch (Exception e) {
        logger.error("Error choosing server using zone aware logic for load
            balancer={}", name, e);
    }
    if (server != null) {
        return server;
    } else {
        logger.debug("Zone avoidance logic is not invoked.");
        return super.chooseServer(key);
    }
}
```

BaseLoadBalancer 对象的 chooseServer 方法实现比较简单，就是直接调用它的 IRule 成员变量的 choose 方法。IRule 是负责实现负载均衡策略的接口，本书会在下一小节进行详细描述，BaseLoadBalancer 的 chooseServer 函数的代码如下所示：

```java
//BaseLoadBalancer.java
public Server chooseServer(Object key) {
    if (counter == null) {
        counter = createCounter();
    }
    counter.increment();
    if (rule == null) {
        return null;
    } else {
        try {
            return rule.choose(key);
        } catch (Throwable t) {
            return null;
        }
    }
}
```

7.3.5 负载均衡策略实现

IRule 是定义 Ribbon 负载均衡策略的接口，你可以通过实现该接口来自定义自己的负载均衡策略，RibbonClientConfiguration 配置类则会给出 IRule 的默认实例。IRule 接口的 choose 方法就是从一堆服务器中根据一定规则选出一个服务器。IRule 有很多默认的实现类，这些实现类根据不同的算法和逻辑来进行负载均衡。

图 7-4 展示了 Ribbon 提供的 IRule 的实现类。

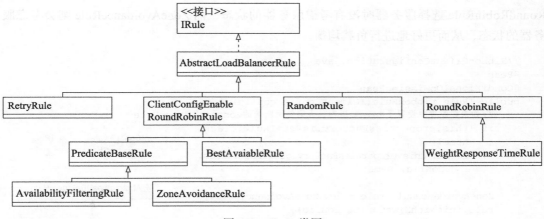

图 7-4　IRule 类图

在大多数情况下，这些默认的实现类是可以满足需求的，如果有特殊需求，可以自己实现。Ribbon 内置的 IRule 子类如下所示。

- BestAvailableRule：选择最小请求数的服务器。
- ClientConfigEnabledRoundRobinRule：使用 RandomRobinRule 随机选择一个服务器。
- RoundRobinRule：以 RandonRobin 方法轮询选择服务器。
- RetryRule：在选定的负载均衡策略上添加重试机制。
- WeightedResponseTimeRule：根据响应时间去计算一个权重（weight），响应时间越长，权重越低，权重越低的服务器，被选择的可能性就越低。
- ZoneAvoidanceRule：根据服务器所属的服务区的整体运行状况来轮询选择。

Ribbon 的负载均衡策略既有 RoundRobinRule 和 RandomRule 这样的不依赖于服务器运行状况的策略，也有 AvailabilityFilteringRule 和 WeightedResponseTimeRule 等多种基于服务器运行状况决策的策略。这些策略既可以依据单个服务器的运行状况，也可以依据整个服务区的运行状况选择具体调用的服务器，适用于各种场景需求。

ZoneAvoidanceRule 是 Ribbon 默认的 IRule 实例，较为复杂，在本小节的后续部分再进行讲解。而 ClientConfigEnabledRoundRobinRule 是比较常用的 IRule 的子类之一，它使用的负载均衡策略是最为常见的 Round Robin 策略，即简单轮询策略。其具体实现如下所示：

```java
public class ClientConfigEnabledRoundRobinRule extends AbstractLoadBalancerRule {
    RoundRobinRule roundRobinRule = new RoundRobinRule();
    @Override
    public Server choose(Object key) {
        return roundRobinRule.choose(key);
    }
}
```

RoundRobinRule 会以轮询的方式依次将选择不同的服务器,从序号为 1 的服务器开始,直到序号为 N 的服务器,下次选择服务器时,再从序号为 1 的服务器开始。但是 RoundRobinRule 选择服务器时没有考虑服务器的状态,而 ZoneAvoidanceRule 则会考虑服务器的状态,从而更好地进行负载均衡。

```java
//RibbonClientConfiguration.java
@Bean
@ConditionalOnMissingBean
public IRule ribbonRule(IClientConfig config) {
    //如果在配置中设置了Rule就返回,否则使用默认的zoneAvoidanceRule
    if (this.propertiesFactory.isSet(IRule.class,
        name)) {
        return this.propertiesFactory.get(IRule.class,
            config, name);
    }
    ZoneAvoidanceRule rule = new ZoneAvoidanceRule();
    rule.initWithNiwsConfig(config);
    return rule;
}
```

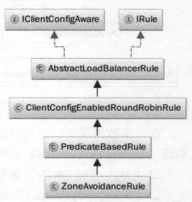

图 7-5 ZoneAvoidanceRule 类图

如上代码所示,ZoneAvoidanceRule 是 Ribbon 默认的 IRule 实例,它使用 CompositePredicate 来根据服务区的运行状况和服务器的可用性选择服务器。ZoneAvoidanceRule 相关的类图如图 7-5 所示。

ZoneAvoidanceRule 是根据服务器所属的服务区的运行状况和可用性来进行负载均衡。PredicateBasedRule 是 ZoneAvoidanceRule 的基类,它选择服务器的策略是先使用 ILoadBalancer 获取服务器列表,再使用 AbstractServerPredicate 进行服务器的过滤,最后使用轮询策略从剩余的服务器列表中选出最终的服务器。PredicateBasedRule 的具体实现如下所示:

```java
public abstract class PredicateBasedRule extends ClientConfigEnabledRoundRobinRule {
    public abstract AbstractServerPredicate getPredicate();
    @Override
    public Server choose(Object key) {
        ILoadBalancer lb = getLoadBalancer();
        Optional<Server> server = getPredicate().chooseRoundRobinAfterFiltering(
            lb.getServerList(false), key);
        if (server.isPresent()) {
            return server.get();
        } else {
```

```
        return null;
    }
}
```

PredicateBasedRule 接口的 getPredicate 抽象接口需要子类实现，不同的子类提供不同的 AbstractServerPredicate 实例来实现不同的服务器过滤策略。而 ZoneAvoidanceRule 使用的是由 ZoneAvoidancePredicate 和 AvailabilityPredicate 组成的复合策略 CompositePredicate，前一个判断判定一个服务区的运行状况是否可用，去除不可用的服务区的所有服务器，后一个用于过滤掉连接数过多的服务器。ZoneAvoidanceRule 的 getPredicate 方法的相关实现如下所示：

```
//ZoneAvoidanceRule.java
public ZoneAvoidanceRule() {
    super();
    ZoneAvoidancePredicate zonePredicate = new ZoneAvoidancePredicate(this);
    AvailabilityPredicate availabilityPredicate = new AvailabilityPredicate(this);
    compositePredicate = createCompositePredicate(zonePredicate, availabilityPredicate);
}

//将两个Predicate组合成一个CompositePredicate
private CompositePredicate createCompositePredicate(ZoneAvoidancePredicate p1,
        AvailabilityPredicate p2) {
    return CompositePredicate.withPredicates(p1, p2)
        .addFallbackPredicate(p2)
        .addFallbackPredicate(AbstractServerPredicate.alwaysTrue())
        .build();
}
//实现getPredicate接口，返回构造方法中生成的compositePredicate
@Override
public AbstractServerPredicate getPredicate() {
    return compositePredicate;
}
```

CompositePredicate 的 chooseRoundRobinAfterFiltering 方法继承父类 AbstractServerPredicate 的实现。它会首先调用 getEligibleServers 方法通过 Predicate 过滤服务器列表，然后使用轮询策略选择出一个服务器进行返回，如下所示：

```
//AbstractServerPredicate.java
//先用Predicate来获取一个可用服务器的集合，然后用轮询算法来选择一个服务器
public Optional<Server> chooseRoundRobinAfterFiltering(List<Server> servers) {
    List<Server> eligible = getEligibleServers(servers);
    if (eligible.size() == 0) {
        return Optional.absent();
    }
    // (i + 1) % n
    return Optional.of(eligible.get(nextIndex.getAndIncrement() % eligible.size()));
}
```

当 loadBalancerKey 为 null 时，getEligibleServers 方法会使用 serverOnlyPredicate 来依

次过滤服务器列表。getEligibleServers 方法的具体实现如下所示：

```java
//AbstractServerPredicate.java
public List<Server> getEligibleServers(List<Server> servers, Object loadBalancerKey) {
    if (loadBalancerKey == null) {
        return ImmutableList.copyOf(Iterables.filter(servers, this.getServerOnlyPredicate()));
    } else {
        //遍历 servers,调用对应 Predicate 的 apply 方法来判断该服务器是否可用
        List<Server> results = Lists.newArrayList();
        for (Server server: servers) {
            if (this.apply(new PredicateKey(loadBalancerKey, server))) {
                results.add(server);
            }
        }
        return results;
    }
}
```

serverOnlyPredicate 则会调用其 apply 方法，并将 Server 对象封装成 PredicateKey 当作参数传入。AbstractServerPredicate 并没有实现 apply 方法，由它的子类来实现从而达到不同子类实现不同过滤策略的目的，serverOnlyPredicate 的代码如下所示：

```java
//AbstractServerPredicate.java
private final Predicate<Server> serverOnlyPredicate =  new Predicate<Server>() {
    @Override
    public boolean apply(@Nullable Server input) {
        return AbstractServerPredicate.this.apply(new PredicateKey(input));
    }
};
```

ZoneAvoidanceRule 类中 CompositePredicate 对象的 apply 方法就会依次调用 ZoneAvoidancePredicate 和 AvailabilityPredicate 的 apply 方法。

ZoneAvoidancePredicate 以服务区（Zone）为单位考察所有服务区的整体运行情况，对于不可用的区域整个丢弃，从剩下服务区中选可用的服务器。并且会判断出最差的服务区，排除掉最差服务区。ZoneAvoidancePredicate 的 apply 方法如下代码所示。

```java
//ZoneAvoidancePredicate.java
public boolean apply(@Nullable PredicateKey input) {
    if (!ENABLED.get()) {
        return true;
    }
    String serverZone = input.getServer().getZone();
    if (serverZone == null) {
        //如果服务器没有服务区相关的信息,则直接返回
        return true;
    }
    //LoadBalancerStats 存储着每个服务器或者节点的执行特征和运行记录。这些信息可供动态负载均
      衡策略使用
    LoadBalancerStats lbStats = getLBStats();
    if (lbStats == null) {
        //如果没有则直接返回
```

```
        return true;
    }
    if (lbStats.getAvailableZones().size() <= 1) {
        // 如果只有一个服务区，那么也是直接返回
        return true;
    }
    //为了效率，先看一下lbStats中记录的服务区列表是否包含当前这个服务区
    Map<String, ZoneSnapshot> zoneSnapshot = ZoneAvoidanceRule.createSnapshot(lbStats);
    if (!zoneSnapshot.keySet().contains(serverZone)) {
        // 如果该serverZone不存在，那么也直接返回
        return true;
    }
    //调用ZoneAvoidanceRule的getAvailableZone方法来获取可用的服务区列表
    Set<String> availableZones = ZoneAvoidanceRule.getAvailableZones(zoneSnapshot,
        triggeringLoad.get(), triggeringBlackoutPercentage.get());
    //判断当前服务区是否在可用服务区列表中
    if (availableZones != null) {
        return availableZones.contains(input.getServer().getZone());
    } else {
        return false;
    }
}
```

服务区是多个服务实例的集合，不同服务区之间的服务实例相互访问会有更大的网络延迟，服务区之内的服务实例访问网络延迟较小。ZoneSnapshot 存储了关于服务区的一些运行状况数据，比如说实例数量、断路器断开数、活动请求数和实例平均负载。ZoneSnapshot 的定义如下所示：

```
//ZoneSnapshot.java
public class ZoneSnapshot {
    final int instanceCount;        //实例数
    final double loadPerServer;     //平均负载
    final int circuitTrippedCount;  //断路器断开数量
    final int activeRequestsCount;  //活动请求数量
}
```

ZoneAvoidanceRule 的 createSnapshot 方法其实就是将所有的服务区列表转化为以其名称为键值的哈希表，供 ZoneAvoidancePredicate 的 apply 方法使用，如下所示：

```
//ZoneAvoidanceRule.java
//将LoadbalancerStats中的availableZones列表转换为Map再返回
static Map<String, ZoneSnapshot> createSnapshot(LoadBalancerStats lbStats) {
    Map<String, ZoneSnapshot> map = new HashMap<String, ZoneSnapshot>();
    for (String zone : lbStats.getAvailableZones()) {
        ZoneSnapshot snapshot = lbStats.getZoneSnapshot(zone);
        map.put(zone, snapshot);
    }
    return map;
}
```

getAvailableZones 方法是用来筛选服务区列表的，首先，它会遍历一遍 ZoneSnapshot

哈希表，在遍历的过程中，它会做两件事情：依据 ZoneSnapshot 的实例数、实例的平均负载时间和实例故障率等指标将不符合标准的 ZoneSnapshot 从列表中删除，它会维护一个最坏 ZoneSnapshot 列表，当某个 ZoneSnapshot 的平均负载时间小于但接近全局最坏负载时间时，就会将该 ZoneSnapshot 加入到最坏 ZoneSnapshot 列表中，如果某个 ZoneSnapshot 的平均负载时间大于最坏负载时间时，它将会清空最坏 ZoneSnapshot 列表，然后以该 ZoneSnapshot 的平均负载时间作为全局最坏负载时间，继续最坏 ZoneSnapshot 列表的构建。在方法最后，如果全局最坏负载数据大于系统设定的负载时间阈值，则在最坏 ZoneSnapshot 列表中随机选择出一个 ZoneSnapshot，将其从列表中删除。

图 7-6 显示了 getAvailableZone 方法的筛选流程。

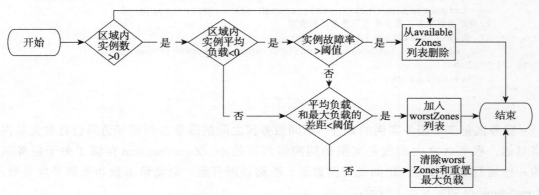

图 7-6　getAvailableZones 的筛选流程

而 getAvailableZones 具体实现如下所示：

```java
//ZoneAvoidanceRule.java
public static Set<String> getAvailableZones(
        Map<String, ZoneSnapshot> snapshot, double triggeringLoad,
        double triggeringBlackoutPercentage) {
    Set<String> availableZones = new HashSet<String>(snapshot.keySet());
    Set<String> worstZones = new HashSet<String>();
    double maxLoadPerServer = 0;
    boolean limitedZoneAvailability = false;
    //遍历所有的服务区来判定
    for (Map.Entry<String, ZoneSnapshot> zoneEntry : snapshot.entrySet()) {
        String zone = zoneEntry.getKey();
        ZoneSnapshot zoneSnapshot = zoneEntry.getValue();
        //判定该服务区中的服务实例数
        int instanceCount = zoneSnapshot.getInstanceCount();
        if (instanceCount == 0) {
            //如果服务区中没有服务实例，那么去除掉该服务区
            availableZones.remove(zone);
            limitedZoneAvailability = true;
        } else {
            double loadPerServer = zoneSnapshot.getLoadPerServer();
            //服务区内实例平均负载小于零，或者实例故障率（断路器断开次数/实例数）大于等于阈值
```

```
                （默认为0.99999），则去掉该服务区
            if (((double) zoneSnapshot.getCircuitTrippedCount()) 
                    / instanceCount >= triggeringBlackoutPercentage
                    || loadPerServer < 0) {
                availableZones.remove(zone);
                limitedZoneAvailability = true;
            } else {
                //如果该服务区的平均负载和最大负载的差距小于一定量，则将该服务区加入到最坏服务
                  区集合中
                if (Math.abs(loadPerServer - maxLoadPerServer) < 0.000001d) {
                    worstZones.add(zone);
                } else if (loadPerServer > maxLoadPerServer) {
                    //否则，如果该zone平均负载还大于最大负载
                    maxLoadPerServer = loadPerServer;
                    //清除掉最坏服务区集合，将该服务区加入
                    worstZones.clear();
                    worstZones.add(zone);
                }
            }
        }
    }
    //如果最大平均负载小于设定的阈值则直接返回
    if (maxLoadPerServer < triggeringLoad && !limitedZoneAvailability) {
        // zone override is not needed here
        return availableZones;
    }
    //如果大于，则从最坏服务区集合中随机剔除一个
    String zoneToAvoid = randomChooseZone(snapshot, worstZones);
    if (zoneToAvoid != null) {
        availableZones.remove(zoneToAvoid);
    }
    return availableZones;
}
```

ZoneAvoidancePredicate 的 apply 方法调用结束之后，AvailabilityPredicate 的 apply 方法也会被调用。该预测规则依据断路器是否断开或者服务器连接数是否超出阈值等标准来进行服务过滤。AvailabilityPredicate 的 apply 方法如下所示：

```
//AvailabilityPredicate.java
public boolean apply(@Nullable PredicateKey input) {
    LoadBalancerStats stats = getLBStats();
    if (stats == null) {
        return true;
    }
    //获得关于该服务器的信息记录
    return !shouldSkipServer(stats.getSingleServerStat(input.getServer()));
}
private boolean shouldSkipServer(ServerStats stats) {
    //如果该服务器的断路器已经打开，或者它的连接数大于预设的阈值，那么就需要将服务器过滤掉
    if ((CIRCUIT_BREAKER_FILTERING.get() && stats.isCircuitBreakerTripped())
            || stats.getActiveRequestsCount() >= activeConnectionsLimit.get()) {
```

```
        return true;
    }
    return false;
}
```

本小节主要讲解了 ClientConfigEnabledRoundRobinRule 和 ZoneAvoidanceRule 这两个负载均衡策略，其他策略的具体实现，有兴趣的读者可以自行了解。

7.4 进阶应用

本节讲解 Ribbon 相关的进阶应用，主要有 Ribbon 的 API 使用、与 Netty 的结合使用和只读数据库负载均衡的实现。

7.4.1 Ribbon API

Ribbon 除了与 RestTemplate 和 OpenFegin 一同使用之外，还可以依靠自己独立的 API 接口来实现特定需求。

比如下面的代码中，使用 LoadBalancerBuilder 的 buildFixedServerListLoadBalancer 创建出 ILoadBalancer 实例，然后使用 LoadBalancerCommand.Builder 的接口生成 LoadBalancerCommand 实例来发送网络请求。通过调用 LoadBalancerCommand 的 submit 方法传入匿名的 ServerOperation 来完成网络请求的发送。LoadBalancerCommand 的 submit 方法会使用你配置的负载均衡策略来选出一个服务器，然后调用匿名的 ServerOperation 的 call 方法，将选出的服务器传入，来进行网络传输的处理和操作。

```
public class URLConnectionLoadBalancer {
    private final ILoadBalancer loadBalancer;
    private final RetryHandler retryHandler = new DefaultLoadBalancerRetryHandler(0,
        1, true);
    public URLConnectionLoadBalancer(List<Server> serverList) {
        //使用LoadBalancerBuilder的接口来创建ILoadBalancer实例
        loadBalancer = LoadBalancerBuilder.newBuilder().buildFixedServerListLoad
            Balancer(serverList);
    }

    public String call(final String path) throws Exception {
        //使用LoadBalancerCommand.Builder接口来配置Command实例，然后在回调方法中使用选中
            的服务器信息发送HTTP请求
        return LoadBalancerCommand.<String>builder()
                .withLoadBalancer(loadBalancer)
                .withRetryHandler(retryHandler)
                .build()
                .submit(new ServerOperation<String>() {
            @Override
            public Observable<String> call(Server server) {
                URL url;
                try {
```

```
            url = new URL("http://" + server.getHost() + ":" + server.
                getPort() + path);
            HttpURLConnection conn = (HttpURLConnection) url.openConnection();
            return Observable.just(conn.getResponseMessage());
        } catch (Exception e) {
            return Observable.error(e);
        }
     }
   })).toBlocking().first();
}

public LoadBalancerStats getLoadBalancerStats() {
    return ((BaseLoadBalancer) loadBalancer).getLoadBalancerStats();
}

public static void main(String[] args) throws Exception {
    URLConnectionLoadBalancer urlLoadBalancer = new URLConnectionLoadBalancer
        (Lists.newArrayList(
            new Server("www.google.com", 80),
            new Server("www.linkedin.com", 80),
            new Server("www.yahoo.com", 80)));
    for (int i = 0; i < 6; i++) {
        System.out.println(urlLoadBalancer.call("/"));
    }
    System.out.println("=== Load balancer stats ===");
    System.out.println(urlLoadBalancer.getLoadBalancerStats());
}
}
```

FeignLoadBalancer 就是使用 Ribbon 的 LoadBalancerCommand 来实现有关 OpenFeign 的网络请求，只不过 FeignLoadBalancer 将网络请求交给其 Client 实例处理，而上边例子中的代码是交给 HttpURLConnection 处理。读者使用 Ribbon 的独立 API 可以在任何项目中使用 Ribbon 所提供的负载均衡机制。

7.4.2 使用 Netty 发送网络请求

Ribbon 除了可以和 RestTemplate、OpenFeign 一起使用之外，还可以与 Netty 进行集成，也就是说，Ribbon 使用负载均衡策略选择完服务器之后，再交给 Netty 处理网络请求。其实，上一小节介绍的 Ribbon 的 LoadBalancerCommand 的 submit 方法可以直接使用 Netty 框架，也就是在 ServerOperation 的 call 方法中使用 Netty 代替 HttpURLConnection 来发送网络请求。但是，Ribbon 已经封装好了与 Netty 进行集成的相关实现。RibbonTransport 封装了生成 LoadBalancingHttpClient<ByteBuf, ByteBuf> 对象的各类工厂函数。使用 LoadBalancingHttpClient<ByteBuf, ByteBuf> 的 submit 方法发送网络请求，底层默认使用 Netty 框架来进行网络请求（其中 ByteBuf 是 Netty 框架内的字符串缓存实例），代码如下所示：

```
public class SimpleGet {
```

```java
@edu.umd.cs.findbugs.annotations.SuppressWarnings
public static void main(String[] args) throws Exception {
    LoadBalancingHttpClient<ByteBuf, ByteBuf> client = RibbonTransport.
        newHttpClient();
    //HttpClientRequest.createGet接口可以直接生成对应的请求
    HttpClientRequest<ByteBuf> request = HttpClientRequest.createGet("http://
        www.google.com/");
    final CountDownLatch latch = new CountDownLatch(1);
    client.submit(request)
        .toBlocking()
        .forEach(new Action1<HttpClientResponse<ByteBuf>>() {
            @Override
            public void call(HttpClientResponse<ByteBuf> t1) {
                System.out.println("Status code: " + t1.getStatus());
                t1.getContent().subscribe(new Action1<ByteBuf>() {

                    @Override
                    public void call(ByteBuf content) {
                        //可以直接对Netty的ByteBuf进行操作
                        System.out.println("Response content: " + content.
                            toString(Charset.defaultCharset()));
                        latch.countDown();
                    }

                });
            }
        });
    latch.await(2, TimeUnit.SECONDS);
}
```

如上述代码所示，使用 LoadBalancingHttpClient 对象实例可以直接发送 Netty 相关的网络请求，并使用 CountDownLatch 实例来通知等待网络请求结果的线程。

7.4.3 只读数据库的负载均衡实现

读者在学习了 FeignLoadBalancer 的原理和 Ribbon 的 API 之后，可以为任何需要负载均衡策略的项目添加 Ribbon 的集成。

比如一个数据库中间件项目，它支持多个读库的数据读取，它希望在对多个读库进行数据读取时可以支持一定的负载均衡策略。那么，读者就可以通过集成 Ribbon 来实现读库之间的负载均衡。

首先，你需要定义 DBServer 类来继承 Ribbon 的 Server 类，用于存储只读数据库服务器的状态信息，比如说 IP 地址、数据库连接数、平均请求响应时间等，然后定义一个 DBLoadBalancer 来继承 BaseLoadBalancer 类。下述示例代码通过 WeightedResponseTimeRule 对 DBServer 列表进行负载均衡选择，然后使用自定义的 DBPing 来检测数据库是否可用。示例代码如下所示：

```java
public DBLoadBalancer buildFixedDBServerListLoadBalancer(List<DBServer> servers) {
    IRule rule = new WeightedResponseTimeRule();
    IPing ping = new DBPing();
    DBLoadBalancer lb = new DBLoadBalancer(config, rule, ping);
    lb.setServersList(servers);
    return lb;
}
```

使用 DBLoadBalancer 的过程也很简单，通过 LoadBalancerCommand 的 withLoadBalancer 来使用它，然后在 submit 的回调函数中使用选出的数据库和 SQL 语句交给其他组件来执行 SQL 操作。DBConnectionLoadBalancer 的具体实现如下所示：

```java
public class DBConnectionLoadBalancer {

    private final ILoadBalancer loadBalancer;
    private final RetryHandler retryHandler = new DefaultLoadBalancerRetryHandler
        (0, 1, true);

    public DBConnectionLoadBalancer(List<DBServer> serverList) {
        loadBalancer = LoadBalancerBuilder.newBuilder().buildFixedDBServerListLo
            adBalancer(serverList);
    }
    public String executeSQL(final String sql) throws Exception {
        //使用LoadBalancerCommand来进行负载均衡，具体策略可以在Builder中进行设置
        return LoadBalancerCommand.<String>builder()
            .withLoadBalancer(loadBalancer)
            .build()
            .submit(new ServerOperation<String>() {
                @Override
                public Observable<String> call(Server server) {
                    URL url;
                    try {
                        return Observable.just(DBManager.execute(server, sql));
                    } catch (Exception e) {
                        return Observable.error(e);
                    }
                }
            }).toBlocking().first();
    }
}
```

ILoadBalancer 通过一定的负载策略从读数据库列表中选出一个数据库来让 DBManager 执行 SQL 语句，通过这种方式，读者就可以实现读数据库的负载均衡机制。

7.5 本章小结

在现代网站架构中，一台服务器的处理能力往往受限于服务器自身的可扩展硬件能力。所以，在需要处理大量用户请求的时候，通常都会引入负载均衡器，将多台普通服务器组

成一个服务集群,来完成高并发的请求处理任务。Ribbon 作为 Spring Cloud 的负载均衡组件,可以为开发者提供开箱即用的负载均衡能力,让开发者免于了解复杂的负载均衡策略,专注于业务逻辑的实现。Ribbon 还为开发者预留了自定义负载均衡策略的接口,让有需求的开发者在 Ribbon 框架内部定制自己的负载均衡策略。读者可以通过了解 Ribbon 来对负载均衡机制拥有全面深入的认识,为之后使用其他负载均衡框架打下基础。

第 8 章 Chapter 8

API 网关：Spring Cloud Gateway

在单体应用程序架构下，客户端（Web 或移动端）通过向服务端发起一次网络调用来获取数据。负载均衡器将请求路由给 N 个相同的应用程序实例中的一个。然后应用程序会查询各种数据库表处理业务逻辑，并将响应返回给客户端。微服务架构下，单体应用被切割成多个微服务，如果将所有的微服务直接对外暴露，势必会出现安全方面的各种问题。

客户端可以直接向每个微服务发送请求，其问题主要包括：

- 客户端需求和每个微服务暴露的细粒度 API 不匹配。
- 部分服务使用的协议不是 Web 友好协议。可能使用 Thrift 二进制 RPC，也可能使用 AMQP 消息传递协议。
- 微服务难以重构。如果合并两个服务，或者将一个服务拆分成两个或更多服务，这类重构非常困难。

针对如上问题，一个常用的解决方案是使用 API 网关。API 网关自身也是一个服务，并且是后端服务的唯一入口。从面向对象设计的角度看，它与外观模式类似。API 网关封装了系统内部架构，为每个客户端提供一个定制的 API。除此之外，它还可以负责身份验证、监控、负载均衡、限流、降级与应用检测等功能。

本章第一部分将会介绍 Spring Cloud Gateway 的相关特性；第二部分为基础应用，通过网关服务和用户服务示例，讲解 Spring Cloud Gateway 的基本功能；第三部分将会结合源码讲解 Spring Cloud Gateway 基本功能；最后是 Spring Cloud Gateway 的进阶学习，介绍如何使用限流机制和熔断降级等功能。

8.1 Spring Cloud Gateway 介绍

Spring Cloud Gateway 基于 Spring Boot 2，是 Spring Cloud 的全新项目，该项目提供

了一个构建在 Spring 生态之上的 API 网关,包括 Spring 5、Spring Boot 2 和 Project Reactor。Spring Cloud Gateway 旨在提供一种简单而有效的途径来转发请求,并为它们提供横切关注点,例如:安全性、监控/指标和弹性。

Spring Cloud Gateway 具有如下特征:

- 基于 Java 8 编码;
- 支持 Spring Framework 5;
- 支持 Spring Boot 2;
- 支持动态路由;
- 支持内置到 Spring Handler 映射中的路由匹配;
- 支持基于 HTTP 请求的路由匹配(Path、Method、Header、Host 等);
- 过滤器作用于匹配的路由;
- 过滤器可以修改下游 HTTP 请求和 HTTP 响应(增加/修改头部、增加/修改请求参数、改写请求路径等);
- 通过 API 或配置驱动;
- 支持 Spring Cloud DiscoveryClient 配置路由,与服务发现与注册配合使用。

在 Finchley 正式版之前,Spring Cloud 推荐的网关是 Netflix 提供的 Zuul(笔者这里指的都是 Zuul 1.x,是一个基于阻塞 I/O 的 API Gateway)。与 Zuul 相比,Spring Cloud Gateway 建立在 Spring Framework 5、Project Reactor 和 Spring Boot 2 之上,使用非阻塞 API。Spring Cloud Gateway 还支持 WebSocket,并且与 Spring 紧密集成,拥有更好的开发体验。Zuul 基于 Servlet 2.5,使用阻塞架构,它不支持任何长连接,如 WebSocket。Zuul 的设计模式和 Nginx 较像,每次 I/O 操作都是从工作线程中选择一个执行,请求线程被阻塞直到工作线程完成,但是差别是 Nginx 用 C++ 实现,Zuul 用 Java 实现,而 JVM 本身会有第一次加载较慢的情况,使得 Zuul 的性能相对较差。Zuul 已经发布了 Zuul 2.x,基于 Netty、非阻塞、支持长连接,但 Spring Cloud 目前还没有整合。Zuul 2.x 的性能肯定会较 Zuul 1.x 有较大提升。在性能方面,根据官方提供的基准(benchmark)测试⊖,Spring Cloud Gateway 的 RPS(每秒请求数)是 Zuul 的 1.6 倍。综合来说,Spring Cloud Gateway 在提供的功能和实际性能方面,表现都很优异。

8.2 基础应用

下面示例启动两个服务:Gateway-Server 和 User-Server。示例的场景为,客户端请求后端服务,网关提供后端服务的统一入口。后端的服务都注册到服务发现 Consul(与 Eureka 一样,同为服务注册中心)。网关通过负载均衡将客户端请求转发到具体的后端服务,如图 8-1 所示。

⊖ 项目地址为 https://github.com/spencergibb/spring-cloud-gateway-bench

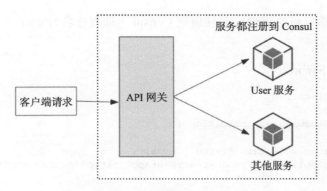

图 8-1 网关示意图

8.2.1 用户服务

用户服务功能较为简单,需要注册到 Consul 上,并提供一个接口 /test。

1. 引入依赖

需要的依赖如下所示:

```xml
<dependency>
    <groupId>org.springframework.cloud</groupId>
    <artifactId>spring-cloud-starter-consul-discovery</artifactId>
</dependency>
<dependency>
    <groupId>org.springframework.boot</groupId>
    <artifactId>spring-boot-starter-web</artifactId>
</dependency>
```

spring-cloud-starter-consul-discovery 用于服务注册与发现,spring-boot-starter-web 为启动类 jar 包,引入 Web MVC 和 Tomcat 等 Web 项目所需要的组件。

2. 配置文件

配置文件如下所示:

```yaml
spring:
  application:
    name: user-service
  cloud:
    consul:
      host: localhost
      port: 8500
      discovery:
        ip-address: ${HOST_ADDRESS:localhost}
        port: ${SERVER_PORT:${server.port}}
        instance-id: user-${server.port}
        service-name: user
server:
  port: 8005
```

配置文件设置端口号为 8005，且注册到 Consul 上的服务名为 user。

3. 暴露接口

暴露的接口如下所示：

```
@SpringBootApplication
@RestController
public class GatewayUserApplication {

    public static void main(String[] args) {
        SpringApplication.run(GatewayUserApplication.class, args);
    }

    @GetMapping("/test")
    public String test() {
        return "ok";
    }
}
```

对外暴露 /test 接口，返回字符串 ok 即可。

8.2.2 网关服务

网关服务提供路由配置、路由断言和过滤器等功能。下面将会在网关中分别实现这些功能。

1. 引入依赖

需要引入的依赖如下所示：

```xml
<!--依赖于WebFlux，必须引入-->
<dependency>
    <groupId>org.springframework.boot</groupId>
    <artifactId>spring-boot-starter-webflux</artifactId>
</dependency>
<dependency>
    <groupId>org.springframework.cloud</groupId>
    <artifactId>spring-cloud-gateway-core</artifactId>
</dependency>
<!--服务发现组件，排除Web依赖-->
<dependency>
    <groupId>org.springframework.cloud</groupId>
    <artifactId>spring-cloud-starter-consul-discovery</artifactId>
</dependency>
<!--Kotlin依赖-->
<dependency>
    <groupId>org.jetbrains.kotlin</groupId>
    <artifactId>kotlin-stdlib</artifactId>
    <optional>true</optional>
</dependency>
<dependency>
    <groupId>org.jetbrains.kotlin</groupId>
```

```xml
        <artifactId>kotlin-reflect</artifactId>
        <optional>true</optional>
</dependency>
```

如上引入了 Kotlin 相关的依赖，这里需要支持 Kotlin 的路由配置。Spring Cloud Gateway 的使用需要排除 Web 相关的配置，且必须引入 WebFlux 的依赖，应用服务启动时会检查 WebFlux 依赖。

2. 路由断言

路由断言有多种类型，根据请求的时间、Host 地址、路径和请求方法等。如下定义的是一个基于路径的路由断言匹配。

```java
@Bean
public RouterFunction<ServerResponse> testFunRouterFunction() {
    RouterFunction<ServerResponse> route = RouterFunctions.route(
            RequestPredicates.path("/testfun"),
            request -> ServerResponse.ok().body(BodyInserters.fromObject("hello")));
    return route;
}
```

当请求的路径为 /testfun 时，直接返回状态码 200，且响应体为 hello 的字符串。

3. 过滤器

网关经常需要对路由请求进行过滤，对符合条件的请求进行一些操作，如增加请求头、增加请求参数、增加响应头和断路器等功能。例如下面的示例代码：

```java
@Bean
public RouteLocator customRouteLocator(RouteLocatorBuilder builder, ThrottleGatewayFilterFactory
    throttle) {
        return builder.routes()
                .route(r -> r.path("/image/webp")
                        .filters(f ->
                                f.addResponseHeader("X-AnotherHeader", "baz"))
                        .uri("http://httpbin.org")
                )
                .build();
}
```

如上代码实现了当请求路径为 /image/webp 时，网关将请求转发到 http://httpbin.org，并对响应进行过滤处理，增加响应的头部 X-AnotherHeader: baz。

4. 自定义路由

除了通过 Gateway 提供的 API 自定义路由，还可以通过配置进行定义，如下所示：

```yaml
spring:
    cloud:
        gateway:
            locator:
                enabled: true
            default-filters:
```

```
        - AddResponseHeader=X-Response-Default-Foo, Default-Bar
      routes:
      - id: default_path_to_http
          uri: blueskykong.com
          order: 10000
          predicates:
          - Path=/test/**
```

如上的配置定义了全局过滤器与一条路由。全局过滤器为所有的响应加上头部 X-Response-Default-Foo: Default-Bar。另外还定义了 id 为 default_path_to_http 的路由，优先级比较低。符合路由断言条件的请求将会转发到 blueskykong.com。

5. Kotlin 自定义路由

Spring Cloud Gateway 支持使用 Kotlin 自定义路由，如下所示：

```
@Configuration
class AdditionalRoutes {
    @Bean
    fun additionalRouteLocator(builder: RouteLocatorBuilder): RouteLocator = builder.
        routes {
          route(id = "test-kotlin") {
              path("/image/png")
              filters {
                  addResponseHeader("X-TestHeader", "foobar")
              }
              uri("http://blueskykong.com")
          }
        }
}
```

当请求的路径是 /image/png，将会转发到 http://blueskykong.com，并设置了过滤器，在其响应头中加上了 X-TestHeader: foobar 头部。

6. WebSocket

还可以配置 WebSocket 的网关路由，如下所示：

```
spring:
    cloud:
        gateway:
            default-filters:
            - AddResponseHeader=X-Response-Default-Foo, Default-Bar
            routes:
            - id: websocket_test
                uri: ws://localhost:9000
                order: 9000
                predicates:
                - Path=/echo
```

网关会将外部的 WebSocket 请求转发到 ws://localhost:9000。

7. 基于服务发现

网关与服务注册与发现组件进行结合，通过 serviceId 转发到具体的服务实例。在前面已经引入了相应的依赖，配置文件如下所示。

```
spring:
    cloud:
        gateway:
            locator:
                enabled: true
            routes:
            - id: service_to_user
              uri: lb://user
              predicates:
              - Path=/user/**
              filters:
              - StripPrefix=1
```

上面的配置开启了 DiscoveryClient 服务发现。路由定义了所有请求路径以 /user 开头的请求，都将会转发到 user 服务，并应用路径过滤器截掉路径的第一部分前缀。即访问 /user/test 的实际请求转换成了 http://172.16.1.100:8005/test。

8.2.3 客户端的访问

在前面两小节中，网关和用户服务实现的功能，读者可以自行下载源码进行尝试。笔者这里只展示访问用户服务的结果。发送一个请求到网关，由于加上了对应服务的前缀，所以请求地址为 http://localhost:9090/user/test。

网关成功负载均衡到 user-server，并返回了 ok。响应的头部中包含了全局过滤器设置的头部 X-Response-Default-Foo: Default-Bar，如图 8-2 所示。

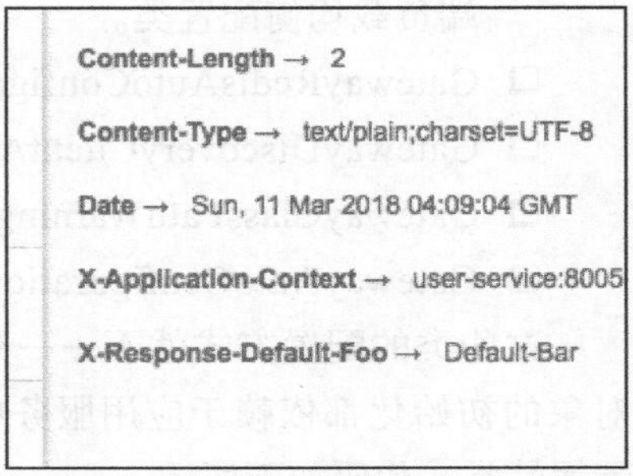

图 8-2 请求头部信息图

8.3 源码解析

作为后端服务的统一入口，API 网关可提供请求路由、协议转换、安全认证、服务鉴权、流量控制与日志监控等服务。使用微服务架构将所有的应用管理起来，那么 API 网关就起到了微服务网关的作用；如果只是使用 REST 方式进行服务之间的访问，使用 API 网关对调用进行管理，那么 API 网关起到的就是 API 服务治理的作用。不管哪一种使用方式，都不影响 API 网关核心功能的实现。当请求到达网关时，网关处理的流程如图 8-3 所示。

具体步骤如下：

1）请求发送到网关，DispatcherHandler 是 HTTP 请求的中央分发器，将请求匹配到相应的 HandlerMapping。

2）请求与处理器之间有一个映射关系，网关将会对请求进行路由，handler 此处会匹配到 RoutePredicateHandlerMapping，以匹配请求所对应的 Route。

3）随后到达网关的 Web 处理器，该 WebHandler 代理了一系列网关过滤器和全局过滤器的实例，如对请求或者响应的头部进行处理（增加或者移除某个头部）。

4）最后，转发到具体的代理服务。

这里比较重要的功能点是路由的过滤和路由的定位，Spring Cloud Gateway 提供了非常丰富的路由过滤器和路由断言。下面将会按照自上而下的顺序分析这部分的源码。

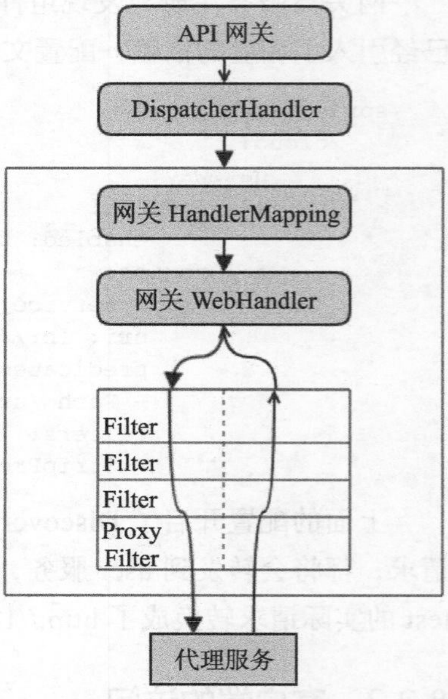

8.3.1 初始化配置

在引入 Spring Cloud Gateway 的依赖后，Starter 的 jar 包将会自动初始化一些类：

图 8-3 网关处理流程图

- GatewayLoadBalancerClientAutoConfiguration，客户端负载均衡配置类。
- GatewayRedisAutoConfiguration，Redis 的自动配置类。
- GatewayDiscoveryClientAutoConfiguration，服务发现自动配置类。
- GatewayClassPathWarningAutoConfiguration，WebFlux 依赖检查的配置类。
- GatewayAutoConfiguration，核心配置类，配置路由规则、过滤器等。

这些类的配置方式就不一一列出讲解了，主要看一下涉及的网关属性配置定义，很多对象的初始化都依赖于应用服务中配置的网关属性，GatewayProperties 是网关中主要的配置属性类，代码如下所示：

```
@ConfigurationProperties("spring.cloud.gateway")
@Validated
public class GatewayProperties {

    //路由列表
    @NotNull
    @Valid
    private List<RouteDefinition> routes = new ArrayList<>();

    private List<FilterDefinition> defaultFilters = new ArrayList<>();

    private List<MediaType> streamingMediaTypes = Arrays.asList(MediaType.TEXT_
        EVENT_STREAM,
            MediaType.APPLICATION_STREAM_JSON);
    ...
}
```

GatewayProperties 中有三个属性，分别是路由、默认过滤器和 MediaType 的配置，在之前的基础应用的例子中演示了配置的前两个属性。routes 是一个列表，对应的对象属性是路由定义 RouteDefinition；defaultFilters 是默认的路由过滤器，会应用到每个路由中；streamingMediaTypes 默认支持两种类型：APPLICATION_STREAM_JSON 和 TEXT_EVENT_STREAM。

8.3.2 网关处理器

请求到达网关之后，会有各种 Web 处理器对请求进行匹配与处理，图 8-4 为 Spring Cloud Gateway 中主要涉及的 WebHandler。

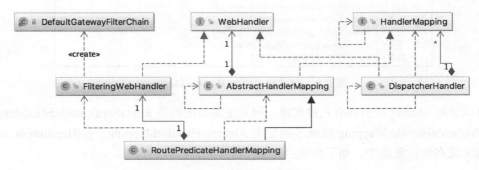

图 8-4　WebHandler 核心类图

这一小节将会按照以下顺序讲解负责请求路由选择和定位的处理器：

```
DispatcherHandler -> RoutePredicateHandlerMapping -> FilteringWebHandler -> DefaultGatewayFilterChain
```

1. 请求的分发器

Spring Cloud Gateway 引入了 Spring WebFlux，DispatcherHandler 是其访问入口，请求分发处理器。在之前的项目中，引入了 Spring MVC，而它的分发处理器是 DispatcherServlet。下面具体看一下网关收到请求后，如何匹配 HandlerMapping，代码如下所示：

```java
public class DispatcherHandler implements WebHandler, ApplicationContextAware {
    @Override
    public Mono<Void> handle(ServerWebExchange exchange) {
        if (this.handlerMappings == null) {
            //不存在handlerMappings则报错
            return Mono.error(HANDLER_NOT_FOUND_EXCEPTION);
        }
        return Flux.fromIterable(this.handlerMappings)
            .concatMap(mapping -> mapping.getHandler(exchange))
            .next()
            .switchIfEmpty(Mono.error(HANDLER_NOT_FOUND_EXCEPTION))
            .flatMap(handler -> invokeHandler(exchange, handler))
            .flatMap(result -> handleResult(exchange, result));
```

```
            }
            ...
        }
```

DispatcherHandler 实现了 WebHandler 接口，WebHandler 接口用于处理 Web 请求。DispatcherHandler 的构造函数会初始化 HandlerMapping。核心处理的方法是 handle (ServerWebExchange exchange)，而 HandlerMapping 是一个定义了请求与处理器对象映射的接口且有多个实现类，如 ControllerEndpointHandlerMapping 和 RouterFunctionMapping。调试网关中的 handler 映射，如图 8-5 所示。

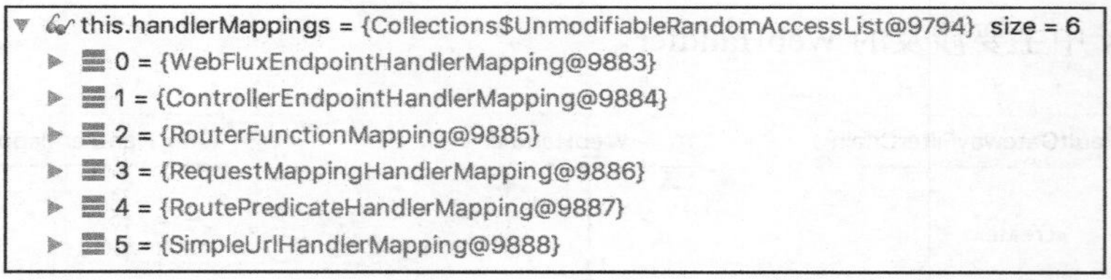

```
▼ 🔍 this.handlerMappings = {Collections$UnmodifiableRandomAccessList@9794} size = 6
    ▶ ≣ 0 = {WebFluxEndpointHandlerMapping@9883}
    ▶ ≣ 1 = {ControllerEndpointHandlerMapping@9884}
    ▶ ≣ 2 = {RouterFunctionMapping@9885}
    ▶ ≣ 3 = {RequestMappingHandlerMapping@9886}
    ▶ ≣ 4 = {RoutePredicateHandlerMapping@9887}
    ▶ ≣ 5 = {SimpleUrlHandlerMapping@9888}
```

图 8-5　调试网关中的 HandlerMapping

可以看到 handler 映射共有六种实现，网关主要关注的是 RoutePredicateHandlerMapping。RoutePredicateHandlerMapping 继承了抽象类 AbstractHandlerMapping，getHandler(exchange) 方法就定义在该抽象类中，如下所示：

```
public abstract class AbstractHandlerMapping extends ApplicationObjectSupport
    implements HandlerMapping, Ordered {
    @Override
    public Mono<Object> getHandler(ServerWebExchange exchange) {
        return getHandlerInternal(exchange).map(handler -> {
            ...
            return handler;
        });
    }

    protected abstract Mono<?> getHandlerInternal(ServerWebExchange exchange);
}
```

可以看出，抽象类在 handler 映射中用于抽取公用的功能，不是我们关注的重点，此处代码省略。具体的实现定义在 HandlerMapping 子类中。

AbstractHandlerMapping#getHandler 返回了相应的 Web 处理器，随后到达 DispatcherHandler#invokeHandler。

从图 8-5 可以看出，mapping#getHandler 返回的是 FilteringWebHandler。DispatcherHandler#invokeHandler 方法调用相应的 WebHandler，获取该 WebHandler 有对应的适配器。在图 8-6 的调试中对应的是 SimpleHandlerAdapter，适配器类的实现较为简单，直接调用了对应的 WebHandler 的处理方法。

```
private Mono<HandlerResult> invokeHandler(ServerWebExchange exchange, Object handler) {    exchange: DefaultServerWebExchange@20207   handler: FilteringWebHandler@12139
    if (this.handlerAdapters != null) {
        for (HandlerAdapter handlerAdapter : this.handlerAdapters) {   handlerAdapters:  size = 3
            if (handlerAdapter.supports(handler)) {
                return handlerAdapter.handle(exchange, handler);    exchange: DefaultServerWebExchange@20207   handler: FilteringWebHandler@12139
            }
        }
    }
    return Mono.error(new IllegalStateException("No HandlerAdapter: " + handler));
}
```

图 8-6　注入 FilteringWebHandler

2. 路由断言的 HandlerMapping

RoutePredicateHandlerMapping 用于匹配具体的 Route，并返回处理 Route 的 FilteringWebHandler，如下所示：

```
public class RoutePredicateHandlerMapping extends AbstractHandlerMapping {

    public RoutePredicateHandlerMapping(FilteringWebHandler webHandler, RouteLocator
            routeLocator) {
        this.webHandler = webHandler;
        this.routeLocator = routeLocator;
        setOrder(1);
    }
    ...
}
```

RoutePredicateHandlerMapping 的构造函数接收两个参数：FilteringWebHandler 网关过滤器和 RouteLocator 路由定位器，setOrder(1) 用于设置该对象初始化的优先级。Spring Cloud Gateway 的 GatewayWebfluxEndpoint 提供的 HTTP API 不需要经过网关转发，它通过 RequestMappingHandlerMapping 进行请求匹配处理，因此需要将 RoutePredicateHandlerMapping 的优先级设置为低于 RequestMappingHandlerMapping。

```
// RoutePredicateHandlerMapping.java
protected Mono<?> getHandlerInternal(ServerWebExchange exchange) {
    //设置GATEWAY_HANDLER_MAPPER_ATTR为 RoutePredicateHandlerMapping
    exchange.getAttributes().put(GATEWAY_HANDLER_MAPPER_ATTR, getClass().getSimpleName());

    return lookupRoute(exchange)
        .flatMap((Function<Route, Mono<?>>) r -> {
            //设置 GATEWAY_ROUTE_ATTR为匹配的Route
            exchange.getAttributes().put(GATEWAY_ROUTE_ATTR, r);
            return Mono.just(webHandler);
        }).switchIfEmpty(Mono.empty().then(Mono.fromRunnable(() -> {
            //logger
        })));
}
//顺序匹配请求对应的Route
protected Mono<Route> lookupRoute(ServerWebExchange exchange) {
    return this.routeLocator.getRoutes()
        .filterWhen(route -> {
            exchange.getAttributes().put(GATEWAY_PREDICATE_ROUTE_ATTR, route.
                getId());
```

```
            return route.getPredicate().apply(exchange);
        })
        .next()
        .map(route -> {
            //校验 Route的有效性
            validateRoute(route, exchange);
            return route;
        });
}
```

如上为获取 handler 的方法,用于匹配请求的 Route,并返回处理 Route 的 FilteringWebHandler。首先设置 GATEWAY_HANDLER_MAPPER_ATTR 为 RoutePredicateHandlerMapping 的类名;然后顺序匹配请求对应的 Route,RouteLocator 接口用于获取在网关中定义的路由,并根据请求的信息,与路由定义的断言进行匹配(路由的定义也有优先级,按照优先级顺序匹配)。最后设置 GATEWAY_ROUTE_ATTR 为匹配的 Route,并返回相应的 handler。

3. 过滤器的 Web 处理器

FilteringWebHandler 通过创建所请求 Route 对应的 GatewayFilterChain,在网关处进行过滤处理,实现代码如下:

```java
public class FilteringWebHandler implements WebHandler {
    private final List<GatewayFilter> globalFilters;

    public FilteringWebHandler(List<GlobalFilter> globalFilters) {
        this.globalFilters = loadFilters(globalFilters);
    }
    private static List<GatewayFilter> loadFilters(List<GlobalFilter> filters) {
        return filters.stream()
            .map(filter -> {
                //适配器模式,用以适配GlobalFilter
                GatewayFilterAdapter gatewayFilter = new GatewayFilterAdapter(filter);
                //判断是否实现Ordered接口
                if (filter instanceof Ordered) {
                    //实现了Ordered接口,则返回的是OrderedGatewayFilter对象
                    int order = ((Ordered) filter).getOrder();
                    return new OrderedGatewayFilter(gatewayFilter, order);
                }
                return gatewayFilter;
            }).collect(Collectors.toList());
    }
    ...
}
```

其中,全局变量 globalFilters 是 Spring Cloud Gateway 中定义的全局过滤器。构造函数通过传入的全局过滤器,对这些过滤器进行适配处理。因为过滤器的定义有优先级,这里的处理主要是判断是否实现 Ordered 接口,如果实现了 Ordered 接口,则返回的是 OrderedGatewayFilter 对象。否则,返回过滤器的适配器,用以适配 GlobalFilter,适配器类

比较简单，不再列出。最后将这些过滤器设置为全局变量 globalFilters，如下所示：

```java
// FilteringWebHandler.java
public Mono<Void> handle(ServerWebExchange exchange) {
    Route route = exchange.getRequiredAttribute(GATEWAY_ROUTE_ATTR);
    List<GatewayFilter> gatewayFilters = route.getFilters();
    //加入全局过滤器
    List<GatewayFilter> combined = new ArrayList<>(this.globalFilters);
    combined.addAll(gatewayFilters);
    //过滤器排序
    AnnotationAwareOrderComparator.sort(combined);
    //按照优先级，对该请求进行过滤
    return new DefaultGatewayFilterChain(combined).filter(exchange);
}

private static class DefaultGatewayFilterChain implements GatewayFilterChain {
    private final int index;
    private final List<GatewayFilter> filters;
    public DefaultGatewayFilterChain(List<GatewayFilter> filters) {
        this.filters = filters;
        this.index = 0;
    }
    private DefaultGatewayFilterChain(DefaultGatewayFilterChain parent, int index) {
        this.filters = parent.getFilters();
        this.index = index;
    }
    public List<GatewayFilter> getFilters() {
        return filters;
    }

    @Override
    public Mono<Void> filter(ServerWebExchange exchange) {
        return Mono.defer(() -> {
            if (this.index < filters.size()) {
                GatewayFilter filter = filters.get(this.index);
                DefaultGatewayFilterChain chain = new DefaultGatewayFilterChain (this,
                    this.index + 1);
                return filter.filter(exchange, chain);
            } else {
                return Mono.empty();
            }
        });
    }
}
```

FilteringWebHandler#handle 方法首先获取请求对应的路由的过滤器和全局过滤器，将两部分组合；然后对过滤器列表排序，AnnotationAwareOrderComparator 是 OrderComparator 的子类，支持 Spring 的 Ordered 接口的优先级排序；最后按照优先级，生成过滤器链，对该请求进行过滤处理。这里过滤器链是通过内部静态类 DefaultGatewayFilterChain 实现，该类实现了 GatewayFilterChain 接口，用于按优先级过滤。

8.3.3 路由定义定位器

RouteDefinitionLocator 是路由定义定位器的顶级接口，具体的路由定义定位器都继承自该接口，其类图如图 8-7 所示。

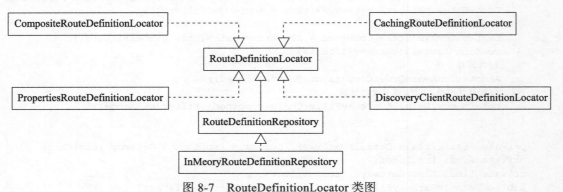

图 8-7　RouteDefinitionLocator 类图

RouteDefinitionLocator 接口定义如下所示：

```
public interface RouteDefinitionLocator {
    Flux<RouteDefinition> getRouteDefinitions();
}
```

可以看到唯一的 getRouteDefinitions 方法，用以获取路由定义。RouteDefinition 对象作为属性定义在 GatewayProperties 中，而网关服务在启动时读取了配置文件中的相关配置。RouteDefinition 的定义如下所示：

```
public class RouteDefinition {
    @NotEmpty
    private String id = UUID.randomUUID().toString();

    @NotEmpty
    @Valid
    private List<PredicateDefinition> predicates = new ArrayList<>();

    @Valid
    private List<FilterDefinition> filters = new ArrayList<>();

    @NotNull
    private URI uri;

    private int order = 0;
}
```

在 RouteDefinition 中，主要有五个属性：路由 id、URI 转发地址、order 优先级、PredicateDefinition 路由断言定义和 FilterDefinition 过滤器的定义。再深入的话，可以看到断言和过滤器属性是一个 Map 数据结构，用以存放多个对应的键值数组。

通过 RouteDefinitionLocator 的类图，可以看出该接口有四个实现类：

- 基于属性配置的（PropertiesRouteDefinitionLocator）
- 基于服务发现的（DiscoveryClientRouteDefinitionLocator）
- 组合方式的（CompositeRouteDefinitionLocator）
- 缓存方式的（CachingRouteDefinitionLocator）

在类图中，还有一个接口 RouteDefinitionRepository 继承自 RouteDefinitionLocator，用于对路由定义进行操作，如保存和删除路由定义。下面我们分别介绍这几种路由定义定位器的实现。

1. 路由定义的数据访问操作

RouteDefinitionRepository 接口中的方法用于对 RouteDefinition 进行增、删、查操作，如下所示：

```
public interface RouteDefinitionRepository extends RouteDefinitionLocator,
    RouteDefinitionWriter {
}

//RouteDefinitionWriter对路由定义进行操作
public interface RouteDefinitionWriter {
    Mono<Void> save(Mono<RouteDefinition> route);
    Mono<Void> delete(Mono<String> routeId);
}
```

RouteDefinitionRepository 继承自 RouteDefinitionWriter，封装了 RouteDefinitionLocator 操作的方法，在实现网关内置的 API 端点时会用到这里的接口。InMemoryRouteDefinitionRepository 实现了 RouteDefinitionRepository 接口，基于内存的路由定义仓库，也是唯一的实现类。当然我们可以根据需要自行扩展，存放在其他存储介质。

2. 基于属性配置的 RouteDefinitionLocator

基于属性配置的路由定义定位器是 PropertiesRouteDefinitionLocator，从类名就可以知道该类是从配置文件读取路由配置，配置文件如 YAML 和 Properties 等类型，如下所示：

```
public class PropertiesRouteDefinitionLocator implements RouteDefinitionLocator {
    public PropertiesRouteDefinitionLocator(GatewayProperties properties) {
        this.properties = properties;
    }

    @Override
    public Flux<RouteDefinition> getRouteDefinitions() {
        return Flux.fromIterable(this.properties.getRoutes());
    }
}
```

PropertiesRouteDefinitionLocator 的实现是通过构造函数传入的 GatewayProperties 对象，读取其中的路由配置信息，关于 GatewayProperties 在前面介绍过。如下为基础应用中的网关属性配置：

```yaml
spring:
    cloud:
        gateway:
            default-filters:
            - AddResponseHeader=X-Response-Default-Foo, Default-Bar
            routes:
            - id: default_path_to_httpbin
              uri: ${test.uri}
              order: 10000
              predicates:
              - Path=/**
```

在如上配置中，增加了默认过滤器 default-filters 和一条路由配置，在基础应用中有讲解，此处不再赘述。

3. 基于服务发现的 RouteDefinitionLocator

基于服务发现的路由定义定位器是 DiscoveryClientRouteDefinitionLocator，该类通过服务发现组件获取注册中心的服务信息，即路由定义的源变成了配置中心。服务注册与发现组件配合提供负载均衡，是微服务架构中很常用的组合，可以多实例部署实现负载均衡，避免单点故障等等。需要注意的是，引入服务注册依赖时需要去除 **spring-boot-starter-web** 的依赖，避免和 Spring Cloud Gateway 依赖的 WebFlux 冲突。该类的代码如下所示：

```java
@ConfigurationProperties("spring.cloud.gateway.discovery.locator")
public class DiscoveryLocatorProperties {
    // 开启服务发现
    private boolean enabled = false;
    // 路由的前缀，默认为discoveryClient.getClass().getSimpleName() + "_".
    private String routeIdPrefix;
    // SpEL表达式，判断网关是否集成一个服务，默认为true
    private String includeExpression = "true";
    // SpEL 表达式，为每个路由创建uri，默认为 'lb://'+serviceId */
    private String urlExpression = "'lb://'+serviceId";
    // 在断言和过滤器中小写serviceId，默认为false
    private boolean lowerCaseServiceId = false;
    // 断言定义
    private List<PredicateDefinition> predicates = new ArrayList<>();
    // 过滤器定义
    private List<FilterDefinition> filters = new ArrayList<>();
    ...
}
```

在 DiscoveryLocatorProperties 中定义了如上的属性，要启用服务必须设置 spring.cloud.gateway.discovery.locator.enabled=true。includeExpression 属性用于判断网关是否集成一个服务，默认为 true，当我们设置为 metadata['edge'] == 'true' 时，则会判断 ServiceInstance 中的元数据 'edge' 属性，如下所示：

```java
// DiscoveryClientRouteDefinitionLocator.java
public Flux<RouteDefinition> getRouteDefinitions() {
    //对includeExpression和urlExpression的表达式处理
```

```
    return Flux.fromIterable(discoveryClient.getServices())
        .map(discoveryClient::getInstances)
        .filter(instances -> !instances.isEmpty())
        .map(instances -> instances.get(0))
        .filter(instance -> {
            // 根据includeExpression表达式，过滤不符合的ServiceInstance
        })
        .map(instance -> {
            String serviceId = instance.getServiceId();
            RouteDefinition routeDefinition = new RouteDefinition();
            routeDefinition.setId(this.routeIdPrefix + serviceId);
            String uri = urlExpr.getValue(evalCtxt, instance, String.class);
            routeDefinition.setUri(URI.create(uri));

            final ServiceInstance instanceForEval = new DelegatingServiceInstance
                (instance, properties);

            for (PredicateDefinition original : this.properties.getPredicates()) {
    // 增加配置的断言表达式
            }
            for (FilterDefinition original : this.properties.getFilters()) {
                FilterDefinition filter = new FilterDefinition();
                filter.setName(original.getName());
                //增加配置中的过滤器
            }
                return routeDefinition;
        });
}
```

从源码可以看出，getRouteDefinitions 方法通过注册发现客户端获取注册服务信息，组装成多个 RouteDefinition 路由定义的数组，并将配置中定义的路由断言和过滤器应用到 RouteDefinition 中。在前面介绍过 includeExpression 属性，根据 includeExpression 表达式，过滤不符合的 ServiceInstance。关于路由决策工厂和网关过滤器工厂，将在下面章节具体介绍。

4. 缓存方式与组合方式的 RouteDefinitionLocator

缓存方式的路由定义定位器 CachingRouteDefinitionLocator，通过传入的代理路由定义定位器来获取路由定义，并缓存到本地。当需要缓存更新时，可以通过触发路由刷新事件 RefreshRoutesEvent，将本地的缓存清空并通过代理 RouteDefinitionLocator 重新请求路由定义信息。

组合方式的路由定义定位器 CompositeRouteDefinitionLocator 是一种组合模式，组合的逻辑很简单，通过传入的 RouteDefinitionLocator 类型的参数作为代理，路由定义实际上是由传入的路由定义定位器产生。

8.3.4 路由定位器

直接获取路由的方法是通过 RouteLocator 接口获取。同样，该顶级接口有多个实现类，如图 8-8 所示为 RouteLocator 的类图。

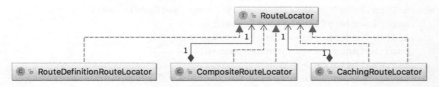

图 8-8 RouteLocator 的类图

```
public interface RouteLocator {
    Flux<Route> getRoutes();
}
```

与上一节介绍的路由定义定位器接口类似，只有一个 getRoutes 方法用以获取路由信息，Route 的定义如下所示：

```
public class Route implements Ordered {
    //路由Id
    private final String id;
    //路由地址
    private final URI uri;
    //路由的优先级
    private final int order;
    //路由断言，以此判断请求路径是否匹配
    private final AsyncPredicate<ServerWebExchange> predicate;
    //网关过滤器
    private final List<GatewayFilter> gatewayFilters;
    ...
}
```

Route 路由定义了路由断言、过滤器、路由地址和路由的优先级等主要信息。请求到达时，会在转发到代理的服务之前，依次经过路由断言进行匹配路由和网关过滤器处理。

通过 RouteLocator 的类图，可以知道 RouteLocator 有三个实现类：

- 基于路由定义方式的（RouteDefinitionRouteLocator）
- 缓存方式的（CachingRouteLocator）
- 组合方式的（CompositeRouteLocator）

1. 基于路由定义方式的 RouteLocator

路由的获取可以通过 RouteDefinitionRouteLocator 获取 RouteDefinition，并将路由定义转换成 Route。图 8-9 是 RouteDefinitionRouteLocator 中获取路由的流程图。

RouteDefinitionRouteLocator 在实例化时先进行初始化的过程，对路由断言和过滤器进行初始化，如下所示：

```
public class RouteDefinitionRouteLocator implements RouteLocator, BeanFactoryAware {
    public RouteDefinitionRouteLocator(RouteDefinitionLocator routeDefinitionLocator,
        List<RoutePredicateFactory> predicates, List<GatewayFilterFactory> gatewayFilterFactories,
        GatewayProperties gatewayProperties) {
        //设置routeDefinitionLocator
        this.routeDefinitionLocator = routeDefinitionLocator;
```

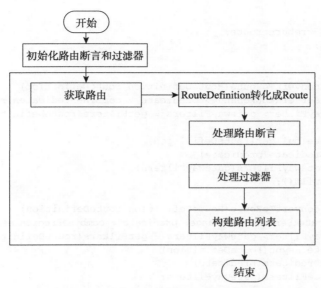

图 8-9 路由定义转换成 Route 流程图

```
        //初始化路由断言
        initFactories(predicates);
        //初始化网关过滤器
        gatewayFilterFactories.forEach(factory -> this.gatewayFilterFactories.
            put(factory.name(), factory));
        this.gatewayProperties = gatewayProperties;
    }

    private void initFactories(List<RoutePredicateFactory> predicates) {
        predicates.forEach(factory -> {
            String key = factory.name();
            ...
            this.predicates.put(key, factory);
        });
    }
    ...
}
```

可以看到 RouteDefinitionRouteLocator 构造函数有多个参数：路由定义定位器、路由断言工厂、网关过滤器以及网关配置。根据传入的参数，设置 routeDefinitionLocator 和网关配置，并初始化路由断言和网关过滤器。

RouteDefinitionRouteLocator 的实现方式是基于路由定义来获取路由，它实现了 RouteLocator 接口，用以获取路由信息，获取路由的方法如下所示：

```
// RouteDefinitionRouteLocator.java
public Flux<Route> getRoutes() {
    return this.routeDefinitionLocator.getRouteDefinitions()
        .map(this::convertToRoute)
        .map(route -> {
```

```
                ...
                return route;
            });
    }
    private Route convertToRoute(RouteDefinition routeDefinition) {
        Predicate<ServerWebExchange> predicate = combinePredicates(routeDefinition);
        List<GatewayFilter> gatewayFilters = getFilters(routeDefinition);

        return Route.builder(routeDefinition)
                .predicate(predicate)
                .gatewayFilters(gatewayFilters)
                .build();
    }
    private Route convertToRoute(RouteDefinition routeDefinition) {
        AsyncPredicate<ServerWebExchange> predicate = combinePredicates(routeDefinition);
        List<GatewayFilter> gatewayFilters = getFilters(routeDefinition);
        return Route.async(routeDefinition)
                .asyncPredicate(predicate)
                .replaceFilters(gatewayFilters)
                .build();
    }
```

获取路由的方法，基于传入的具体 RouteDefinitionLocator 获取路由定义，map 方法再将每个 RouteDefinition 转换为 Route。RouteDefinitionLocator#convertToRoute 是具体的转换方法，在该方法中，涉及路由断言和网关过滤器的处理，处理完之后就可以将这些属性构建为 Route 对象。

网关过滤器来自两部分：网关配置默认的过滤器以及路由定义中的过滤器。首先获取网关配置默认的过滤器，根据过滤器名称获取到对应的过滤器并生成元组，注意生成的过滤器还会转换成有优先级的 OrderedGatewayFilter；然后，对路由定义中的过滤器进行同样的操作，这里不再赘述；最后，根据过滤器的优先级排序，返回排序后的过滤器。

2. 缓存方式和组合方式的 RouteLocator

CachingRouteLocator 是缓存路由的 RouteLocator 实现类。想要获取路由，只需要调用 RouteLocator#getRoutes 方法即可。这里和缓存路由定义定位器的实现很相似。根据传入的代理路由定位器获取路由信息，并将获取的路由信息设置到缓存的路由中。通过监听 RefreshRoutesEvent（缓存路由定义定位器也是监听该事件），进而刷新缓存的路由信息。Spring Cloud Gateway 提供了内置端点触发，这里就不详细讲解了。

CompositeRouteLocator 是组合路由的 RouteLocator 实现类，可以组合多种 RouteLocator 的实现类，将具体实现类获取的路由进行组合，提供了统一访问路由的入口。

8.3.5 路由断言

Spring Cloud Gateway 创建 Route 对象时，使用 RoutePredicateFactory 创建 Predicate 对象。Predicate 对象可以赋值给 Route。简单来说，路由断言用于匹配请求对应的 Route。路

由决策工厂 RoutePredicateFactory 的定义如下所示：

```java
public interface RoutePredicateFactory<C> extends ShortcutConfigurable, Configurable<C> {
    String PATTERN_KEY = "pattern";

    default Predicate<ServerWebExchange> apply(Consumer<C> consumer) {
        C config = newConfig();
        consumer.accept(config);
        beforeApply(config);
        return apply(config);
    }
    //还有另一种apply 实现，返回AsyncPredicate,借助reactor实现异步

    default void beforeApply(C config) {}

    default String name() {
        return NameUtils.normalizeRoutePredicateName(getClass());
    }
}
```

该接口继承自 ShortcutConfigurable 接口，该接口将在后面的多个实现类中出现，基于传入的具体 RouteDefinitionLocator 获取路由定义时，已经用到该接口中的默认方法。断言的种类很多，不同的断言需要的配置参数不一样，所以每种断言和过滤器的实现会实现 ShortcutConfigurable 接口，指定自身参数个数和顺序。下面具体看一下实现：

```java
public interface ShortcutConfigurable {

    static String normalizeKey(String key, int entryIdx, ShortcutConfigurable argHints,
        Map<String, String> args) { // 1
        if (key.startsWith(NameUtils.GENERATED_NAME_PREFIX) && !argHints.shortcutFieldOrder().
            isEmpty()
                && entryIdx < args.size() && entryIdx < argHints.shortcutFieldOrder().
                    size()) {
            key = argHints.shortcutFieldOrder().get(entryIdx);
        }
        return key;
    }
    static Object getValue(SpelExpressionParser parser, BeanFactory beanFactory,
        String entryValue) { // 2
        Object value;
        String rawValue = entryValue;
        if (rawValue != null) {
            rawValue = rawValue.trim();
        }
        if (rawValue != null && rawValue.startsWith("#{") && entryValue.endsWith("}"))
            { // 3
            StandardEvaluationContext context = new StandardEvaluationContext();
            context.setBeanResolver(new BeanFactoryResolver(beanFactory));
            Expression expression = parser.parseExpression(entryValue, new
                TemplateParserContext());
            value = expression.getValue(context);
        } else {
```

```
                value = entryValue;
            }
            return value;
        }
        default ShortcutType shortcutType() { // 默认的ShortcutType为DEFAULT
            return ShortcutType.DEFAULT;
        }
        default List<String> shortcutFieldOrder() { // 4
            return Collections.emptyList();
        }
        ...
}
```

上述代码主要做了以下工作：

1）对键进行标准化处理，因为键有可能是自动生成，当键以 _genkey_ 开头时，表明是自动生成的。

2）获取真实值，需要传入 Spring EL 解析器、Bean 工厂等工具类。

3）对传入的 entryValue 是一个表达式的情况进行处理，这里默认是 Spring EL 表达式。

4）返回有关参数数量和快捷方式解析顺序的提示。

ShortcutConfigurable 接口中提供了如上的默认方法，主要用于对构造的过滤器和断言参数进行标准化处理，将表达式和生成的键进行转换。在该接口中，还提供了枚举类型的 ShortcutType：DEFAULT 和 GATHER_LIST，如下所示：

```
enum ShortcutType {
    DEFAULT {
        //…
    },

    GATHER_LIST {
        ...
    };

    public abstract Map<String, Object> normalize(Map<String, String> args,
        ShortcutConfigurable shortcutConf, SpelExpressionParser parser, BeanFactory
        beanFactory);
}
```

ShortcutType 中定义了一个抽象方法 normalize，该方法传入了多个参数，这些参数都是键值对标准化时所必需的。这里是工厂模式的应用，具体的工厂类根据获取的 shortcutType 类型，只需要调用该抽象方法即可。DEFAULT 方式是普通的 map 组合方式，map 里面有多组键值对，键值对都经过标准化处理；GATHER_LIST 方式是将值进行聚合，存放到一个列表中，该种方式会检测 fieldOrder 的大小是否为 1，这里因为聚合为 map 中的一个键值对，所以必须为 1。

路由决策工厂 RoutePredicateFactory 包含的主要实现类如图 8-10 所示。可以看到该接口具体有多个实现类，抽象类 AbstractRoutePredicateFactory 实现了路由断言工厂，没有实际的方法。具体的路由决策工厂实现类都是继承自抽象类 AbstractRoutePredicateFactory，包括 Datetime、请求的远端地址、路由权重、请求头部、Host 地址、请求方法、请求 URL 中的路径和请求参数等类型的路由断言。下面分别对这些类型的路由断言工厂进行解析。

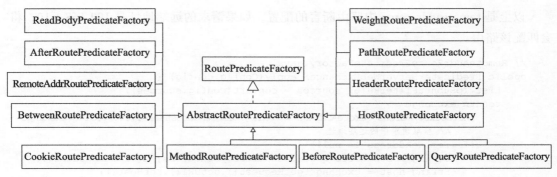

图 8-10　RoutePredicateFactory 类图

1. Datetime 类型的断言工厂

Datetime 类型的断言工厂有三种，分别为：

- AfterRoutePredicateFactory：接收一个日期参数判断请求时间是否配置时间之后。
- BeforeRoutePredicateFactory：接收一个日期参数，判断请求日期是否在指定日期之前。
- BetweenRoutePredicateFactory：接收两个日期参数，判断请求日期是否在这两个指定日期之间。

以 AfterRoutePredicateFactory 为例，介绍 Datetime 类型的断言工厂的应用，如下所示：

```
spring:
  cloud:
    gateway:
      routes:
      - id: after_route
        uri: http://example.org
        predicates:
        - After=2018-03-20T10:42:47.789+08:00[Asia/Shanghai]
```

上面的配置文件指定了路由的断言。关键字是 After，表示晚于指定时间，如上的配置使得请求的时间必须晚于上海时间 2018 年 3 月 20 号 10:42。

2. 基于远端地址的断言工厂

RemoteAddrRoutePredicateFactory 属于根据请求 IP 进行路由决策的类型，接收 CIDR（无类别域间路由）表示法（IPv4 或 IPv6）的字符串列表（列表最小长度为 1）作为参数，例如 192.168.0.1/16，其中 192.168.0.1 是 IP 地址，16 是子网掩码。

```
spring:
  cloud:
```

```yaml
gateway:
  routes:
  - id: remoteaddr_route
    uri: http://example.org
    predicates:
    - RemoteAddr=192.168.1.1/24
```

以上是 remoteaddr_route 中路由断言的配置，如果请求的远端地址是 192.168.1.10，将会匹配该路由。

```java
// RemoteAddrRoutePredicateFactory.java
public Predicate<ServerWebExchange> apply(Config config) {
    List<IpSubnetFilterRule> sources = convert(config.sources);
    return exchange -> {
        InetSocketAddress remoteAddress = config.remoteAddressResolver.resolve(exchange);
            // 获取请求中的远端地址
        if (remoteAddress != null) {
            String hostAddress = remoteAddress.getAddress().getHostAddress();
            String host = exchange.getRequest().getURI().getHost();
            for (IpSubnetFilterRule source : sources) {
                // 遍历配置好的RemoteAddr列表，如果远端地址在其列表之中，则返回匹配
                if (source.matches(remoteAddress)) {
                    return true;
                }
            }
        }
        return false;
    };
}
private List<IpSubnetFilterRule> convert(List<String> values) {
    List<IpSubnetFilterRule> sources = new ArrayList<>();
    for (String arg : values) { // 遍历配置文件中指定的RemoteAddr数组
        addSource(sources, arg);
    }
    return sources;
}
private void addSource(List<IpSubnetFilterRule> sources, String source) {
    if (!source.contains("/")) { //当RemoteAddr没有子网掩码时，默认为 /32
        source = source + "/32";
    }
    String[] ipAddressCidrPrefix = source.split("/",2);
    // 分别获取RemoteAddr中的ip地址和子网掩码
    String ipAddress = ipAddressCidrPrefix[0];
    int cidrPrefix = Integer.parseInt(ipAddressCidrPrefix[1]);
    //根据ip和子网，确定RemoteAddr的范围，并加入到sources中
    sources.add(new IpSubnetFilterRule(ipAddress, cidrPrefix, IpFilterRuleType.ACCEPT));
}
```

如上为基于请求 IP 的路由断言具体实现，首先获取配置文件中的 RemoteAddr 列表；然后将配置的 RemoteAddr 列表转换成 sources 列表，主要是根据 IP 地址和子网掩码确定地址的范围；最后判断请求的远端地址是否在设置的 IP 列表中。

3. 路由权重

Spring Cloud Gateway 提供了基于路由权重的断言工厂，配置时指定分组和权重值即可。Nginx upstream 也可以指定路由权重，Nginx 配置文件中指定的该后端的权重值是固定不变的。WeightRoutePredicateFactory 同时实现了权重的功能，按照路由权重选择同一个分组中的路由。下面我们看一下在 Spring Cloud Gateway 中如何配置带有权重的路由：

```
spring:
    cloud:
        gateway:
            locator:
                enabled: true
            routes:
            - id: weight_route1
              uri: http://blueskykong.com
              order: 6000
              predicates:
              - Weight=group3, 1
              - Path=/weight/**
              filters:
              - StripPrefix=2
            - id: weight_route2
              uri: http://baidu.com
              order: 6000
              predicates:
              - Path=/weight/**
              - Weight=group3, 9
              filters:
              - StripPrefix=1
```

如上配置了两个对于"/weight/**"路径转发的路由定义。这两个路由属于同一个权重分组，且 weight_route1 的权重为 1，weight_route2 的权重为 9。对于 10 个访问 /weight/** 路径的请求来说，将会有 9 个路由到 weight_route2，1 个路由到 weight_route1。

下面介绍下该权重的算法实现过程。配置中有两个路由，如下所示：

```
weight_route1: group3, 1
weight_route2: group3, 9
```

实现过程为：

1）首先构造 weights（group3）数组：weights=[1,9]

2）规范化（Normalize）：weights = weights/sum(weights) = [0.1,0.9]

3）计算区间范围：ranges = weights.collect(0, (s,w) -> s + w) = [0, 0.1, 1.0]

4）生成随机数：r = random()

5）搜索随机数所在的区间：i = integer s.t. r>=ranges[i] && r <ranges[i+1]

6）选择相应的路由：routes[i]

网关应用服务在启动时会发布 WeightDefinedEvent，而在 WeightCalculatorWebFilter 过滤器中定义了事件的监听器，当接收到事件 WeightDefinedEvent 时，会自动添加 WeightConfig

到权重配置中。请求在经过 WeightCalculatorWebFilter 时会生成一个随机数，根据随机数所在的区间选择对应分组的路由。

```java
// WeightRoutePredicateFactory.java
public Predicate<ServerWebExchange> apply(WeightConfig config) {
    return exchange -> {
        Map<String, String> weights = exchange.getAttributeOrDefault(WEIGHT_ATTR,
                Collections.emptyMap());
        String routeId = exchange.getAttribute(GATEWAY_PREDICATE_ROUTE_ATTR);
        String group = config.getGroup();
        if (weights.containsKey(group)) {
            String chosenRoute = weights.get(group);
            return routeId.equals(chosenRoute);
        }
        return false;
    };
}
```

如上，当应用到配置的路由断言 WeightRoutePredicate 时，会根据 ServerWebExchange 中的 WEIGHT_ATTR 值，判断当前的 routeId 与对应分组的 routeId 是否一致。

4. 其他断言工厂

（1）基于 Cookie 的断言工厂

CookieRoutePredicateFactory 是 Cookie 类型的路由断言工厂，接收两个参数：cookie 的名字和一个正则表达式。此断言匹配具有给定名称并且值与正则表达式匹配的 cookie。以下是 cookie_route 中路由断言的配置示例：

```yaml
spring:
  cloud:
    gateway:
      routes:
      - id: cookie_route
        uri: http://example.org
        predicates:
        - Cookie=chocolate, ch.p
```

如果请求的 cookie 名为 chocolate，且其值与正则表达式 ch.p 相匹配，则该请求与该路由匹配。

（2）基于头部的断言工厂

HeaderRoutePredicateFactory 是头部类型的路由断言工厂，接收两个参数：头部名和一个正则表达式。以下是 header_route 中路由断言配置的示例：

```yaml
spring:
  cloud:
    gateway:
      routes:
      - id: header_route
        uri: http://example.org
```

```
predicates:
- Header=X-Request-Id, \d+
```

如果请求的头部中有 X-Request-Id，并且该头部值匹配 \d+ 正则表达式，则与该路由匹配。

（3）基于 Host 地址的断言工厂

HostRoutePredicateFactory 是 Host 地址类型的路由断言工厂，接收一个参数：主机名的模式串。该模式是一个以"."作为分隔符的 Ant 风格的模式。这个断言与 Host 头部匹配模式串的请求匹配。以下是 host_route 中路由断言的配置示例：

```
spring:
    cloud:
        gateway:
            routes:
            - id: host_route
              uri: http://example.org
              predicates:
              - Host=**.blueskykong.com
```

如果请求有一个 Host 头部拥有值 www.blueskykong.com 或 beta.blueskykong.com，将会匹配该路由。

（4）基于请求方法的断言工厂

MethodRoutePredicateFactory 是请求类型的路由断言工厂，接收 HTTP 请求方法作为参数。以下是 method_route 中路由断言的配置示例：

```
spring:
    cloud:
        gateway:
            routes:
            - id: method_route
              uri: http://example.org
              predicates:
              - Method=GET
```

如果请求的 Method 为 GET，则匹配该路由。

（5）基于请求路径的断言工厂

PathRoutePredicateFactory 是基于请求路径的路由断言工厂，接收一个参数：Spring 的 PathMatcher 模式串。以下是 path_route 中路由断言的配置示例：

```
spring:
    cloud:
        gateway:
            routes:
            - id: path_route
              uri: http://example.org
              predicates:
              - Path=/foo/{segment}
```

如果请求路径是 /foo/1 或者 /foo/bar，将会匹配该路由。

（6）基于请求参数的断言工厂

QueryRoutePredicateFactory 是请求参数的路由断言工厂，接收两个参数：一个必须的请求 param 和一个可选的正则表达式。以下是 query_route 中路由断言的配置示例：

```
spring:
    cloud:
        gateway:
            routes:
            - id: query_route
              uri: http://example.org
              predicates:
              - Query=baz, ba.
```

如果请求中包含 baz 查询参数，且其值匹配 ba. 正则表达式，将会匹配该路由。

（7）基于请求体内容的断言工厂

ReadBodyPredicateFactory 可以根据请求体的内容进行路由判断。网关中增加如下的路由定义：

```
@Bean
public RouteLocator customRouteLocator(RouteLocatorBuilder builder) {
    //@formatter:off
    return builder.routes()
            .route("read_body_pred", r -> r.path("/test/**").and().readBody(String.
                class,
                    s -> s.trim().equalsIgnoreCase("hello"))
                    .filters(f ->
                            f.addResponseHeader("X-TestHeader", "read_body_pred")
                    ).uri("http://example.org")
            )
            .build();
    //@formatter:on
}
```

该路由定义了基于请求体内容的断言工厂。该过滤器用法比较灵活，需要传入两个参数：inClass 转换类型和断言表达式。在约定好的情况下，请求体内容可以强转成指定类型的对象。由于网关与业务解耦，直接转换成对象很困难。我们还可以通过将请求体中的字符串转换成 JSONObject 等方式来获取请求体中指定字段的内容。在如上的配置中，请求体对象类型强转为 String，定义了断言表达式，匹配请求体的内容为"hello"（忽略大小写的情况下）。

8.3.6 网关过滤器

GatewayFilter 网关过滤器用于拦截和链式处理 Web 请求，可以实现横切的、与应用无关的需求，比如安全、访问超时的设定等。GatewayFilter 的定义如下所示：

```
public interface GatewayFilter {
    Mono<Void> filter(ServerWebExchange exchange, GatewayFilterChain chain);
}
```

接口中定义了唯一的方法 filter，用于处理 Web 请求，并且可以通过给定的过滤器链传递到下一个过滤器。该接口有多个实现类，其类图如图 8-11 所示。

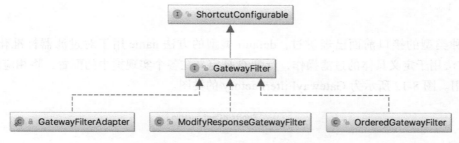

图 8-11　GatewayFilter 类图

从类图可以看到，GatewayFilter 有三个实现类，ModifyResponseGatewayFilter 是一个内部类，用于修改响应体；OrderedGatewayFilter 是一个有序的网关过滤器；GatewayFilterAdapter 是一个适配器类，是定义在 Web 处理器中的内部类。除此之外，在 GatewayFilterFactory 实现类中内部实际上是创建了一个 GatewayFilter 的匿名类。下面分别介绍网关过滤器的实现类。

1. 有序的 GatewayFilter 和 GatewayFilterAdapter

OrderedGatewayFilter 顾名思义，有序的网关过滤器。很多情况下，过滤器都是有优先级的，因此有序的网关过滤器使用的场景很多。

在实现过滤器接口的同时，有序的网关过滤器类还实现了 Ordered 接口。构造函数中传入需要代理的网关过滤器以及优先级，就可以构造一个有序的网关过滤器。具体的过滤功能实现，只需要调用代理的过滤器即可。

GatewayFilterAdapter 用于网关过滤器的适配。在网关过滤器链 GatewayFilterChain 中应用 GatewayFilter 过滤请求，需要通过 GatewayFilterAdapter 将全局过滤器 GlobalFilter 适配成 GatewayFilter。

2. GatewayFilterFactory 的内部类

网关过滤器工厂接口有多个实现类，在每个 GatewayFilterFactory 实现类的 apply(T config) 方法里，都声明了一个实现 GatewayFilter 的内部类。

路由过滤器允许以某种方式修改传入的 HTTP 请求或传出的 HTTP 响应。路由过滤器的作用域是一个特定的路由。Spring Cloud Gateway 包含许多内置的 GatewayFilter 工厂。首先看一下 GatewayFilterFactory 接口中定义的方法，如下所示：

```
public interface GatewayFilterFactory<C> extends ShortcutConfigurable, Configurable<C> {
```

```
default GatewayFilter apply(Consumer<C> consumer) {
    C config = newConfig();
    consumer.accept(config);
    return apply(config);
}
default String name() {
    return NameUtils.normalizeFilterName(getClass());
}
}
```

这种类型的接口前面已经讲过，default 类型的方法 name 用于对过滤器标准化命名，apply 方法用于定义具体的过滤操作，泛型 C 是定义在各个实现类中的配置，将相应的配置进行应用。图 8-12 所示为 GatewayFilterFactory 的类图。

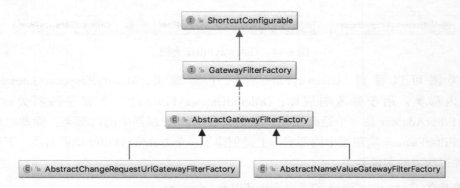

图 8-12　GatewayFilterFactory 的类图

抽象类 AbstractGatewayFilterFactory 实现了 GatewayFilterFactory 接口，和之前讲的断言工厂抽象类功能一样，没有具体的方法，只接收一个参数，如 StripPrefix 过滤器。有两个抽象类继承自 AbstractGatewayFilterFactory：AbstractChangeRequestUriGatewayFilterFactory 和 AbstractNameValueGatewayFilterFactory。

第一个抽象类，改变请求的 URI 过滤器，也是接收一个参数。该过滤器通过方法 determineRequestUri(ServerWebExchange, T) 实现改变请求 URI 的逻辑；另一个抽象类 AbstractNameValueGatewayFilterFactory 继承自 AbstractGatewayFilterFactory，从命名就可以看出，该抽象类用于提供通用的方法给键值对参数类型的网关过滤器，接收两个参数，如 AddResponseHeader 这种类型的过滤器。

网关过滤器有 20 多个实现类，包括头部过滤器、路径类过滤器、Hystrix 过滤器和变更请求 URL 的过滤器，还有参数和状态码等其他类型的过滤器。

3. 头部过滤器

头部类型的过滤器有如下几种：

❑ AddRequestHeaderGatewayFilterFactory：增加请求的头部信息，并将头部传递到下游。

- AddResponseHeaderGatewayFilterFactory：将会为匹配的请求增加响应头部，传递到下游的相应头部。
- RemoveRequestHeaderGatewayFilterFactory：在请求发送到下游之前，将会为匹配的请求移除设置的头部信息。
- RemoveResponseHeaderGatewayFilterFactory：在返回结果给客户端之前，将会移除设置的响应头部信息。
- SetRequestHeaderGatewayFilterFactory：当请求经过网关转发时，该过滤器将会用给定的名字替换所有的头部，而不是增加头部信息。
- SetResponseHeaderGatewayFilterFactory：将会用给定的名字替换所有的头部，而不是增加头部信息。因此，如果下游的服务响应为 X-Response-Foo:1234，网关客户端接收到的响应将会被替换为 X-Response-Foo:Bar。

头部类型的过滤器可以对请求和响应的头部进行处理，实现原理较为简单。下面以 AddRequestHeaderGatewayFilterFactory 为例，讲解该过滤器的应用。

AddRequestHeaderGatewayFilterFactory 过滤器接收两个参数：头部名和值，配置如下所示：

```
spring:
    cloud:
        gateway:
            routes:
            - id: add_request_header_route
              uri: http://example.org
              filters:
              - AddRequestHeader=X-Request-Foo, Bar
```

add_request_header_route 路由将会为匹配的请求，增加 X-Request-Foo:Bar 头部，并将头部传递到下游。

4. 路径过滤器

路径类型的过滤器有如下四种：
- PrefixPathGatewayFilterFactory：将所有匹配的请求的路径加上设置的前缀。
- RewritePathGatewayFilterFactory：使用 Java 正则表达式来灵活地重写请求路径。
- SetPathGatewayFilterFactory：接收的参数为路径的模板。它提供了一种通过允许路径的模板化分段来操纵请求路径的简单方法。这使用了 Spring Framework 的 uri 模板，并允许多个匹配段。如下所示：

```
spring:
    cloud:
        gateway:
            routes:
            - id: setpath_route
              uri: http://example.org
```

```
            predicates:
            - Path=/foo/{segment}
            filters:
            - SetPath=/{segment}
```

一个匹配的请求的路径为 /foo/bar，在构造发送到下游的请求之前，setpath_route 的路由配置会为请求设置请求路径 /bar。

- StripPrefixGatewayFilterFactory：接收一个参数：序号。该参数指明在发送到下游之前，将会移除路径中的分段数量。如下所示：

```
spring:
    cloud:
        gateway:
            routes:
            - id: nameRoot
              uri: http://user-service
              predicates:
              - Path=/user/**
              filters:
              - StripPrefix=2
```

当一个请求 /user/bar/foo 通过网关请求 user-service 时，最终的请求路径将会变成 http://172.16.1.100:8080/foo（假设负载均衡到 user-service 的实例地址为 172.16.1.100:8080）。

下面以 RewritePathGatewayFilterFactory 为例，介绍路径类的过滤器的应用和实现。RewritePathGatewayFilterFactory 接收两个参数：路径的正则表达式和替换字符串。它使用 Java 正则表达式来灵活地重写请求路径。

```
spring:
    cloud:
        gateway:
            routes:
            - id: rewritepath_route
              uri: http://example.org
              predicates:
              - Path=/foo/**
              filters:
              - RewritePath=/foo/(?<segment>.*), /$\{segment}
```

一个请求路径为 /foo/bar，在构造下游请求时，将会设置请求路径为 /bar。注意由于 YAML 规范，$\ 被替换为 $。

```
public class RewritePathGatewayFilterFactory extends AbstractGatewayFilterFactory
    <RewritePathGatewayFilterFactory.Config> {

    @Override
    public GatewayFilter apply(Config config) {
        //替换$\\
        String replacement = config.replacement.replace("$\\", "$");
        return (exchange, chain) -> {
            ServerHttpRequest req = exchange.getRequest();
```

```java
            addOriginalRequestUrl(exchange, req.getURI());
            //获取路径
            String path = req.getURI().getRawPath ();
            //替换路径值
            String newPath = path.replaceAll(config.regexp, replacement);
            //设置新的请求
            ServerHttpRequest request = req.mutate()
                    .path(newPath)
                    .build();
            //在传递到下游之前,保存了原始的路径
            exchange.getAttributes().put(GATEWAY_REQUEST_URL_ATTR, request.getURI());

            return chain.filter(exchange.mutate().request(request).build());
        };
    }

}
```

如上所示,重写路径的过滤器实现比较简单,首先由于 YAML 规范, $\\\\$ 被替换为 $;其次获取路径并用配置的字符串替换;最后在传递到下游之前,保存了原始的路径,这里就是替换后的路径。

5. Hystrix 过滤器

属于熔断类型的过滤器,接收一个参数:HystrixCommand 的命令。

```yaml
spring:
  cloud:
    gateway:
      routes:
      - id: hytstrix_route
        uri: http://example.org
        filters:
        - name: Hystrix
          args:
            name: fallbackcmd
            fallbackUri: forward:/fallbackcontroller
```

如上的配置将会用一个 HystrixCommand 名为 fallbackcmd 包装剩余的过滤器。Hystrix 过滤器接收一个可选的 fallbackUri 参数。当前情况下,只有 forward 模式的 URIs 支持。如果服务被降级,该请求将会转发到该 URI 对应的控制器。

```java
public class HystrixGatewayFilterFactory extends AbstractGatewayFilterFactory
    <HystrixGatewayFilterFactory.Config> {
    public HystrixGatewayFilterFactory(DispatcherHandler dispatcherHandler) {
        super(Config.class);
        this.dispatcherHandler = dispatcherHandler;
    }
    @Override
    public GatewayFilter apply(Config config) {
        if (config.setter == null) {
        // 设置groupKey,默认使用当前类名
```

```
            HystrixCommandGroupKey groupKey = HystrixCommandGroupKey.Factory.asKey
                (getClass().getSimpleName());
            // 设置commandKey，使用配置文件中的name属性
            HystrixCommandKey commandKey = HystrixCommandKey.Factory.asKey(config.
                name);
            // 设置setter，HystrixObservableCommand构造器的流接口，组成为commandKey和
                groupKey
            config.setter = Setter.withGroupKey(groupKey)
                    .andCommandKey(commandKey);
        }

        return (exchange, chain) -> {
        // 构造HystrixObservableCommand，传入过滤器链，具体请求，fallbackUri等
            RouteHystrixCommand command = new RouteHystrixCommand(config.setter,
                config.fallbackUri, exchange, chain);

            return Mono.create(s -> {
                Subscription sub = command.toObservable().subscribe(s::success,
                    s::error, s::success); // 1
                s.onCancel(sub::unsubscribe);
            }).onErrorResume((Function<Throwable, Mono<Void>>) throwable -> {
                if (throwable instanceof HystrixRuntimeException) {
                    HystrixRuntimeException e = (HystrixRuntimeException) throwable;
                    if (e.getFailureType() == TIMEOUT) { setResponseStatus (exchange,
                        HttpStatus.GATEWAY_TIMEOUT);
                        return exchange.getResponse().setComplete();
                    }
                }
                return Mono.empty();
            }).then(); // 2
        };
    }
    private class RouteHystrixCommand extends HystrixObservableCommand<Void> {
        ...
        @Override
        protected Observable<Void> construct() {
            return RxReactiveStreams.toObservable(this.chain.filter(exchange));
        }
    }
}
```

- 使用 Hystrix Command Observable 订阅，因为调用 toObservable() 时，将会执行 HystrixObservableCommand 中的 construct()，获得过滤器链执行的 Observable。
- 当 Hystrix Command 执行超时，设置响应 504 状态码，并回写客户端响应。发生其他异常时，例如断路器打开，最终返回客户端 200 状态码，内容为空。

Hystrix 过滤器的实现中，构造函数接收了一个全局变量 DispatcherHandler 处理器，即 webflux 中的分发处理器。如上面的配置所示，Hystrix 过滤器除了接收配置文件中定义的 HystrixCommand 命令外，还有一个可选的 fallbackUri 参数，其只支持 forward 模式的 URI。

根据配置文件中的信息，构造 RouteHystrixCommand 对象，这是一个内部类，继

承自 HystrixObservableCommand，用于网关过滤器的 HystrixCommand 命令包装。HystrixObservableCommand 用于打包执行潜在风险功能的代码，通常是通过网络进行服务调用，具有统计和隔离等功能。

6. 变更请求 URL 的过滤器

从图 8-12 的网关过滤器工厂类图可见，抽象类 AbstractChangeRequestUriGatewayFilterFactory 也是继承自 AbstractGatewayFilterFactory。该抽象类用于改变请求的 URL。有多种方式可以改变请求的 URL，如根据请求头部的字段、请求的参数或者请求体中的具体字段。在 Spring Cloud Gateway 的过滤器中有一个基于请求头部的实现方式 RequestHeaderToRequestUriGatewayFilterFactory。如下的配置为具体用法：

```
spring:
    cloud:
        gateway:
            routes:
            - id: request_header
              uri: http://baidu.com
              predicates:
              - Path=/header/**
              - Header=X-Next-Url, .+
              filters:
              - RequestHeaderToRequestUri=X-Next-Url
```

对于符合要求的请求，应用 RequestHeaderToRequestUri 过滤器。如果使用这种方式的过滤器，建议在路由断言中增加头部的判断，确保需要应用的头部存在于请求中，且符合头部的规则（应用正则表达式）。在如上的路由定义中，将会应用头部 X-Next-Url 的值作为新的转发 URL。下面介绍如何实现改变请求的 URL 地址：

```
public abstract class AbstractChangeRequestUriGatewayFilterFactory<T>
        extends AbstractGatewayFilterFactory<T> {
    private final int order;

    public AbstractChangeRequestUriGatewayFilterFactory(Class<T> clazz, int order) {
        super(clazz);
        this.order = order;
    }

    public AbstractChangeRequestUriGatewayFilterFactory(Class<T> clazz) {
        this(clazz, RouteToRequestUrlFilter.ROUTE_TO_URL_FILTER_ORDER + 1);
    }

    protected abstract Optional<URI> determineRequestUri(ServerWebExchange exchange,
        T config);

    public GatewayFilter apply(T config) {
        return new OrderedGatewayFilter((exchange, chain) -> {
            Optional<URI> uri = this.determineRequestUri(exchange, config);
            uri.ifPresent(u -> {
```

```
            Map<String, Object> attributes = exchange.getAttributes();
            attributes.put(GATEWAY_REQUEST_URL_ATTR, u);
        });
        return chain.filter(exchange);
    }, this.order);
}
```

如上为改变请求 URL 的抽象类，其中定义了一个抽象方法 determineRequestUri，由具体的继承类实现。从其构造函数中可以看到，该过滤器的优先级低于 RouteToRequestUrlFilter（一个全局过滤器，是路由到请求的 URL 过滤器，会设置请求的 URL 属性值，后面会具体讲解），这是为了保证最终的改变请求 URL 过滤器生效。实际应用时会构造一个有序的 OrderedGatewayFilter，内部的实现其实就是将抽象方法 determineRequestUri 获取的地址放进请求的 GATEWAY_REQUEST_URL_ATTR 属性中。最后，放入过滤器链继续向下传递，如下所示：

```
// RequestHeaderToRequestUriGatewayFilterFactory.java
protected Optional<URI> determineRequestUri(ServerWebExchange exchange,
        NameConfig config) {
    String requestUrl = exchange.getRequest().getHeaders().getFirst(config.getName());
    return Optional.ofNullable(requestUrl).map(url -> {
        try {
            return new URL(url).toURI();
        }
        catch (MalformedURLException | URISyntaxException e) {
            ...
            return null;
        }
    });
}
```

如上为根据头部值改变请求 URL 的实现，取出相应的值并转换成 URI 返回。我们还可以通过请求参数中定义的值来设置新的请求 URL。例如定义如下的路由：

```
@Bean
public RouteLocator customRouteLocator(RouteLocatorBuilder builder) {
    //@formatter:off
    return builder.routes()
            .route(r -> r.path("/test/**").and().query("newurl")
                .filters(f -> f.changeRequestUri(e -> Optional.of(URI.create(e.
                    getRequest().getQueryParams().getFirst("newurl "))))) .uri
                    ("http://example.com"))
            .build();
    //@formatter:on
}
```

其中的配置将请求的参数中的 newurl 的值取出来，设置为新的请求 URL。实际上是实现了抽象方法 AbstractChangeRequestUriGatewayFilterFactory#determineRequestUri，读者可以自行验证如上的实现。

7. 其他过滤器

除了上述过滤器，还有其他类型的过滤器，下面简单介绍这些过滤器的用法。

- AddRequestParameterGatewayFilterFactory：属于请求参数类的过滤器，该过滤器接收两个参数：请求参数和值。将会为匹配的请求增加设置的请求参数，并传递到下游。
- RequestRateLimiterGatewayFilterFactory：属于限流类型的过滤器，该过滤器接收三个参数：令牌桶上限、平均填充速度和关键字 Bean 名称。基于令牌桶算法实现的限流，新请求来临时，会各自拿走一个令牌，如果没有令牌可拿了就阻塞或者拒绝服务。
- ModifyRequestBodyGatewayFilterFactory：修改请求体内容的过滤器，接收三个参数，inClass 源类型、outClass 目标类型和重写函数。
- ModifyResponseBodyGatewayFilterFactory：修改响应体内容的过滤器，和上一个过滤器一样，接收三个参数。

我们可在项目中配置使用修改请求体和响应体这两种过滤器，如下所示：

```
@Bean
public RouteLocator customRouteLocator(RouteLocatorBuilder builder) {
    return builder.routes()
        .route("modify_body", r -> r.path("/modify/**")
            .filters(f -> f.modifyRequestBody(String.class, String.class,
                (exchange, s) -> Mono.just(s.toUpperCase())))
                .modifyResponseBody(String.class, String.class, (exchange, s)
                    -> Mono.just(s.toUpperCase())))
            .uri("http://localhost:8005/body"))
        .build();
}
```

当请求路径的模式串符合 "/modify/**" 时，将会应用到名为 modify_body 的路由，代理服务地址为 http://localhost:8005/body。上面的配置应用了两个过滤器，第一个修改请求体的内容，将请求体的字符串全部转为大写；第二个是将请求返回的响应体，全部转换成小写。

- RetryGatewayFilterFactory：重试过滤器，当转发到代理服务时，遇到指定的服务端错误，如 httpStatus 为 500，我们可以设定请求重试次数。除了对指定的异常重试之外，还可以指定请求的方法，GET 或 POST。通常可以将重试过滤器应用到 default-filters 中。
- PreserveHostHeaderGatewayFilterFactory：PreserveHostHeader 过滤器没有参数。该过滤器设置一个请求属性，将会检测并决定是否发送原始 host 头部，而不是由 HTTP 客户端确定的 host 头部，如下所示：

```
...
filters:
- PreserveHostHeader
```

- RedirectToGatewayFilterFactory：属于转发类型的过滤器，该过滤器接收状态码和一个 URL 参数。该状态码应该是 300 系列的 HTTP 转发码，比如 301。URL 应该是有效的，这将是用于定位头部的值，如下所示：

```
spring:
  cloud:
    gateway:
      routes:
      - id: prefixpath_route
        uri: http://example.org
        filters:
        - RedirectTo=302, http://acme.org
```

上面 prefixpath_route 的路由配置，将会为匹配的请求发送一个头部为 Location:http://acme.org，且状态码为 302。

❑ SaveSessionGatewayFilterFactory：属于会话类型的过滤器，在转发到调用的下游时，该过滤器强制执行 WebSession::save 操作。这是一种特殊的用途，当使用类似于 Spring Session MongoDB 这种惰性数据存储时可使用这种操作，并且需要确保在转发之前会话的数据已经被存储，如下所示：

```
filters:
- SaveSession
```

❑ SecureHeadersGatewayFilterFactory：属于头部类型的过滤器，该过滤器将会增加一些响应的头部[⊖]。如下为默认增加的头部及其值：

```
X-Xss-Protection:1; mode=block
Strict-Transport-Security:max-age=631138519
X-Frame-Options:DENY
X-Content-Type-Options:nosniff
Referrer-Policy:no-referrer
Content-Security-Policy:default-src 'self' https:; font-src 'self' https:
    data:; img-src 'self' https: data:; object-src 'none'; script-src https:;
    style-src 'self' https: 'unsafe-inline'
X-Download-Options:noopen
X-Permitted-Cross-Domain-Policies:none
```

想要改变默认值，可以通过在 spring.cloud.gateway.filter.secure-headers 中配置适合的属性。可以改变的属性如下所示：

```
xss-protection-header
strict-transport-security
frame-options
content-type-options
referrer-policy
content-security-policy
download-options
permitted-cross-domain-policies
```

❑ SetStatusGatewayFilterFactory：属于状态码类的过滤器，该过滤器接收一个状态码

⊖ 详细信息可参考 https://blog.appcanary.com/2017/http-security_headers.html。

作为参数,该参数应该是一个合理的状态码。可以是整型值 404,也可以是枚举类型的字符串 NOT_FOUND。

```
spring:
  cloud:
    gateway:
      routes:
      - id: setstatusstring_route
        uri: http://example.org
        filters:
        - SetStatus=BAD_REQUEST
      - id: setstatusint_route
        uri: http://example.org
        filters:
        - SetStatus=401
```

如上的配置,不管哪一种情况,都会将请求的响应状态码设置为 401。

8.3.7 全局过滤器

GlobalFilter 接口与 GatewayFilter 具有相同的方法定义。全局过滤器是一系列特殊的过滤器,将会根据条件应用到所有的路由中,该接口的设计和用法在将来的版本中可能会发生变化。全局过滤器用于拦截链式的 Web 请求,可以实现横切的、与应用无关的需求,比如安全、访问超时的设定等等。前面章节也讲了过滤器,定制的过滤器的粒度更细,然而有些过滤器需要全局应用,Spring Cloud Gateway 中也有提供全局过滤器的定义与实现。

全局过滤器的接口定义与网关过滤器 GatewayFilter 是一样的,下面通过类图看一下全局网关过滤器有哪些实现类,如图 8-13 所示。

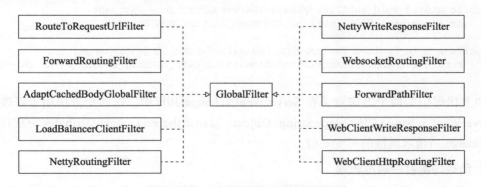

图 8-13　GlobalFilter 类图

从图中可以看到 GlobalFilter 有十个实现类,包括路由转发、负载均衡、ws 路由、Netty 路由等全局过滤器。下面将会介绍这些全局过滤器的实现。

1. 转发路由过滤器

ForwardRoutingFilter 在交换属性 ServerWebExchangeUtils.GATEWAY_REQUEST_URL_

ATTR 中查找 URL，如果 URL 为转发模式即 forward: /// localendpoint，它将使用 Spring DispatcherHandler 来处理请求。未修改的原始 URL 将追加到 GATEWAY_ORIGINAL_REQUEST_URL_ATTR 属性的列表中。ForwardRoutingFilter 的实现如下所示：

```java
public class ForwardRoutingFilter implements GlobalFilter, Ordered {
    public ForwardRoutingFilter(DispatcherHandler dispatcherHandler) {
        this.dispatcherHandler = dispatcherHandler;
    }

    @Override
    public Mono<Void> filter(ServerWebExchange exchange, GatewayFilterChain chain) {
        URI requestUrl = exchange.getRequiredAttribute(GATEWAY_REQUEST_URL_ATTR);
        //获取请求的URI的格式
        String scheme = requestUrl.getScheme();
        //该请求已被路由处理或者URI的格式不是"forward"
        if (isAlreadyRouted(exchange) || !"forward".equals(scheme)) {
            return chain.filter(exchange);
        }
        setAlreadyRouted(exchange);
        //DispatcherHandler进行处理
        return this.dispatcherHandler.handle(exchange);
    }
}
```

转发路由过滤器比较简单，构造函数传入请求的分发处理器 DispatcherHandler。过滤器具体处理时，首先获取请求地址的 URL 前缀；然后进行判断，如果该请求已被路由处理或者 URL 的前缀不是 "forward"，则继续在过滤器链传递；否则设置路由状态位并交由 DispatcherHandler 进行处理。涉及的请求状态位设置与检查如下所示：

```java
public static void setAlreadyRouted(ServerWebExchange exchange) {
    exchange.getAttributes().put(GATEWAY_ALREADY_ROUTED_ATTR, true);
}
public static boolean isAlreadyRouted(ServerWebExchange exchange) {
    return exchange.getAttributeOrDefault(GATEWAY_ALREADY_ROUTED_ATTR, false);
}
```

如上代码中的两个方法定义在 ServerWebExchangeUtils 中，这两个方法用于修改与查询 ServerWebExchange 中的 Map<String, Object> getAttributes()，#getAttributes 方法返回当前 exchange 所请求属性的可变映射。

2. 负载均衡客户端过滤器

LoadBalancerClientFilter 在交换属性 GATEWAY_REQUEST_URL_ATTR 中查找 URL，如果 URL 有一个 lb 前缀，即 lb: // myservice，它使用 LoadBalancerClient 将名称解析为实际的主机和端口，如示例中的 myservice。未修改的原始 URL 将追加到 GATEWAY_ORIGINAL_REQUEST_URL_ATTR 属性的列表中。该过滤器还将查看 GATEWAY_SCHEME_PREFIX_ATTR 属性，以判断它是否等于 lb，然后应用相同的规则。

```java
// LoadBalancerClientFilter.java
```

```java
public Mono<Void> filter(ServerWebExchange exchange, GatewayFilterChain chain) {
    URI url = exchange.getAttribute(GATEWAY_REQUEST_URL_ATTR);
    String schemePrefix = exchange.getAttribute(GATEWAY_SCHEME_PREFIX_ATTR);
    if (url == null || (!"lb".equals(url.getScheme()) && !"lb".equals(schemePrefix))) {
        return chain.filter(exchange);  //继续在过滤器链传递
    }
    addOriginalRequestUrl(exchange, url);  //保存原始的url
    //负载均衡到具体的服务实例
    final ServiceInstance instance = loadBalancer.choose(url.getHost());

    URI uri = exchange.getRequest().getURI();
    //如果没有提供前缀的话,则会使用默认的'<scheme>',否则使用'lb:<scheme>'机制。
    String overrideScheme = null;
    if (schemePrefix != null) {
        overrideScheme = url.getScheme();
    }
    //根据获取的服务实例信息,重新组装请求的url
    URI requestUrl = loadBalancer.reconstructURI(new DelegatingServiceInstance(instance,
        overrideScheme), uri);
    // Routing 相关的GatewayFilter 会通过GATEWAY_REQUEST_URL_ATTR属性,发起请求。
    exchange.getAttributes().put(GATEWAY_REQUEST_URL_ATTR, requestUrl);
    return chain.filter(exchange);
}
```

从中可以看到,负载均衡客户端过滤器的实现步骤如下:构造函数传入负载均衡客户端,依赖中添加 Spring Cloud Netflix Ribbon 即可注入该 Bean。获取请求的 URL 及其前缀;如果 URL 不为空且前缀为 lb 或者网关请求的前缀是 lb,则保存保存原始的 URL,负载到具体的服务实例并根据获取的服务实例信息,重新组装请求的 URL。最后,添加请求 URL 到 GATEWAY_REQUEST_URL_ATTR,并提交过滤器链继续过滤;否则继续在过滤器链传递。另外,在组装请求的地址时,如果 loadbalancer 没有提供前缀的话,则会使用默认的 <scheme>,即 overrideScheme 为 null,否则使用 lb:<scheme> 机制。

3. 基于 Netty 的路由和响应过滤器

如果位于 ServerWebExchangeUtils.GATEWAY_REQUEST_URL_ATTR 请求属性中的 URL 具有 http 或 https 前缀,Netty 路由过滤器将运行。它使用 Netty HttpClient 进行下游代理请求。响应放在 ServerWebExchangeUtils.CLIENT_RESPONSE_ATTR 请求属性中,在过滤器链中传递,如下所示:

```java
// NettyRoutingFilter.java
public Mono<Void> filter(ServerWebExchange exchange, GatewayFilterChain chain) {
    URI requestUrl = exchange.getRequiredAttribute(GATEWAY_REQUEST_URL_ATTR);
    String scheme = requestUrl.getScheme();
    if (isAlreadyRouted(exchange) || (!"http".equals(scheme) && !"https".equals
        (scheme))) {
        return chain.filter(exchange);
    }
    ...
    return this.httpClient.request(method, url, req -> {
```

```
            final HttpClientRequest proxyRequest = req.options(NettyPipeline.SendOptions::
                flushOnEach)
                    .headers(httpHeaders)
                    .chunkedTransfer(chunkedTransfer)
                    .failOnServerError(false)
                    .failOnClientError(false);
            if (preserveHost) {
                String host = request.getHeaders().getFirst(HttpHeaders.HOST);
                proxyRequest.header(HttpHeaders.HOST, host);
            }
            return proxyRequest.sendHeaders()
                    .send(request.getBody().map(dataBuffer ->
                        ((NettyDataBuffer)dataBuffer).getNativeBuffer())); 
    }).doOnNext(res -> {
        ServerHttpResponse response = exchange.getResponse();
        HttpHeaders headers = new HttpHeaders();
        res.responseHeaders().forEach(entry -> headers.add(entry.getKey(), entry.
            getValue()));
        HttpHeaders filteredResponseHeaders = HttpHeadersFilter.filter(
                this.headersFilters.getIfAvailable(), headers, exchange, Type.RESPONSE);
        response.getHeaders().putAll(filteredResponseHeaders);
        // 设置状态码，代码简化了
        response.setStatusCode(HttpStatus.valueOf(res.status().code()));
        exchange.getAttributes().put(CLIENT_RESPONSE_ATTR, res);
    }).then(chain.filter(exchange));
}
```

NettyRoutingFilter 过滤器的构造函数有两个参数：一个是基于 Netty 实现的 HttpClient，通过该属性请求后端的 Http 服务。另一个是 ObjectProvider 类型的 headersFilters，用于头部过滤。ObjectProvider 接口是一种专为注入点设计的 ObjectFactory 变体，返回该工厂维护的对象实例。实际的过滤处理和客户端负载均衡方式的流程类似：首先获取请求的 URL 及前缀，判断前缀是不是 http 或者 https，如果该请求已经被路由或者前缀不合法，则调用过滤器链直接向后传递；否则正常对头部进行过滤操作，具体的头部过滤方法如下所示：

```
// HttpHeadersFilter.java
static HttpHeaders filterRequest(List<HttpHeadersFilter> filters,
                     ServerWebExchange exchange) {
    HttpHeaders headers = exchange.getRequest().getHeaders();
    return filter(filters, headers, exchange, Type.REQUEST);
}

static HttpHeaders filter(List<HttpHeadersFilter> filters, HttpHeaders input,
        ServerWebExchange exchange, Type type) {
    HttpHeaders response = input;
    if (filters != null) {
        HttpHeaders reduce = filters.stream()
                .filter(headersFilter -> headersFilter.supports(type))
                .reduce(input,
                    (headers, filter) -> filter.filter(headers, exchange),
                    (httpHeaders, httpHeaders2) -> {
                        httpHeaders.addAll(httpHeaders2);
```

```
                    return httpHeaders;
                });
        return reduce;
    }
    return response;
}
```

filterRequest 用于对请求头部的信息进行处理,是定义在接口 HttpHeadersFilter 中的默认方法,该接口有三个实现类,请求头部将会经过这三个头部过滤器,并最终返回修改之后的头部。三种过滤器的源码实现就不列出了,功能如下:

- ❑ ForwardedHeadersFilter:增加 Forwarded 头部,头部值为协议类型、host 和目的地址。
- ❑ XForwardedHeadersFilter:增加 X-Forwarded-For、X-Forwarded-Host、X-Forwarded-Port 和 X-Forwarded-Proto 头部。代理转发时,用以自定义的头部信息向下游传递。
- ❑ RemoveHopByHopHeadersFilter:为了定义缓存和非缓存代理的行为,我们将 HTTP 头字段分为两类:端到端头部字段和逐跳头部字段。端到端的标题字段,将其发送给请求或响应的最终收件人;逐跳标题字段,仅对单个传输级别连接有意义,并且不由缓存存储或由代理转发。

所以该头部过滤器会移除逐跳头部字段[⊖],包括如下 9 个字段:

```
Proxy-Authenticate
Proxy-Authorization
TE
Trailer
Transfer-Encoding
Upgrade
proxy-connection
content-length
```

利用上面准备好的请求客户端信息,请求具体的服务地址,将 Netty Response 赋值给响应。同时将头部和状态码放到 HTTP 头部中,后续的过滤器能够修改响应。注意这里是推迟提交响应,直到所有路由过滤器都运行完毕。将客户端响应设置为 ServerWebExchange 属性,并稍后写入响应 NettyWriteResponseFilter。

NettyWriteResponseFilter 与 NettyRoutingFilter 成对使用。"预"过滤阶段没有任何内容,因为 CLIENT_RESPONSE_ATTR 在 WebHandler 运行之前不会被添加。

```java
// NettyWriteResponseFilter.java
public Mono<Void> filter(ServerWebExchange exchange, GatewayFilterChain chain) {
    return chain.filter(exchange).then(Mono.defer(() -> {
        HttpClientResponse clientResponse = exchange.getAttribute(CLIENT_RESPONSE_
            ATTR);
        if (clientResponse == null) {
            return Mono.empty();
        }
        ServerHttpResponse response = exchange.getResponse();
```

⊖ 详细信息可参考 https://tools.ietf.org/html/。

```
            //将响应写到客户端
            NettyDataBufferFactory factory = (NettyDataBufferFactory) response.bufferFactory();

            final Flux<NettyDataBuffer> body = clientResponse.receive()
                    .retain()
                    .map(factory::wrap);
            MediaType contentType = response.getHeaders().getContentType();
            return (isStreamingMediaType(contentType) ?
                response.writeAndFlushWith(body.map(Flux::just)) : response.writeWith(body));
        }));
    }
```

如果 CLIENT_RESPONSE_ATTR 请求属性中存在 Netty HttpClientResponse，则会应用 NettyWriteResponseFilter。它在其他过滤器完成后运行，并将代理响应写回网关客户端响应。

成对出现的 WebClientHttpRoutingFilter 和 WebClientWriteResponseFilter 过滤器，与基于 Nettty 的路由和响应过滤器执行相同的功能，但不需要使用 Netty。目前该功能处于实验阶段，这里暂时不做讲解。

4. 路由到指定请求 URL 的过滤器

如果 ServerWebExchangeUtils.GATEWAY_ROUTE_ATTR 请求属性中有 Route 对象，则会运行 RouteToRequestUrlFilter 过滤器。它会根据请求 URI 创建一个新的 URI，但会使用 Route 对象的 URI 属性进行更新。新的 URI 位于 ServerWebExchangeUtils.GATEWAY_REQUEST_URL_ATTR 请求属性中。该过滤器会组装成发送到代理服务的 URL 地址（有一个特殊情况是应用了变更请求的 URL 过滤器之后，其优先级低于 RouteToRequestUrlFilter，请求的地址还会变更），继续向后传递到路由转发的过滤器，如下所示：

```
// RouteToRequestUrlFilter.java
public Mono<Void> filter(ServerWebExchange exchange, GatewayFilterChain chain) {
    Route route = exchange.getAttribute(GATEWAY_ROUTE_ATTR);
    if (route == null) {
        return chain.filter(exchange);
    }
    URI uri = exchange.getRequest().getURI();
    boolean encoded = containsEncodedQuery(uri);
    URI routeUri = route.getUri();

    if (hasAnotherScheme(routeUri)) {
        exchange.getAttributes().put(GATEWAY_SCHEME_PREFIX_ATTR, routeUri.getScheme());
        routeUri = URI.create(routeUri.getSchemeSpecificPart());
    }
    //拼接requestUrl
    URI requestUrl = UriComponentsBuilder.fromUri(uri)
            .uri(routeUri)
            .build(encoded)
            .toUri();
    exchange.getAttributes().put(GATEWAY_REQUEST_URL_ATTR, requestUrl);
```

```
            return chain.filter(exchange);
    }
```

首先获取请求中的 Route，如果为空则直接提交过滤器链；否则获取 routeUri，并判断 routeUri 是否特殊，如果是则需要处理 URL，保存前缀到 GATEWAY_SCHEME_PREFIX_ATTR，并将 routeUri 替换为 schemeSpecificPart；然后拼接 requestUrl，将原有请求的 URI 转化为路由中定义的 routeUri；最后，提交到过滤器链继续传递。

拼接 requestUrl 时要注意，如果 Route.uri 属性配置带有 Path，则会覆盖请求的 Path。比如：Route.uri 为 http://bin.org:80/123，请求的 URI 为 http://127.0.0.1:8080/test/segment，则最后拼接的 URL 为 http://httpbin.org:80/123。与之关联的是 UriComponentsBuilder#uri 方法。

5. Websocket 路由过滤器

如果请求中的 ServerWebExchangeUtils.GATEWAY_REQUEST_URL_ATTR 属性对应的值 URL 前缀为 ws 或 wss，则应用 Websocket 路由过滤器。它使用 Spring Web Socket 作为底层通信组件向下游转发 WebSocket 请求。Websocket 可以通过在前缀前加上前缀 lb 来实现负载平衡，如 lb:ws://serviceid。

```java
// WebsocketRoutingFilter.java
public Mono<Void> filter(ServerWebExchange exchange, GatewayFilterChain chain) {
changeSchemeIfIsWebSocketUpgrade(exchange);//检查WebSocket是否upgrade
    URI requestUrl = exchange.getRequiredAttribute(GATEWAY_REQUEST_URL_ATTR);

    String scheme = requestUrl.getScheme();
    if (isAlreadyRouted(exchange) || (!"ws".equals(scheme) && !"wss".equals
        (scheme))) {   //判断是否处理
        return chain.filter(exchange);
    }
    setAlreadyRouted(exchange);

    HttpHeaders headers = exchange.getRequest().getHeaders();
    HttpHeaders filtered = HttpHeadersFilter.filter(getHeadersFilters(),
            headers);

    List<String> protocols = headers.get(SEC_WEBSOCKET_PROTOCOL);
    if (protocols != null) {
        //用户定义的字符串，客户端支持的子协议
        protocols = headers.get(SEC_WEBSOCKET_PROTOCOL).stream()
                .flatMap(header -> Arrays.stream(commaDelimitedListToStringArray
                    (header)))
                .map(String::trim)
                .collect(Collectors.toList());
    }
    //将请求代理转发
    return this.webSocketService.handleRequest(exchange,
            new ProxyWebSocketHandler(requestUrl, this.webSocketClient,
                filtered, protocols));
}
```

如上所示，Websocket 路由过滤器进行处理时，首先获取请求的 URL 及其前缀，判断是否能够进行过滤处理；对于未被路由且请求前缀为 ws 或 wss 的请求，设置路由状态位，构造过滤后的头部，这里的头部过滤处理和之前 Netty 路由过滤器处理一样，不再赘述；最后将请求通过代理转发。ProxyWebSocketHandler 是 WebSocketHandler 的实现类，处理客户端 WebSocket Session。下面看一下代理 WebSocket 处理器的具体实现：

```java
// ProxyWebSocketHandler.java
    public Mono<Void> handle(WebSocketSession session) {
        return client.execute(url, this.headers, new WebSocketHandler() {
            @Override
            public Mono<Void> handle(WebSocketSession proxySession) {
                Mono<Void> proxySessionSend = proxySession
                    .send(session.receive().doOnNext(WebSocketMessage::retain));
                Mono<Void> serverSessionSend = session
                    .send(proxySession.receive().doOnNext(WebSocketMessage::retain));
                return Mono.when(proxySessionSend, serverSessionSend).then();
            }
            @Override
            public List<String> getSubProtocols() {
                return ProxyWebSocketHandler.this.subProtocols;
            }
        });
    }
}
```

WebSocketClient#execute 方法连接后端被代理的 WebSocket 服务。连接成功后，回调 WebSocketHandler 实现的内部类的 handle(WebSocketSession session) 方法。WebSocketHandler 实现的内部类进行消息的转发：客户端 => 具体业务服务 => 客户端；然后合并代理服务的会话信息 proxySessionSend 和业务服务的会话信息 serverSessionSend。

6. 其他过滤器

除此之外，还有如下两种全局过滤器：

- AdaptCachedBodyGlobalFilter——用于缓存请求体的过滤器，在全局过滤器中的优先级较高。
- ForwardPathFilter——请求中的 gatewayRoute 属性对应 Route 对象，当 Route 中的 URI scheme 为 forward 模式时，该过滤器用于设置请求的 URI 路径为 Route 对象中的 URI 路径。

8.3.8 API 端点

Spring Cloud Gateway 提供了内置的端点，用于提供路由相关的操作，如过滤器列表、路由列表、单个路由信息等等。Spring Cloud Gateway 的内置端点纳管到 Spring Boot-Actuator 中。从网关服务的启动日志，可以看到网关的内置端点。Spring Cloud Gateway 中提供了多个内置端点。如下为 Spring Cloud Gateway 的主要 API 端点。

- /actuator/gateway/routes/{id},methods=[DELETE]，删除单个路由。
- /actuator/gateway/routes/{id},methods=[POST]，增加单个路由。
- /actuator/gateway/routes/{id},methods=[GET]，查看单个路由。
- /actuator/gateway/routes],methods=[GET]，获取路由列表。
- /actuator/gateway/refresh,methods=[POST]，路由刷新。
- /actuator/gateway/globalfilters,methods=[GET]，获取全局过滤器列表。
- /actuator/gateway/routefilters,methods=[GET]，路由过滤器工厂列表。
- /actuator/gateway/routes/{id}/combinedfilters,methods=[GET]，获取单个路由的联合过滤器。

可以看到如上的 API 端点包括三类：路由操作、获取过滤器和路由刷新。路由操作包括增加、删除单个路由，以及获取路由列表；获取过滤器有三个端点：全局过滤器列表、过滤器工厂列表和单个路由的联合过滤器；路由缓存刷新的实现如下所示：

```
@PostMapping("/refresh")
public Mono<Void> refresh() {
    this.publisher.publishEvent(new RefreshRoutesEvent(this));
    return Mono.empty();
}
```

其实就是发布一个更新路由的事件，然后监听器触发缓存路由的更新。CachingRouteLocator 中实现了相关的刷新操作。

8.4 应用进阶

网关作为服务端对外的唯一入口，除了基本的路由转发、请求和响应过滤功能之外，还有一些高级的功能。下面将会介绍其中的限流机制、熔断降级和请求的重试。

8.4.1 限流机制

分布式系统中经常会提及限流和降级的概念。所谓限流，可以认为是服务降级的一种，限流就是限制系统的输入和输出流量，以达到保护系统的目的。系统上线之前，一般都会进行压测，压测之后吞吐量是可以被测算的，为了保证系统的稳定运行，一旦达到了设定限制的阈值，就需要限制流量并采取一些措施以完成限制流量的目的。常见的限流方案为：延迟处理、拒绝处理和部分拒绝处理等。

一般高并发系统常见的限流有：限制总并发数（比如数据库连接池、线程池）、限制瞬时并发数（如 nginx 的 limit_conn 模块，用来限制瞬时并发连接数）、限制时间窗口内的平均速率（如 Guava 的 RateLimiter、nginx 的 limit_req 模块，限制每秒的平均速率）；其他还有如限制远程接口调用速率、限制 MQ 的消费速率。另外还可以根据网络连接数、网络流量、CPU 或内存负载等来限流。

1. 限流算法

限流的时候，如果要准确地控制 TPS，简单的做法是维护一个单位时间内的计数器，如判断单位时间已经过去，则将计数器重置零。此做法被认为没有很好地处理单位时间的边界，比如在前一秒的最后一毫秒里和下一秒的第一毫秒都触发了最大的请求数，也就是在两毫秒内发生了两倍的 TPS。常用的平滑限流算法有两种：漏桶算法和令牌桶算法。

漏桶算法——思路很简单，水（请求）先进入到漏桶里，漏桶以一定的速度出水（接口有响应速率），当水流入速度过大会直接溢出（访问频率超过接口响应速率），然后就拒绝请求，可以看出漏桶算法能强行限制数据的传输速率。

这里有两个变量，一个是桶的大小，即流量突发增多时可以存多少的水（burst），另一个是水桶漏洞的大小（rate）。因为漏桶的漏出速率是固定的参数，所以，即使网络中不存在资源冲突（没有发生拥塞），漏桶算法也不能使流突发（burst）到端口速率。因此，漏桶算法对于存在突发特性的流量来说缺乏效率。

令牌桶算法——和漏桶算法效果一样但方向相反，更加容易理解。随着时间流逝，系统会按恒定时间间隔（如果 QPS=100，则间隔是 10 毫秒）往桶里加入令牌（Token，想象和漏洞漏水相反，有个水龙头在不断的加水），如果桶已经满了就不再加了。新请求来临时，会各自拿走一个令牌，如果没有令牌可拿了就阻塞或者拒绝服务。

令牌桶算法的另外一个好处是可以方便地改变速度。一旦需要提高速率，则按需提高放入桶中的令牌的速率。一般会定时（比如 100 毫秒）往桶中增加一定数量的令牌，有些变种算法令实时计算应该增加的令牌的数量。

过滤器的应用有两种方式，一种方式是基于自定义过滤器接口；另一种方式是基于内置的限流过滤器配置实现。下面我们分别介绍这两种实现。

2. 基于内置的限流过滤器配置实现

在 8.3.6 节，介绍过限流过滤器，该限流过滤器基于漏桶算法是实现。下面我们具体看一下如何在项目中使用限流过滤器。

引入的依赖代码如下所示：

```
<dependency>
    <groupId>org.springframework.boot</groupId>
    <artifactId>spring-boot-starter-data-redis-reactive</artifactId>
</dependency>
```

限流的配置存放在 Redis 缓存中，需要引入 Redis 的 Starter。

路由配置代码如下所示：

```
spring:
    cloud:
        gateway:
            routes:
                - id: throttle_filter_route
                  uri: http://blueskykong.com
```

```yaml
order: 10000
predicates:
- Path=/rate/limit/**
filters:
- name: RequestRateLimiter
  args:
      key-resolver: "#{@userKeyResolver}"
      redis-rate-limiter.replenishRate: 10
      redis-rate-limiter.burstCapacity: 20
```

key-resolver 对应自定义限流键，实现 KeyResolver 接口即可。使用 SpEL 表达式 #{@beanName} 从 Spring 容器中获取 Bean 对象。默认情况下，使用 PrincipalNameKeyResolver，即请求认证的 java.security.Principal 作为限流键，如下所示：

```java
@Bean
public KeyResolver userKeyResolver() {
    return exchange -> Mono.just(exchange.getRequest().getQueryParams().getFirst("user"));
}
```

该限流键的定义很简单，获取请求中的 user 参数。burstCapacity 对应令牌桶容量，replenishRate 对应令牌桶每秒填充平均速率。配置执行的过滤器效果是：每个用户的请求每秒限制为 10 次，令牌桶的上限为 20。

3. 基于过滤器接口的自定义实现

（1）实现 GatewayFilterFactory 接口

网关过滤器的核心接口为 GatewayFilterFactory，实现该接口可以自定义过滤器的实现。限流过滤器的实现，是基于 https://github.com/bbeck/token-bucket 提供的令牌桶算法。首先需要引入令牌桶算法的依赖，如下所示：

```xml
<dependency>
    <groupId>org.isomorphism</groupId>
    <artifactId>token-bucket</artifactId>
    <version>1.6</version>
</dependency>
```

token-bucket 库提供了令牌桶算法的实现，该算法对于提供对部分代码的限速访问非常有用。所提供的实现方法是"漏斗"，即桶的容量有限，任何超过此容量的令牌将"溢出"该桶并被丢弃。在这个实现中，RefillStrategy 实例中封装了重新填充桶的规则。在每次尝试消费令牌之前，都会根据 RefillStrategy 填充策略向桶中添加相应的令牌数。

```java
public class ThrottleGatewayFilterFactory implements GatewayFilter {
    private static final Log log = LogFactory.getLog(ThrottleGatewayFilterFactory.
        class);

    private TokenBucket tokenBucket;

    public ThrottleGatewayFilterFactory(int capacity, int refillTokens,
                                         int refillPeriod, TimeUnit refillUnit) {
```

```java
        this.tokenBucket = TokenBuckets.builder()
            .withCapacity(capacity)
            .withFixedIntervalRefillStrategy(refillTokens, refillPeriod, refillUnit)
            .build();
    }

    @Override
    public Mono<Void> filter(ServerWebExchange exchange, GatewayFilterChain chain) {
        log.info("TokenBucket capacity: " + tokenBucket.getCapacity());
        boolean consumed = tokenBucket.tryConsume();
        if (consumed) {
            return chain.filter(exchange);
        }
        exchange.getResponse().setStatusCode(HttpStatus.TOO_MANY_REQUESTS);
        return exchange.getResponse().setComplete();
    }
}
```

节流过滤器的定义如上所示，该过滤器有四个必须的参数：令牌桶的容量、填充的令牌数、周期和周期的单位。应用节流过滤器，将会构造一个令牌桶，并尝试消费该请求。如果消费成功，则过滤器链继续传递；否则，直接拒绝该请求。

（2）路由定义

```java
@Bean
public RouteLocator customRouteLocator(RouteLocatorBuilder builder, ThrottleGatewayFilterFactory
    throttle) {
        return builder.routes()
            .route(r -> r.order(-1)
            .path("/customer/**").filters(f -> f.stripPrefix(2).filter(new
                ThrottleGatewayFilterFactory(1, 1, 5, TimeUnit.SECONDS))).uri("http://
                blueskykong.com")
            .id("ThrottleGatewayFilterFactory_test")).build();
}
```

路由定义将会使用节流过滤器，所以需要注入该过滤器。如上的路由定义，将访问路径为"/customer/**"的请求，转发的 URI 设为 http://blueskykong.com，并应用节流过滤器。这里并没有设定具体的限流键，根据请求的访问时间进行限流。创建了一个容量为 1 的令牌桶，以每 5 秒 1 个令牌的速率填充。

至此，自定义限流过滤器已经完成，应用到路径为 /customer/** 的请求，读者可以根据上述步骤自行测试。

8.4.2 熔断降级

熔断和降级在第 6 章讲解 Hystrix 时有深入讲解，Spring Cloud Gateway 支持 Hystrix 过滤器的应用。这里示例如何使用 Hystrix 过滤器。场景和基础应用时一样，有网关服务和用户服务。

1. 加入依赖

代码如下所示：

```xml
<dependency>
    <groupId>org.springframework.cloud</groupId>
    <artifactId>spring-cloud-starter-netflix-hystrix</artifactId>
    <optional>true</optional>
</dependency>
```

这里加入了必需的 Hystrix 依赖。

2. 定义路由配置

代码如下所示：

```yaml
spring:
  cloud:
    gateway:
      locator:
        enabled: true
      routes:
      - id: hytstrix_route
        uri: lb://user
        predicates:
        - Path=/user/**
        filters:
        - StripPrefix=1
        - Hystrix=myCommandName
```

所有的请求访问路径为 /user/**，将会基于服务发现和负载均衡路由到用户服务，路由定义中使用了两个过滤器：路径类和 Hystrix 类。第一个过滤器将会去掉请求中的第一段；第二个过滤器则会应用 Hystrix 熔断与降级，将请求包装成名为 myCommandName 的路由指令 RouteHystrixCommand，RouteHystrixCommand 继承于 HystrixObservableCommand，其内包含了 Hystrix 中断路、资源隔离等诸多断路器核心功能，当转发服务出现问题时，网关能对此进行快速失败，执行特定的失败逻辑，保护网关中的线程池安全，减少同步等待。

3. 定义 fallback 接口

Spring Cloud Gateway 支持定义 fallback 接口，首先需要定义 fallback 接口，如下所示：

```java
@RequestMapping("/fallbackcontroller")
public Map<String, String> fallbackcontroller(@RequestParam("a") String a) {
    return Collections.singletonMap("from", "fallbackcontroller");
}
```

然后我们将过滤器的配置做如下修改：

```yaml
filters:
    - StripPrefix=1
    - name: Hystrix
      args:
        name: fallbackcmd
```

```
fallbackUri: forward:/fallbackcontroller?a=test
```

当前，Spring Cloud Gateway 的 Hystrix 过滤器只支持 forward: 格式的 URI。如果 fallback 被调用，请求将会被转发到匹配的接口。所以当关闭用户服务时，将会得到如图 8-14 的响应结果。

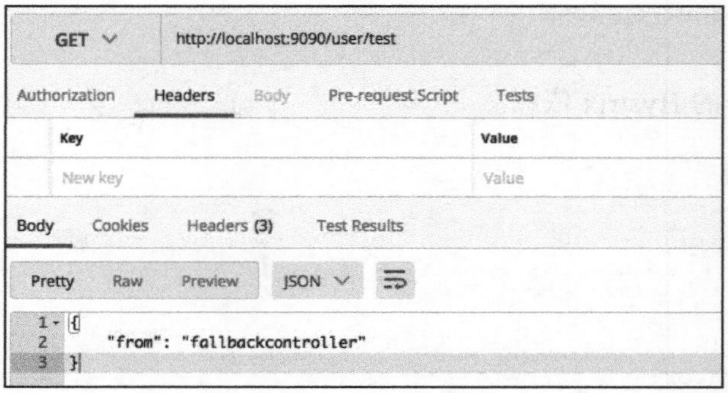

图 8-14　请求返回 fallback 的响应

8.4.3　网关重试过滤器

当转发到代理服务时，遇到指定的服务端错误，如 HTTP 状态码为 500 时，我们可以设定请求重试一定数量。除了对指定的异常重试之外，还可以指定请求的方法，GET 或 POST。

实验场景涉及：网关服务和用户服务。客户端请求经过网关，请求用户服务的 API 接口，遇到指定的异常时，进行重试。

1. 网关服务

网关服务中，新增一个路由的定义 retry_java，请求的判定是路径以 /test 为前缀的请求，并将请求转发到用户服务。当遇到内部服务错误（状态码为 500）时，设定重试的次数为 2。当然该路由也可以通过网关服务的配置文件进行配置，效果是一样的。代码如下所示：

```
@Bean
public RouteLocator retryRouteLocator(RouteLocatorBuilder builder) {
    return builder.routes()
        .route("retry_java", r -> r.path("/test/**")
            .filters(f -> f.stripPrefix(1)
                .retry(config -> config.setRetries(2).setStatuses(HttpStatus.INTERNAL_
                    SERVER_ERROR)))
            .uri("lb://user"))
        .build();
}
```

2. 用户服务

用户服务增加一个 API 接口，请求中传入参数 key 和 count，代码如下所示：

```java
//Controller.java
ConcurrentHashMap<String, AtomicInteger> map = new ConcurrentHashMap<>();

@GetMapping("/exception")
public String testException(@RequestParam("key") String key, @RequestParam(name =
    "count", defaultValue = "3") int count) {
    AtomicInteger num = map.computeIfAbsent(key, s -> new AtomicInteger());
    int i = num.incrementAndGet();
    log.warn("Retry count: "+i);
    if (i < count) {
        throw new RuntimeException("temporarily broken");
    }
    return String.valueOf(i);
}
```

这里主要是演示如何配置网关请求次数，count 是指定的重试次数，默认为 3，第一次和第二次都会抛出运行时异常（状态码为 500），变量 i 是 key 对应的值，初始为 0，每重试一次，i 会递增，直到 i 大于等于 count 的值。

3. 测试结果

根据上面的实现，访问的地址为 http://localhost:9090/test/exception?key=abc&count=2。按照用户服务实现的逻辑，用户服务将会重试一次后成功。图 8-15 为用户服务的控制台日志信息。

```
Retry count: 1

java.lang.IllegalArgumentException: temporarily broken] with root cause
...
Retry count: 2
```

图 8-15　测试 retry 过滤器的响应

从控制台的信息和最后的响应结果可以看出，请求的重试执行成功。

8.5　本章小结

API 网关负责服务请求的路由、组合及协议转换。客户端的所有请求都首先经过 API

网关，然后由它将请求路由到合适的微服务。API 网关经常会调用多个微服务合并结果，以此来处理一个请求，它可以在 Web 协议（如 HTTP 与 WebSocket）与内部使用的非 Web 友好协议之间转换，是微服务架构中一个重要的基础服务。Spring Cloud Gateway 基于最新的 Spring Boot 2.x 开发，与 Spring Cloud 其他组件完美融合；Spring Cloud Gateway 提供了多种类型的路由断言和过滤器，并支持使用 Netty 转发 HTTP 请求。不管是性能还是功能的易用性方面，Spring Cloud Gateway 表现都很卓越，是一个极具潜力的项目。

第 9 章

配置中心:Spring Cloud Config

应用服务除了实现系统功能,还需要连接资源和其他应用,经常有很多需要在外部配置的数据用于调整应用的行为,如切换不同的数据库,设置功能开关等。随着微服务数量的不断增加,需要系统具备可伸缩和可扩展性,除此之外就是能够管理相当多的服务实例的配置数据。在应用的开发阶段,配置信息由各个服务自治管理,但是到了生产环境之后会给运维带来很大的麻烦,特别是微服务的规模比较大,配置的更新更为麻烦。为此,系统需要建立一个统一的配置管理中心,常见的配置中心的实现方法有:

1)硬编码,缺点是需要修改代码,风险大。
2)放在 xml 等配置文件中,和应用一起打包,缺点是更新需要重新打包和重启。
3)文件系统,缺点是依赖操作系统等。
4)读取系统的环境变量,缺点是有大量的配置需要人工设置到环境变量中,不便于管理,且依赖平台。
5)云端存储,缺点是与其他应用耦合。

业界关于分布式配置中心有多种开源的组件,如携程开源的 Apollo、百度的 Disconf、淘宝的 Diamond 等。Spring Cloud 中提供了分布式配置中心 Spring Cloud Config,为外部配置提供了客户端和服务器端的支持。基于 Config 服务器,就可以集中管理各种环境下的各种应用的配置信息。客户端和服务器端与 Spring 中的 Environment 和 PropertySource 概念相匹配,所以这不仅适用于所有的 Spring 应用,而且对于任意语言的应用都能够适用。一个应用可能有多个环境,从开发到测试,再到生产环境,开发者可以管理这些不同环境下的配置,而且能够确保应用在环境迁移后有完整的配置能够正常运行。Config 服务端默认的存储实现是 Git,这能够很容易地支持配置环境的标签版本,而且有各种工具方便地管理这些配置内容。Config 配置服务还支持多种仓库的实现方式,如本地文件系统、SVN 等。

本章第一小节将会实现一个简单的配置中心，包括配置服务器、客户端和 Git 仓库，以基础应用的案例，引出微服务配置中心的主要功能；第二小节首先讲解配置中心的工作，然后介绍配置中心的两个部分：配置客户端和配置服务器，结合源码介绍这两部分主要功能的实现；最后是配置中心的应用进阶，介绍配置中心的其他高级特性，如安全保护、加密解密、多配置仓库等。

9.1 基础应用

在配置中心的基础应用案例中，将会包括两个部分：配置服务器和配置客户端。Config Server 即配置服务器，为配置客户端提供其对应的配置信息，配置信息的来源为配置仓库，启动时需要拉取配置仓库的信息，缓存到本地仓库中；Config Client 即配置客户端，只会在本地配置必要的信息，如指定获取配置的 Config Server 地址，启动时从配置服务器获取配置信息，并支持动态刷新配置仓库中的属性值。

9.1.1 配置客户端

1. 依赖和入口类

```xml
<dependencies>
    <dependency>
        <groupId>org.springframework.cloud</groupId>
        <artifactId>spring-cloud-starter-config</artifactId>
    </dependency>
    <dependency>
        <groupId>org.springframework.cloud</groupId>
        <artifactId>spring-cloud-starter-consul-discovery</artifactId>
    </dependency>
</dependencies>
```

如上所示，依赖中引入了配置客户端的 Starter。由于需要从配置服务器获取配置信息，引入了服务发现组件 Consul 的 Starter。入口类没有特殊配置，此处略。

2. 启动类配置文件

```yaml
server:
    port: 8000
spring:
    application:
        name: config-client
    cloud:
        config:
            label: master
            discovery:
                enabled: true
                service-id: config-server
            enabled: true
```

```
            fail-fast: true
            profile: dev
        consul:
            host: localhost
            port: 8500
            discovery:
                ip-address: localhost
                port: ${server.port}
                instance-id: config-client-${server.port}
                service-name: config-client
```

如上所示，启动类 bootstrap 的配置信息相对比较简单，可以将除了服务发现之外的配置都放到配置中心。在启动类配置文件中，首先需要指定 Consul 注册中心的配置；其次，spring.cloud.config 的相关配置包括 label、profile 和 service-id（即配置服务器的服务名称）。

9.1.2 配置仓库

在建立配置服务器之前，需要先建一个配置仓库。这里选用了 Git 作为配置仓库，当然还有其他种类的仓库选择，这里主要强调的是配置的规则。

在配置仓库 ConfigRepo 中，新建两个环境的配置文件夹：dev 和 test。文件夹中分别存放 Config 客户端的配置文件，目录结构如下所示：

```
.
├── dev
│   └── config-client-dev.yml
└── test
    └── config-client-test.yml
```

1. 配置规则

配置客户端的请求地址和资源文件映射，如下所示：

```
/{application}/{profile}[/{label}]
/{application}-{profile}.yml
/{label}/{application}-{profile}.yml
/{application}-{profile}.properties
/{label}/{application}-{profile}.properties
```

可以看到，配置资源文件同时支持 YAML 和 Properties。这些端点都可以映射到配置文件 {application}-{profile}.yml（或 {application}-{profile}.properties）。YAML 是一种非常方便的格式，用于指定分层配置数据。{application} 对应客户端的应用名 spring.application.name，{profile} 对应不同的 profile 值 spring.cloud.config.profile，{label} 对应配置仓库的分支 spring.cloud.config.label，默认为 master。

2. 客户端提交配置文件

在 dev 目录下的 config-client-dev.yml 加入如下配置：

```
cloud:
    version: Camden SR7
```

在 test 目录下的 config-client-test.yml 加入如下配置：

```
cloud:
    version: Camden SR7 for test
```

最后，将上述配置提交到配置仓库。

9.1.3 服务端

1. 依赖和入口类

```xml
<dependencies>
    <dependency>
        <groupId>org.springframework.cloud</groupId>
        <artifactId>spring-cloud-config-server</artifactId>
    </dependency>
    <dependency>
        <groupId>org.springframework.cloud</groupId>
        <artifactId>spring-cloud-starter-consul-discovery</artifactId>
    </dependency>
</dependencies>
```

如上依赖中引入了服务发现 Consul 和 spring-cloud-config-server。

配置服务器端为外部配置（键值对或者与 YAML 相似的内容格式）提供了 HTTP API 接口。入口类添加 @EnableConfigServer 注解，启用 Config Server，如下所示：

```java
@SpringBootApplication
@EnableConfigServer
public class ConfigServerApplication {
    public static void main(String[] args) {
        SpringApplication.run(CloudApplication.class, args);
    }
}
```

2. 启动类配置文件

```yaml
server:
    port: 8888
spring:
    application:
        name: config-server
    cloud:
        consul:
            host: localhost
            port: 8500
            discovery:
                ip-address: localhost
                port: ${server.port}
                instance-id: config-server-${server.port}
                service-name: config-server
---
spring:
```

```yaml
cloud:
    config:
        server:
            git:
                uri: https://gitee.com/keets/Config-Repo.git
                searchPaths: ${APP_LOCATE:dev}
                username: user
                password: password
```

如上代码中，服务器端的配置信息主要有服务发现和配置仓库的配置信息。uri 对应配置 Git 仓库地址（前缀为 spring.cloud.config.server.git，下文同）；searchPaths 对应配置仓库路径，这里指定了 dev 文件夹，区分了不同的部署环境；username 对应访问 Git 仓库的用户名；password 对应访问 Git 仓库的用户密码；如果是私有配置仓库的话，需要配置用户名和密码，也支持 ssh 的安全秘钥模式，否则不需要添加。上面的配置中使用了私有库，并且使用了用户名密码登录的模式。

3. 验证服务器端配置

配置服务器端启动后，可以看到其拉取的 Git 仓库中的配置信息。目前在 Git 仓库只有 config-client 的配置，对应为 config-client-dev.yml，启动时的日志信息如下所示：

```
Located property source: CompositePropertySource [name='configService', propertySour
    ces=[MapPropertySource {name='configClient'}, MapPropertySource {name='https://
    gitee.com/keets/Config-Repo.git/dev/config-client-dev.yml#dev'}]]
2018-01-10 20:01:35.367  INFO 25897 --- [nio-8000-exec-1]
```

Spring Cloud Config 服务器端负责将 Git 中存储的配置文件发布成 REST 接口，所以在建好配置仓库和配置服务器之后，就可以验证服务器端能否正常提供接口。根据上面端点的对应规则，请求 http://localhost:8888/config-client/dev，得到如下结果：

```
{
    "name": "config-client",
    "profiles": [
        "dev"
    ],
    "label": "master",
    "version": "c5b6f3f78a0b3492a9ad01df212347839197e2e2",
    "state": null,
    "propertySources": [
        {
            "name": "https://gitee.com/keets/Config-Repo.git/dev/config-client-dev.
                yml#dev",
            "source": {
                "spring.profiles": "dev",
                "cloud.version": "Camden SR7"
            }
        }
    ]
}
```

上面返回的结果显示了应用名、profile、Git 版本、配置文件的 URL 以及配置内容等信息。根据我们前面的讲解，配置内容还可以通过请求 http://localhost:8888/config-client-dev.yml 获取。

9.1.4　配置验证

下面验证客户端是否能够正常获取配置。这需要客户端增加 API 接口，并且提交客户端的配置文件到配置仓库中。

1. 客户端增加 API 接口

在上节客户端的基础上，我们增加一个 API 端点接口，端点很简单，用来获取 cloud.version 的值，如下所示：

```
@RestController
@RequestMapping("/cloud")
public class TestController {

    @Value("${cloud.version}")
    private String version;

    @GetMapping("/version")
    public String version() {
        return version;
    }
}
```

2. 获取配置

启动配置服务器端和客户端，可以使用 POSTMAN 或者浏览器等工具验证客户端是否正确获取了配置信息，如图 9-1 所示。

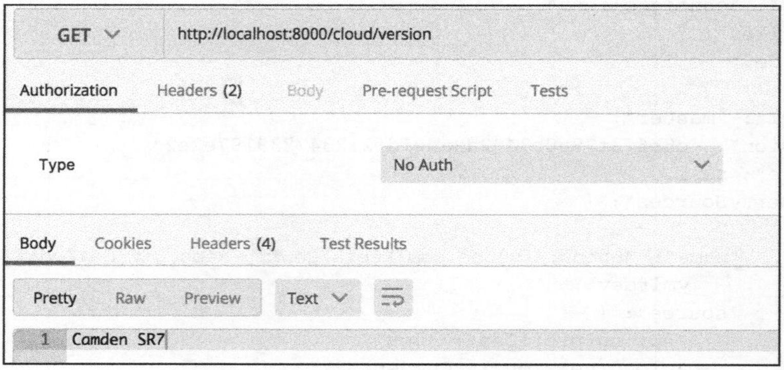

图 9-1　请求客户端获取配置

从图中可以看出，config-client 已经正常获取了配置信息。当我们提交了配置文件之后，最简单的方法就是重启服务以重新获取配置。客户端并不能主动感知到配置的变化，因此需要重启去获取新的配置。

9.1.5 配置动态更新

1. 引入依赖

需要引入 actuator 监控模块，用于动态刷新相关的配置信息，如下所示：

```xml
<dependency>
    <groupId>org.springframework.boot</groupId>
    <artifactId>spring-boot-starter-actuator</artifactId>
</dependency>
```

增加了 spring-boot-starter-actuator 包，spring-boot-starter-actuator 具有监控的功能，actuator 创建了多个监控端点，如 /beans、/health 等，可以监控程序在运行时状态，其中也包括 /refresh 动态刷新的功能。

2. 开启更新机制

还需要给加载变量的类标注 @RefreshScope，使用该注解的类，在客户端执行 /refresh 的时候就会更新此类的变量值，如下所示：

```java
@RestController
@RequestMapping("/cloud")
@RefreshScope
public class TestController {

    @Value("${cloud.version}")
    private String version;

    @GetMapping("/version")
    public String getVersion() {
        return version;
    }
}
```

3. 更改配置

我们更改配置并提交到 Git 仓库，如下所示：

```
cloud:
    version: Edgware.RELEASE
```

另外，Spring Boot 2.0.x 以上版本的默认配置中，Actuator 没有暴露所有 HTTP 端口。这里需要将 /actuator/refresh 端点暴露出来，在配置文件添加如下配置：

```
management:
    endpoints:
        web:
            exposure:
                include: "*"
```

4. 测试

打开命令行或者其他工具,输入命令"curl -X POST http://localhost:8000/actuator/refresh",执行 POST 到刷新端点的请求。返回如下的结果:

```
[
    "config.client.version",
    "cloud.version"
]
```

当我们再次访问 /cloud/version 接口,返回的结果则是 Edgware.RELEASE。这说明在不需要重启的状态下,cloud.version 的配置值已经更新好了,具体原理将会在下一节进行源码解析。每次手动刷新客户端也很麻烦,有没有办法使得提交代码后就自动调用客户端来更新呢? Git 的 webhook 可以实现该功能,每次 push 代码后,都会给远程 HTTP URL 发送一个 POST 请求,这个 HTTP URL 就是我们上面请求的 /actuator/refresh API 端点,具体操作比较简单,读者可以自己尝试一下。

9.2 源码解析

通过上一节的基础应用,我们可以知道配置中心包括如下元素:
- 配置服务器:为配置客户端提供其对应的配置信息,配置信息的来源为配置仓库,启动时即拉取配置仓库的信息,缓存到本地仓库中。
- 配置客户端:除了配置服务器之外的应用服务,启动时从配置服务器拉取其对应的配置信息。
- 配置仓库:为配置服务器提供配置源信息,配置仓库的实现可以支持多种方式。

配置中心的应用架构如图 9-2 所示。

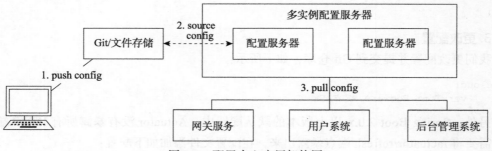

图 9-2 配置中心应用架构图

在部署环境之前,需要将相应的配置信息推送到配置仓库;先启动配置服务器,启动之后,将配置信息拉取并同步至本地仓库;然后,配置服务器对外提供 REST 接口,其他所有的配置客户端启动时根据 spring.cloud.config 配置的 {application}/{profile}/{label} 信息去配置服务器拉取相应的配置。配置仓库支持多样的源,如 Git、SVN、JDBC 数据库和本地文件系统等。最后，其他应用启动，从配置服务器拉取配置，如图 9-3 所示。配置中心还支持动态刷新配置信息，不需要重启应用，结合消息总线提供的刷新 API，webhook 调用该端点 API，达到动态刷新的效果。

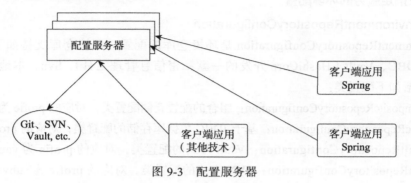

图 9-3　配置服务器

总的来说，Spring Cloud Config 具有如下特性：
1）提供配置的服务器端和客户端支持。
2）集中管理分布式环境下的应用配置。
3）基于 Spring 环境，可以无缝地与 Spring 应用集成。
4）可用于任何语言开发的程序，为其管理与提供配置信息。
5）默认实现基于 Git 仓库，可以进行版本管理。

下面我们分别来对配置服务器和客户端的主要功能进行源码解析。

9.2.1　配置服务器

从上一节的基础应用，可以总结出配置服务器的主要功能有：连接配置仓库、拉取远端的配置并缓存到本地、对外提供配置信息的 REST 接口。本节将会围绕配置服务器的这几个主要功能进行讲解。

1. Config Server 配置类

在讲解配置服务器前，首先介绍 Config Server 配置类。注解 @EnableConfigServer 可以开启应用服务对配置中心的支持。

当启用了 Config Server 之后，配置服务器在启动时就需要对 Config Server 进行自动配置，在 ConfigServerAutoConfiguration 中引入了多个配置类，主要包括：

- EnvironmentRepositoryConfiguration：环境变量存储相关的配置。
- CompositeConfiguration：组合方式的环境仓储配置。
- ResourceRepositoryConfiguration：资源仓储相关的配置。
- ConfigServerEncryptionConfiguration：加密端点的配置。
- ConfigServerMvcConfiguration：对外暴露 MVC 端点控制器的配置。
- TransportConfiguration：配置 clone 或 fetch 传输命令的回调。

@Import 注解导入了这些配置类，这些类也是 Config Server 的主要功能实现。下面重点介绍核心的配置类，通过配置类来引入 Config Server 核心功能。

2. 获取指定服务的环境配置

（1）EnvironmentRepositoryConfiguration

EnvironmentRepositoryConfiguration 是环境仓库的配置，环境仓库支持如下的配置仓库形式：JDBC、Vault（HashiCorp 开发的一款私密信息管理工具）、Svn、本地文件系统、Git 库，包括如下配置类：

- CompositeRepositoryConfiguration：组合的配置仓储配置类，对应的 profie 为 composite。
- JdbcRepositoryConfiguration：基于 JDBC 数据库存储的配置类，对应的 profie 为 jdbc。
- VaultRepositoryConfiguration：Vault 方式的配置类，对应的 profie 为 vault。
- SvnRepositoryConfiguration：SVN 方式的配置类，对应的 profie 为 subversion。
- NativeRepositoryConfiguration：本地文件方式的配置类，对应的 profie 为 native。
- GitRepositoryConfiguration：Git 方式的配置类，对应的 profie 为 git。
- DefaultRepositoryConfiguration：默认方式的配置类，即不指定 profile 的情况，与 profie 为 git 的方式相同。

环境仓库支持多种方式，每种方式的实现原理都是相同的，笔者这里将以其中最常用的默认方式进行讲解，即 Git 方式配置环境仓库。

既然环境仓库支持多种方式，那么怎么指定配置服务器启动时使用哪种方式？回忆上一节基础应用时的配置为 spring.cloud.config.server.git，但这里的配置并没看出指定哪种配置仓库的方式，所以看一下该配置类的具体实现，如下所示：

```
@Configuration
@ConditionalOnMissingBean(EnvironmentRepository.class)
// 默认的环境仓库，在Config Server没有指定时，将会默认初始化
class DefaultRepositoryConfiguration {
    ...
    @Bean
    public MultipleJGitEnvironmentRepository defaultEnvironmentRepository() {
        MultipleJGitEnvironmentRepository repository = new MultipleJGitEnvironme
            ntRepository(
                this.environment); // 使用的是Git仓库的方式
        repository.setTransportConfigCallback(this.transportConfigCallback);
        if (this.server.getDefaultLabel() != null) {
        // label对应于配置仓库的标签，参数缺失时，设置默认label为master；
            repository.setDefaultLabel(this.server.getDefaultLabel());
        }
        return repository;
    }
}
```

EnvironmentRepositoryConfiguration 中还声明了多个其他的配置类，上面只展示了默认方式实现代码。可以看出，每种配置仓库的实现都对应声明的配置类，用 @Profile 注解来

进行激活相应的配置类，并在配置服务器的 application.properties 或者 application.yml 中指定 spring.profiles.active=jdbc。当我们没有设置的时候，使用的是默认的环境仓库为 Git 方式，所以看到 DefaultRepository 配置的注解有 @ConditionalOnMissingBean(EnvironmentRepository.class)，当上下文中不存在 EnvironmentRepository 对象时才会实例化该默认的环境仓库配置类。

（2）EnvironmentRepository 接口和 SearchPathLocator 接口

各类环境仓库都实现了顶级接口 EnvironmentRepository 和 SearchPathLocator。前者的定义如下所示：

```java
public interface EnvironmentRepository {
    Environment findOne(String application, String profile, String label);
}
```

该接口定义了一个获取指定应用服务环境信息的方法，返回的是 Environment 对象。需要传入的参数也很熟悉，有 application、profile 和 label，对应于客户端应用的信息。该方法根据传入的客户端服务的信息，用于获取对应的配置信息。如下为我们获取到的一个 Environment 对象：

```json
{
    "name": "config-client",
    "profiles": [
        "dev"
    ],
    "label": null,
    "version": "1e3e73d366cb37b99c918cd9b6f3ac471da17211",
    "propertySources": [
        {
            "name": "https://gitee.com/keets/Config-Repo.git/dev/config-client-dev.yml",
            "source": {
                "spring.profiles": "dev",
                "cloud.version": "Camden SR4"
            }
        }
    ]
}
```

除了传入的客户端应用的信息，还有 version 对应于 Git 提交的 CommitId，propertySources 对应环境变量的源和具体的值。

另一个接口是 SearchPathLocator，根据传入的客户端应用信息，获取对应的配置环境文件的位置信息，该接口的定义如下所示：

```java
public interface SearchPathLocator {

    Locations getLocations(String application, String profile, String label);

    class Locations {
        private final String application;
```

```
    private final String profile;
    private final String label;
    private final String[] locations;
    private final String version;
    ...
}
```

图 9-4 为基础应用中获取指定资源的截图。

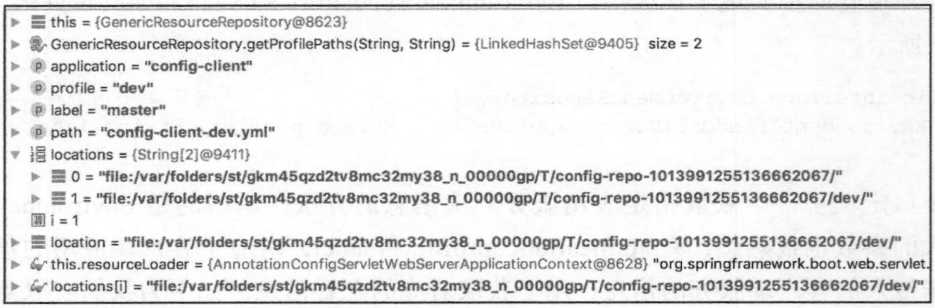

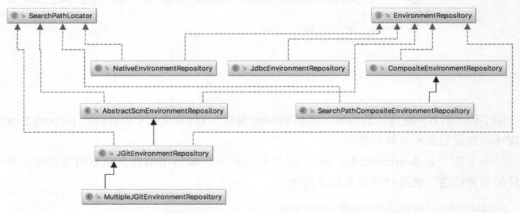

图 9-4　获取指定应用服务的 Locations

locations 对应的是本地仓库的位置，由于根据 profile 区分，所以这里返回的两个位置为根目录和 dev 目录。该接口用于定位资源搜索路径，比如配置信息存储于文件系统或者类路径中，获取配置信息时需要准确定位到这些配置的位置。内部类 Locations 定义了应用服务配置存储信息，和 findOne 方法的传参一样，方法 getLocations 也需要应用服务的相关信息。

图 9-5 为接口 EnvironmentRepository 和 SearchPathLocator 的类图。

图 9-5　EnvironmentRepository 和 SearchPathLocator 的类图

（3）JGit 方式实现配置仓库

MultipleJGitEnvironmentRepository 继承自 JGitEnvironmentRepository，类图如图 9-6 所

示。JGit 也是一种环境仓储的方式①，我们首先需要了解下 JGit 是什么？

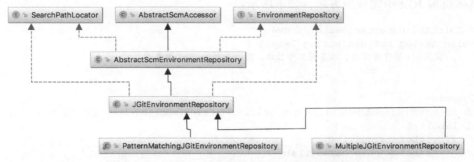

图 9-6　MultipleJGitEnvironmentRepository 类图

从类图可看出，JGitEnvironmentRepository 继承自抽象类 AbstractScmEnvironmentRepository，而该抽象类又继承自 AbstractScmAccessor，AbstractScmAccessor 是 SCM（软件配置管理，Source Control Management）实现的父类，定义了基础的属性组成以获取 SCM 的资源。我们从基类 AbstractScmAccessor 看起，在基类中定义了 SCM 的配置属性和基本方法，包括远端仓库的地址、用户名、密码等属性，这些在使用 Git 仓库时都需要配置。还有一些属性，如 basedir（即本地拷贝之后的工作仓库地址）和 passphrase（即 SSH 的私钥）等，也都是我们在设置仓库时的可选项。

在抽象类 AbstractScmEnvironmentRepository 中，实现了 EnvironmentRepository 接口中获取配置信息的方法，代码如下所示：

```
// AbstractScmEnvironmentRepository.java
public synchronized Environment findOne(String application, String profile, String label) {
    NativeEnvironmentRepository delegate = new NativeEnvironmentRepository(getEnvironment(),
        new NativeEnvironmentProperties());
    Locations locations = getLocations(application, profile, label);
    delegate.setSearchLocations(locations.getLocations());
    Environment result = delegate.findOne(application, profile, "");
    result.setVersion(locations.getVersion());
    result.setLabel(label);
    return this.cleaner.clean(result, getWorkingDirectory().toURI().toString(),
        getUri());
}
```

在获取应用服务的配置信息时，新建了一个本地的环境仓库，作为代理的环境仓库。首先是获取本地仓库中指定应用的位置（一般是一个 tmp 目录，也可以自行指定）；在获取到本地的搜索路径之后，将会根据该路径搜索应用服务的配置信息；最后将得到的结果进行处理，设置应用的 profile 和标签等。由于 delegate#findOne 返回的 Environment 对象会包含 workingDir 和 JSON 键的相关信息，所以这里还调用了 EnvironmentCleaner#clean 方法，对

---

① JGit 是一个用 Java 写成的功能相对健全的 Git 的实现，广泛用在 Java 社区中。JGit 项目由 Eclipse 维护，它的主页在 http://www.eclipse.org/jgit。

结果进行处理。该抽象类基于 SCM 对配置信息进行管理的子类为 JGitEnvironmentRepository。下面具体讲解 JGit 中实现的方法，如下所示：

```java
// JGitEnvironmentRepository.java
public String refresh(String label) {
    //调用git操作的方法，准备好工作目录，返回最新的HEAD版本号
}

@Override
public synchronized Locations getLocations(String application, String profile, String
    label) {
    if (label == null) {
        label = this.defaultLabel;
    }
    String version = refresh(label);
    return new Locations(application, profile, label, version,
        getSearchLocations(getWorkingDirectory(), application, profile, label));
}
```

JGitEnvironmentRepository 继承抽象类 AbstractScmEnvironmentRepository，在 SCM 的基础上加入了 git 操作相关的方法。获取具体的配置文件地址时，需要传入应用名、profile 和标签信息，根据最新的版本号返回 Locations 定位到资源的搜索路径。refresh 方法用于刷新本地仓库的配置状态，保证每次都能拉取到最新的配置信息。

（4）JGit 获取最新的远端仓库配置

JGit 方式作为 AbstractScmEnvironmentRepository 子类，并没有覆写 findOne，其获取配置的逻辑是：获取指定应用在本地仓库中的路径；根据获取的本地位置作为搜索路径；最后，获取到本地仓库中的配置信息。

其中，第一步的逻辑极为重要，因为这关乎获取的配置是否是最新的。在获取到仓库路径之前，需要检查 Git 仓库的状态，代码如下所示：

```java
// JGitEnvironmentRepository.java
public String refresh(String label) {
    Git git = null;
    try {
        git = createGitClient();
        if (shouldPull(git) && fetchStatus != null) {
            FetchResult fetchStatus = fetch(git, label);
            if(deleteUntrackedBranches) {
                deleteUntrackedLocalBranches(fetchStatus.getTrackingRefUpdates(), git);
            }
            checkout(git, label);
            if (isBranch(git, label)) {
                // 将结果merge
                merge(git, label);
                if (!isClean(git, label)) {
                    logger.warn("The local repository is dirty...");
                    resetHard(git, label, LOCAL_BRANCH_REF_PREFIX + label);
                }
```

```java
            }
        }
        else {
            // 如果没有更新,则直接checkout
            checkout(git, label);
        }
        // 返回Head Version
        return git.getRepository().findRef("HEAD").getObjectId().getName();
    }
    ...
}
protected boolean shouldPull(Git git) throws GitAPIException {
    boolean shouldPull;//判断是否需要拉取
    if (this.refreshRate > 0 && System.currentTimeMillis() - this.lastRefresh <
        (this.refreshRate * 1000)) {
        return false;
    } //检测Git的刷新频率
    Status gitStatus = git.status().call();//获取远端git库的状态
    boolean isWorkingTreeClean = gitStatus.isClean();//状态是否有过提交
    String originUrl = git.getRepository().getConfig().getString("remote", "origin",
        "url");
    if (this.forcePull && !isWorkingTreeClean) {
        shouldPull = true;
        logDirty(gitStatus);
    }
    else {
        shouldPull = isWorkingTreeClean && originUrl != null;
    }
    if (!isWorkingTreeClean && !this.forcePull) {
        this.logger.info("Cannot pull from remote...");
    }
    return shouldPull;
}
```

通过检查远端仓库的Git状态,进而判断本地仓库是否需要刷新。refresh依赖于shouldPull的状态,当有新的提交时或者配置了强制拉取,Git客户端将会fetch所有的更新,并merge到所在分支或tag,更新本地环境仓库。refresh最终返回的是最新一次提交的HEAD Version。

(5)多JGit仓库实现

Config Server中还可以配置多个Git仓库,服务启动时会自动读取这些配置属性,MultipleJGitEnvironmentRepository是JGit实现的子类。MultipleJGitEnvironmentRepository用于处理一个或多个Git仓库的环境仓储,不同之处在于MultipleJGitEnvironmentRepository会遍历所有的仓库。具体实现和JGit方式差异并不是很大,这里不再展开描述。

3. 获取指定服务的资源文件

ResourceRepositoryConfiguration是资源的配置类,在SearchPathLocator对象存在时,将ResourceRepository加入到Spring的上下文中,如下所示:

```java
@Bean
@ConditionalOnBean(SearchPathLocator.class)
public ResourceRepository resourceRepository(SearchPathLocator service) {
    return new GenericResourceRepository(service);
}
```

前面讲到 EnvironmentRepository，相比而言 ResourceRepository 用来定位一个应用的资源，返回的是某一个具体的资源文件，最后将其内容转换成文本格式；而 EnvironmentRepository 返回的信息更加全面，是一种键值对的格式，包括应用的基本信息和指定应用的配置源（可能来自多个源或者共享配置文件），这些键值对可以替换资源文本中的占位符。ResourceRepository 接口的定义如下所示：

```java
public interface ResourceRepository {
    Resource findOne(String name, String profile, String label, String path);
}
```

返回的 Resource 是一个资源描述符的接口，用于抽象底层资源的实际类型，如文件或类路径资源。

```yaml
spring:
    profiles: dev
cloud:
    version: Camden SR4
```

上面是示例项目中客户端服务获取其对应的资源文件的结果，Resource 流转化成 String 文本。ResourceRepository 接口的实现类为 GenericResourceRepository，覆写了 findOne 方法，返回配置数据，如下所示：

```java
// GenericResourceRepository.java
public GenericResourceRepository(SearchPathLocator service) {
    this.service = service;
}
@Override
public synchronized Resource findOne(String application, String profile, String
    label, String path) {
    String[] locations = this.service.getLocations(application, profile, label).
        getLocations();
    try {
        for (int i = locations.length; i-- > 0;) {
            String location = locations[i];
            for (String local : getProfilePaths(profile, path)) {
                Resource file = this.resourceLoader.getResource(location)
                    .createRelative(local);
                if (file.exists() && file.isReadable()) {
                    return file;
                }
            }
        }
        ...
    }
```

从上面实现来看，主要是通过构造函数设置 SearchPathLocator 对象，传入应用名等参数调用 getLocations 方法得到配置资源的具体路径，SearchPathLocator 的实现类会保证是最新的配置仓库。因为 profile 可以有默认值 default，在创建资源文件之前，先调用 getProfilePaths 方法根据 profile 值对 path 进行处理，然后由 resourceLoader#getResource 方法创建绝对路径的配置资源。

4. Config Server 提供的端点

ConfigServerMvcConfiguration 对控制器的端点进行配置。配置服务器对外提供的 API 端点包括三类：Environment、Resource 以及加密解密的端点。我们将会介绍前两类常用的 API 端点。

```
@Configuration
@ConditionalOnWebApplication
public class ConfigServerMvcConfiguration extends WebMvcConfigurerAdapter {
    @Bean
    public EnvironmentController environmentController(EnvironmentRepository
        envRepository, ConfigServerProperties server) {
        EnvironmentController controller = new EnvironmentController(encrypted(e
            nvRepository, server), this.objectMapper);
        controller.setStripDocumentFromYaml(server.isStripDocumentFromYaml());
        return controller;
    }
    ...
    private EnvironmentRepository encrypted(EnvironmentRepository envRepository,
        ConfigServerProperties server) {
        EnvironmentEncryptorEnvironmentRepository encrypted = new EnvironmentEnc
            ryptorEnvironmentRepository(
                envRepository, this.environmentEncryptor);
        encrypted.setOverrides(server.getOverrides());
        return encrypted;
    }
}
```

如上所示，配置文件 ConfigServerMvcConfiguration 将 EnvironmentController 加入了 Spring 的上下文中，并自动注入 EnvironmentRepository 和 ConfigServerProperties 对象。在将配置信息返回给客户端服务之前，远端加密的属性值将会被解密（以 {cipher} 开头的字符串）。这里的 encrypted 方法，就是将给定的 EnvironmentRepository 再次封装，返回一个代理类用以解密属性。Overrides 属性，即为在配置服务器设置的属性，用以强制覆写客户端对应的环境变量，这里一起封装到代理类中。stripDocumentFromYaml 属性用来标识那些不是 map 类型的 YAML 文档，应该去掉 Spring 增加的文档的前缀，默认为 true。

（1）获取 Environment 的端点

Environment 控制器提供了如下的端点：

```
/{application}/{profile}[/{label}]
/{application}-{profile}.yml
```

```
/{label}/{application}-{profile}.yml
/{application}-{profile}.properties
/{label}/{application}-{profile}.properties
```

如上的格式都可以获取指定应用的配置信息，如 URI 为 /config-client/dev 和 /config-client-dev.yml 等。配置仓库采用 Git 的方式，获取指定应用的 Environment 过程如图 9-7 所示。

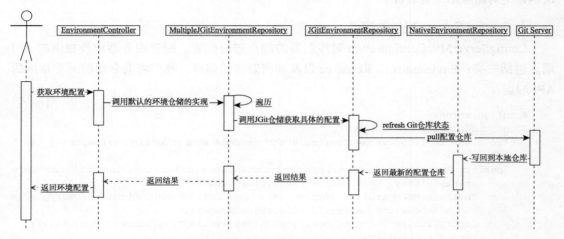

图 9-7 获取指定应用的 Environment 时序图

通过请求的时序图，可以清楚地知道客户端每次拉取的配置都是本地仓库复制的那一份，通过 NativeEnvironmentRepository 代理获取指定应用的配置；Config Server 每次都会检查指定 Git 仓库的状态，当远端仓库有更新时，则会 fetch 到本地进行更新。如下所示为 EnvironmentController 的实现：

```java
public class EnvironmentController {
    //根据参数应用名和profile，返回Environment对象
    @RequestMapping("/{name}/{profiles:.*[^-].*}")
    public Environment defaultLabel(@PathVariable String name,
            @PathVariable String profiles) {
        return labelled(name, profiles, null);
    }
    @RequestMapping("/{label}/{name}-{profiles}.properties")
    public ResponseEntity<String> labelledProperties(@PathVariable String name,
            @PathVariable String profiles, @PathVariable String label,
            @RequestParam(defaultValue = "true") boolean resolvePlaceholders)
            throws IOException {
        validateProfiles(profiles);
        Environment environment = labelled(name, profiles, label);
        Map<String, Object> properties = convertToProperties(environment);
        String propertiesString = getPropertiesString(properties);
        if (resolvePlaceholders) {
            propertiesString = resolvePlaceholders(prepareEnvironment(environment),
                    propertiesString);
        }
```

```
            return getSuccess(propertiesString);
        }
        ...
}
```

控制器调用相应环境仓库实现的 findOne 方法。既可以返回完整的 Environment 对象，包括 name、profiles 和 propertySources 等信息，也可以直接返回配置仓库中的源配置数据，还可以直接返回处理之后的 JSON 对象。在 /{label}/{name}-{profiles}.properties 这个接口对应的实现中，首先调用 labelled 方法返回 Environment 对象，然后将对象转换成 map，由 map 转成 string，最后替换掉系统的环境变量的占位符，返回文本对象。

（2）获取 Resource 的端点

Resource 提供的 API 端点实现其实有些类似 Environment 端点的实现，在此我们仅简单介绍。如下所示为获取 Resource 的 API 端点：

- /{name}/{profile}/{label}/**
- /{name}/{profile}/**

获取资源文件的过程和获取环境变量的过程类似，不同的是，当返回的是一个指定的配置文件时，Resource 控制器默认会将文件中的占位符替换，最后将替换后的资源文件返回给客户端。

Resource 控制器和 EnvironmentController 的初始化过程类似，ResourceController 控制器依赖的获取资源文件的 EnvironmentRepository 也是"Encryptor"代理类，会将解密后的属性值返回给客户端应用。获取具体应用的资源文件的方法，需要应用名、profile、label 和文件名，根据这些参数便可以得到相应的资源文件，Resource 流转换成字符串返回给客户端。默认会将资源文件中的系统环境变量占位符进行替换。

9.2.2 配置客户端

Spring Boot 的应用能够立刻体验 Spring Cloud Config Server 带来的优势，而且还能够收获与环境变化事件有关的特性。当在依赖中添加了 spring-cloud-config-client 依赖，即可对 Spring Cloud Config 进行初始化配置。客户端应用在启动时有两种配置方式：

- 通过 HTTP URI 指定 Config Server。
- 通过服务发现指定 Config Server。

与之相关的配置类为在 spring.factories 文件中定义的启动上下文，如下所示：

```
# Bootstrap components
org.springframework.cloud.bootstrap.BootstrapConfiguration=\
org.springframework.cloud.config.client.ConfigServiceBootstrapConfiguration,\
org.springframework.cloud.config.client.DiscoveryClientConfigServiceBootstrapConfiguration
```

配置客户端应用在引入 spring-cloud-starter-config 依赖后，其配置的 Bean 都会在 SpringApplicatin 启动前加入到它的上下文里去。下面分别介绍这两种配置方式。

1. 通过 HTTP URI 指定 Config Server

通过 HTTP URI 指定 Config Server，是每一个客户端应用默认的启动方式，当 Config Client 启动时，通过 spring.cloud.config.uri（如果不配置，默认为 http://localhost:8888）属性绑定到 Config Server，并利用获取到的远端环境属性初始化 Spring 的环境。客户端应用想要获取 Config Server 中配置信息，需要在环境变量中配置 spring.cloud.config.uri 的值。

ConfigServiceBootstrapConfiguration 进行了两个 Bean 的初始化：ConfigClientProperties 和 ConfigServicePropertySourceLocator。ConfigClientProperties 是对 ConfigClient 的属性进行配置，而 ConfigServicePropertySourceLocator 用于从远程服务器上请求对应服务的配置，并注册到 Spring 容器的 Environment 对象中，如下所示：

```java
public class ConfigClientProperties {
    //配置的前缀
    public static final String PREFIX = "spring.cloud.config";
    //用来标识是否允许获取远端的配置，默认为true
    private boolean enabled = true;
    // 从远端获取配置的默认profile，默认为default
    private String profile = "default";
    // 获取远端配置的应用名，对应于环境变量spring.application.name，默认为application
    @Value("${spring.application.name:application}")
    private String name;
    // 拉取远端配置的标签名，默认的是在Config Server中设置（如git默认为master）
    private String label;
    //远端配置服务器的地址，默认为http://localhost:8888
    private String uri = "http://localhost:8888";
    ...
}
```

从上述代码可以看到，ConfigClientProperties 中定义了 profile、应用名、标签、远端服务器的地址等属性。这些都是 Config Client 启动时必需的信息，如果没有这些配置，客户端将不能正确地从 Config Server 获取其对应的配置信息。

另一个属性资源的定位器类 ConfigServicePropertySourceLocator 依赖于客户端应用配置的属性信息，从远程服务器上请求该应用的配置。下面我们具体看一下该实现类，如下所示：

```java
// ConfigServicePropertySourceLocator.java
@Retryable(interceptor = "configServerRetryInterceptor")
public PropertySource<?> locate(Environment environment) {
    ConfigClientProperties properties = this.defaultProperties.override(environment);
    CompositePropertySource composite = new CompositePropertySource("configService");
    RestTemplate restTemplate = this.restTemplate == null ? getSecureRestTemplate
        (properties) : this.restTemplate;
    ...
    try {
        String[] labels = new String[] { "" };
        if (StringUtils.hasText(properties.getLabel())) {
        // 将传入的label转换成labels数组，label的格式诸如dev,test
            labels = StringUtils.commaDelimitedListToStringArray(properties.
                getLabel());
```

```java
        }
        //保存请求头部的X-Config-State
        String state = ConfigClientStateHolder.getState();
        // 尝试labels数组，通过指定的配置服务器信息，循环调用获取远端的环境配置信息
        for (String label : labels) {
            Environment result = getRemoteEnvironment(restTemplate,
                properties, label.trim(), state);
            if (result != null) {
                //当使用XML时，result.getPropertySources()可能为空
                if (result.getPropertySources() != null) {
                    for (PropertySource source : result.getPropertySources()) {
                        Map<String, Object> map = (Map<String, Object>) source
                            .getSource();
                        composite.addPropertySource(new MapPropertySource(source
                            .getName(), map));
                    }
                }
                // 其他信息的设置，客户端的状态以及版本号等
                if (StringUtils.hasText(result.getState()) || StringUtils.hasText
                    (result.getVersion())) {
                    HashMap<String, Object> map = new HashMap<>();
                    putValue(map, "config.client.state", result.getState());
                    putValue(map, "config.client.version", result.getVersion());
                    composite.addFirstPropertySource(new MapPropertySource("conf
                        igClient", map));
                }
                return composite;
            }
        }
    }
    if (properties.isFailFast()) { // FailFast，如果设置快速响应失败，失败时抛出异常
        throw new IllegalStateException("Could not locate..", error);
    }
    return null;
}
```

重试注解指定了拦截器的配置 configServerRetryInterceptor，该对象的初始化在前面已经讲过。ConfigServicePropertySourceLocator 实质是一个属性资源定位器，其主要方法是 locate(Environment environment)。首先用当前运行环境的 application、profile 和 label 替换 configClientProperties 中的占位符并初始化 RestTemplate，然后遍历 labels 数组直到获取到有效的配置信息，最后还会根据是否快速失败进行重试。

属性资源定位时调用 getRemoteEnvironment 方法，通过 HTTP 的方式获取远程服务器上的配置数据。实现过程为，首先替换请求路径中的占位符，然后进行头部组装，组装好了就可以发送请求，最后返回结果。

在上面的实现中，我们看到获取到的配置信息存放在 CompositePropertySource，那是如何使用它的呢？这里补充另一个重要的类 PropertySourceBootstrapConfiguration，它实现了 ApplicationContextInitializer 接口，该接口会在应用上下文刷新之前回调 refresh()，从而

执行初始化操作,应用启动后的调用栈如下所示:

```
SpringApplicationBuilder.run() -> SpringApplication.run() -> SpringApplication.
    createAndRefreshContext() -> SpringApplication.applyInitializers() -> Proper
    tySourceBootstrapConfiguration.initialize()
```

上述 ConfigServicePropertySourceLocator#locate 方法会在 PropertySourceBootstrapConfiguration#initialize 中被调用,从而保证上下文在刷新之前能够拿到必要的配置信息。initialize 方法如下所示:

```
// PropertySourceBootstrapConfiguration.java
public void initialize(ConfigurableApplicationContext applicationContext) {
    CompositePropertySource composite = new CompositePropertySource(
            BOOTSTRAP_PROPERTY_SOURCE_NAME);
    AnnotationAwareOrderComparator.sort(this.propertySourceLocators);
    boolean empty = true;
    ConfigurableEnvironment environment = applicationContext.getEnvironment();
    for (PropertySourceLocator locator : this.propertySourceLocators) {
        PropertySource<?> source = null;
        source = locator.locate(environment);
        if (source == null) {
            continue;
        }
        composite.addPropertySource(source);
        empty = false;
    }
    if (!empty) {
        MutablePropertySources propertySources = environment.getPropertySources();
        String logConfig = environment.resolvePlaceholders("${logging.config:}");
        LogFile logFile = LogFile.get(environment);
        if (propertySources.contains(BOOTSTRAP_PROPERTY_SOURCE_NAME)) {
            propertySources.remove(BOOTSTRAP_PROPERTY_SOURCE_NAME);
        }
        insertPropertySources(propertySources, composite);
        reinitializeLoggingSystem(environment, logConfig, logFile);
        setLogLevels(environment);
        handleIncludedProfiles(environment);
    }
}
```

在 initialize 方法中进行了如下的操作:

- 根据默认的 AnnotationAwareOrderComparator 排序规则对 propertySourceLocators 数组进行排序。
- 获取运行的环境上下文 ConfigurableEnvironment。
- 遍历 propertySourceLocators 时:
    - 调用 locate 方法,传入获取的上下文 environment。
    - 将 source 添加到 PropertySource 的链表中。
    - 设置 source 是否为空的标识变量 empty。
- source 不为空的情况,才会设置到 environment 中:

- 返回 Environment 的可变形式，可进行的操作有 addFirst 和 addLast。
- 移除 propertySources 中的 bootstrapProperties。
- 根据 Config Server 覆写的规则，设置 propertySources。
- 处理多个活跃（active）profiles 的配置信息。

通过如上过程，可实现将指定的 Config Server 拉取配置信息应用到我们的客户端服务中。

2. 通过服务发现指定 Config Server

Config Client 在启动时，首先会通过服务发现找到 Config Server，然后从 Config Server 拉取其相应的配置信息，并用这些远端的属性资源初始化 Spring 的环境。

如果启用了服务发现，如 Eureka、Consul，则需要设置 spring.cloud.config.discovery.enabled=true，因为默认的是 HTTP URI 的方式，这会导致客户端应用不能利用服务注册。

所有的客户端应用需要配置正确的服务发现信息。比如使用 Spring Cloud Netflix，你需要指定 Eureka 服务器的地址 eureka.client.serviceUrl.defaultZone。在启动时定位服务注册，这样做的开销是需要额外的网络请求，而优点是 Config Server 能够实现高可用，避免单点故障。配置的 Config Server 的 serviceId 默认是"configserver"，可以通过在客户端的 spring.cloud.config.discovery.serviceId 属性来更改。服务发现的客户端实现支持多种类型的元数据 map，如 Eureka 的 eureka.instance.metadataMap。Config Server 的一些额外属性，需要配置在服务注册的元数据中，这样客户端才能正确连接。如果 Config Server 使用了基本的 HTTP 安全，则可以配置证书的用户名和密码；或者是 Config Server 有一个上下文路径，就可以设置 configPath。客户端中的 config 信息可以配置如下：

```
eureka:
    instance:
        ...
        metadataMap:
            user: osufhalskjrtl
            password: lviuhlszvaorhvlo5847
            configPath: /config
```

DiscoveryClientConfigServiceBootstrapConfiguration 中主要配置了 Config Client 通过服务发现组件寻找 Config Server 服务，除此之外还配置了两种事件的监听器：上下文刷新事件和心跳事件，如下所示：

```
// DiscoveryClientConfigServiceBootstrapConfiguration.java
@EventListener(ContextRefreshedEvent.class)
public void startup(ContextRefreshedEvent event) {
    refresh();
}

@EventListener(HeartbeatEvent.class)
public void heartbeat(HeartbeatEvent event) {
    if (monitor.update(event.getValue())) {
```

```
        refresh();
    }
}
```

在 Config Client 获取到 Config Server 中的配置信息之后,剩余的过程与指定 HTTP URI 方式获取 Config Server 是一样的,在上一小节已经讲解。下面将会具体讲解获取配置服务器和事件监听器。

(1) 获取配置服务器

ConfigServerInstanceProvider 用于获取配置的服务器的地址,对其实例化,需要服务发现的客户端 DiscoveryClient。其提供的主要方法 getConfigServerInstance,可以通过传入的 serviceId 参数,获取对应的服务实例,如下所示:

```
public class ConfigServerInstanceProvider {
    public ConfigServerInstanceProvider(DiscoveryClient client) {
        this.client = client;
    }

    @Retryable(interceptor = "configServerRetryInterceptor")
    public ServiceInstance getConfigServerInstance(String serviceId) {
        logger.debug("Locating configserver (" + serviceId + ") via discovery");
        List<ServiceInstance> instances = this.client.getInstances(serviceId);
        if (instances.isEmpty()) {
            throw new IllegalStateException(
                "No instances found of configserver (" + serviceId + ")");
        }
        ServiceInstance instance = instances.get(0);
        return instance;
    }
}
```

上述代码很清晰,主要依赖前面的 DiscoveryClientConfigServiceBootstrapConfiguration 注入的对象 DiscoveryClient,通过 Client 获取对应 serviceId 的实例。

(2) 事件监听器

下面看一下涉及的两个事件监听器:环境上下文刷新和心跳事件。

环境上下文刷新事件,ContextRefreshedEvent 的父类继承自抽象类 ApplicationEvent,当 ApplicationContext 被初始化或者刷新时会唤起该事件,如下所示:

```
public class ContextRefreshedEvent extends ApplicationContextEvent {
    // 创建了一个新的上下文刷新事件,参数是初始化了的ApplicationContext
    public ContextRefreshedEvent(ApplicationContext source) {
        super(source);
    }
}
```

心跳事件定义在 discovery client 中,如果支持来自 discovery server 心跳,则在 DiscoveryClient 的实现中进行广播,并提供给监听器一个基本的服务目录状态变更的指示。当目录更新了,该状态值也需要更新,如同一个版本计数器一样简单,心跳事件的代码如下所示:

```java
public class HeartbeatEvent extends ApplicationEvent {

    private final Object state;
public HeartbeatEvent(Object source, Object state) {
// 创建一个新的事件，参数通常为discovery client和状态值
        super(source);
        this.state = state;
    }

    //代表服务目录的状态值
    public Object getValue() {
        return this.state;
    }
}
```

介绍完这两个事件，我们发现其监听器都依赖于 refresh 方法，下面我们具体了解一下 refresh 方法的功能，如下所示：

```java
// DiscoveryClientConfigServiceBootstrapConfiguration.java
private void refresh() {
    try {// 获取serviceId的服务实例。
        String serviceId = this.config.getDiscovery().getServiceId();
        ServiceInstance server = this.instanceProvider
                .getConfigServerInstance(serviceId);
        String url = getHomePage(server);
        if (server.getMetadata().containsKey("password")) {
        //刷新获取到的服务实例的元数据信息
            String user = server.getMetadata().get("user");
            user = user == null ? "user" : user;
            this.config.setUsername(user); // 更新用户名、密码
            String password = server.getMetadata().get("password");
            this.config.setPassword(password);
        }
        if (server.getMetadata().containsKey("configPath")) {
        // 更新configPath
        ...
        }
        this.config.setUri(url);
    }
}
```

refresh 方法根据上下文环境和心跳事件，刷新服务实例 ConfigClientProperties 中的元数据信息，包括配置的用户名、密码和 configPath。

9.3 应用进阶

在介绍完配置服务中心的基础应用之后，本小节将会介绍一些进阶功能，包括多个 repos 的模式匹配、配置仓库的加密解密、属性覆盖、SVN 配置（本地）仓库和安全保护等功能。

9.3.1 为 Config Server 配置多个 repo

之前的例子中，Config Server 的 git repo 配置信息如下所示：

```
spring.cloud.config.server.git.uri: https://gitee.com/keets/Config-Repo.git
```

它也支持更复杂的需求，通过应用名与 profile 进行模式匹配。模式格式是带有通配符的 {application}/{profile} 名称的逗号分隔列表（其中可能需要引用以通配符开头的模式）。

repo 中的 pattern 属性是一个数组，因此可以使用一个 YAML 数组来绑定多个 pattern。默认情况下，Config Server 只有在第一次请求时配置才会从远端克隆（clone）代码。Config Server 可以设置为在启动时即从远端仓库克隆。

```yaml
spring:
  cloud:
    config:
      server:
        git:
          uri: https://github.com/spring-cloud-samples/config-repo
          repos:
            simple: https://github.com/simple/config-repo
            development:
              pattern:
                - '*/development'
                - '*/staging'
              uri: https://github.com/development/config-repo
            staging:
              pattern: staging*
              cloneOnStart: true
              uri: https://github.com/staging/config-repo
```

如上的配置方式，repos 配置了多个仓库：simple、development 和 staging，下面通过三个不同的 repos 配置方式，讲解多个 repos 的模式匹配。

- simple 仓库：匹配的模式是 simple/*，意为只匹配应用名为 simple 的所有 profile。simple 中使用的快捷配置方式，只能用于设置 URI 的地址，如要设置其他信息（身份验证信息或模式等），则必须要使用完整的形式。
- development 仓库："*/staging" 是 "["*/staging","*/staging,*"]" 的简写，匹配以 staging 开头的 profile 的应用。同理，"*/development" 匹配所有以 development 开头的 profile 的应用。
- staging：pattern 对应于 "staging*"，staging repo 匹配了所有以 staging 开头应用名，并且不限 profile。"/*" 后缀会自动添加到没有 profile 匹配的任何模式。

9.3.2 客户端覆写远端的配置属性

应用的配置源通常都是远端的 Config Server 服务器，默认情况下，本地的配置优先级低于远端配置仓库。如果想用本地应用的系统变量和 config 文件覆盖远端仓库中的属性值，可以通过如下设置：

```
spring:
    cloud:
        config:
            allowOverride: true
            overrideNone: true
            overrideSystemProperties: false
```

- overrideNone：当 allowOverride 为 true 时，overrideNone 设置为 true，外部的配置优先级更低，而且不能覆盖任何存在的属性源。默认为 false。
- allowOverride：标识 overrideSystemProperties 属性是否启用。默认为 true，设置为 false 意为禁止用户的设置。
- overrideSystemProperties：用来标识外部配置是否能够覆盖系统属性，默认为 true。

客户端服务通过如上配置，可以使本地配置优先级更高，且不能被远端仓库中的配置覆盖。

9.3.3 属性覆盖

Config Server 有一个属性覆盖 overrides 特性，允许将配置的属性强制应用到所有的应用服务，而且不能被使用普通 Spring Boot 钩子的应用程序意外地改变。声明属性覆盖，是一种 map 的格式，在 spring.cloud.config.server.overrides 后面增加键值对，如下面的配置所示：

```
spring:
    cloud:
        config:
            server:
                overrides:
                    cloud: Edgware.RELEASE
```

这将会使得所有的配置客户端独立于自己的配置读取 cloud: Edgware.RELEASE。当然，应用程序可以以任何方式使用 Config Server 中的数据，这种情况下覆盖不可执行，但是如果它们是 Spring Cloud Config 客户端，上述配置是有效的。

9.3.4 安全保护

Config Server 中还可以配置 HTTP Basic 安全，拒绝非法用户的配置请求信息。Spring Security 结合 Spring Boot 可以轻松地完成这个需求。使用 Spring Boot 默认配置的 HTTP Basic 安全，只需要引入 Spring Security 的依赖，如下所示：

```
<dependency>
    <groupId>org.springframework.boot</groupId>
    <artifactId>spring-boot-starter-security</artifactId>
</dependency>
```

在 Config Server 中，增加安全的配置，如下所示：

```yaml
spring:
  security:
    user:
      name: aoho
      password: 123456
```

此时,如果客户端应用直接访问对应的 Config Server 将会报错,如下所示:

```
Caused by: org.springframework.web.client.HttpClientErrorException: 401 null
```

因此,客户端需要增加相应的身份验证信息才能正确获取其对应的配置信息,配置如下所示:

```yaml
spring:
  cloud:
    config:
      username: aoho
      password: 123456
```

访问配置服务器需要相应的输入用户名和密码,客户端配置好如上的 Credentials 信息就可以正常访问配置服务器。

9.3.5 加密解密

如果远端配置源包含了加密的内容(以 {cipher} 开头),在客户端 HTTP 请求之前,这些加密了的内容将会先被解密。这样做的好处是,配置仓库中的属性值不再是纯文本。如果一个值不能被解密,将会被移出属性源并且一个额外的属性值将会被加入到该键,只不过加上前缀 "invalid.",表示这个值不可用。这主要是为了防止密码文本被用作密码而意外泄漏。

1. JCE 环境安装

默认情况下 JRE 中自带了 JCE(Java Cryptography Extension),但是默认是一个有限长度的版本,我们这里需要一个不限长度的 JCE,可以从 Oracle 官网下载(如 jce8,http://www.oracle.com/technetwork/java/javase/downloads/jce8-download-2133166.html)。下载之后,解压的目录结构如下所示:

```
.
├── README.txt
├── US_export_policy.jar
└── local_policy.jar
```

上述链接下载解压后拷贝到 JDK/jre/lib/security 目录下覆盖文件。

2. 对称加密与解密

所谓对称,就是采用这种加密方法的双方使用同样的密钥进行加密和解密。密钥是控制加密及解密过程的指令。算法是一组规则,规定如何进行加密和解密。加密的安全性不仅取决于加密算法本身,密钥管理的安全性更是重要。因为加密和解密都使用同一个密钥,如何把密钥安全地传递到解密者手上就成了必须要解决的问题。

对称加解密方式比较简单，从定义可以知道，我们需要配置一个密钥。在 Config Server 中配置一个秘钥的方法如下所示：

```
encrypt:
    key: secret
```

启动 Config Server，通过其提供的多个端点验证我们的配置是否正确，加密解密是否能够生效。

- /encrypt/status 端点验证 Encryptor 的安装状态。如 curl -i "http://localhost:8888/encrypt/status"。
- /encrypt 端点提供了对字符串进行加密的功能，返回密文。如 curl -X POST -d "user" "http://localhost:8888/encrypt/"。
- /decrypt 端点提供了对加密后的密文进行解密的功能，返回明文。如 curl -X POST -d "9f034a63c87496b19f86ab80b4cb0b2f463d116cbb172df0b85286a179e3afb3" "http://localhost:8888/decrypt/"。

通过将需要加密的字符串进行替换，并加上前缀 {cipher}，Config Server 在获取到这个值之后会先对值进行解密，解密之后才会返回给客户端使用。

3. 非对称加密与解密

非对称加密算法需要两个密钥：公开密钥和私有密钥。公开密钥与私有密钥是一对，如果用公开密钥对数据进行加密，只有用对应的私有密钥才能解密；如果用私有密钥对数据进行加密，那么只有用对应的公开密钥才能解密。因为加密和解密使用的是两个不同的密钥，所以这种算法叫作非对称加密算法。

相比于对称加密算法，非对称加密算法更加安全。使用非对称加密，需要生成密钥对，JDK 中自带了 keytool 工具，执行如下命令：

```
keytool -genkeypair -alias config-server -keyalg RSA -keystore config-server.keystore
```

需要注意的是，设置秘钥口令，长度不能小于 6。执行完成之后，会在当前目录生成一个文件 server.keystore，拷贝到 Config Server 中的 src\main\resources 目录下，Config Server 进行如下配置：

```
encrypt:
    key-store:
        location: config-server.keystore
        alias: config-server
        password: 123456
        secret: 123456
```

我们可以使用上面对称算法中列出的端点进行验证，比如执行 curl -X POST -d "user" "http://localhost:8888/encrypt/"，得到的结果明显更加复杂了，如下所示：

```
AQAZL4yLLYh0CAEQKMPkg5WRvjb7Urz+7F2aeruGyG9WYCgKa1/D39DNmzrPgKmoBvCrUJT1a/O/ft8MY8
d1qB8qtlG86wOhopaoiFih1kLxMnqXNH/Q4/fI/b4muOBS+OF0ChodLPUjCtwTUN6KT6ZN/9fkrFI6PCiU
rHd8AZBX80LtpCoy4Ws6C20j/0Fpie6UPOn4Tdpzx1sHkFG/8itcJnWqOaNdM6FpOlKElOOIYbVdeGtEbr
Z0av3xEKUPmBdkFRTwM/7VwdIcPr1qwmsBGLYLVBVHZ0YfVUJpPgBEmaVD7b9WVMP/eyEInvaSCB75qGWa
qc1UVbKtS9U7KTL21mmlr1P9TMfobsG8vwHINfv+PKeOmfcoy47va/NkqHU
```

9.3.6 快速响应失败与重试机制

1. 快速响应失败

在某些情况下，当客户端不能连接到 Config Server 时，可能让服务启动失败更为适合。如果这种行为更为合适，设置启动配置的属性 spring.cloud.config.failFast=true，则客户端会遇到异常而终止。

2. 重试机制

当客户端应用启动时，Config Server 可能因为网络抖动等问题请求超时造成不可用，我们可以使用重试机制进行不断尝试。当然，这是建立在上面快速响应失败关闭的情况下，我们需要添加如下依赖：

```xml
<dependency>
    <groupId>org.springframework.retry</groupId>
    <artifactId>spring-retry</artifactId>
</dependency>
<dependency>
    <groupId>org.springframework.boot</groupId>
    <artifactId>spring-boot-starter-aop</artifactId>
</dependency>
```

spring-retry 默认会尝试 6 次，初始回退间隔为 1000 毫秒，后续回退的指数为 1.1。我们可以自定义这些配置，如下所示：

```yaml
spring:
  cloud:
    config:
      retry:
        max-attempts: 6
        multiplier: 1.1
        initial-interval: 1000
        max-interval: 2000
```

9.4 本章小结

微服务架构中，每个微服务的业务单位进一步细化，服务经过拆分，微服务的数量变得庞大。当这么多的服务需要测试并上产线，可能会存在多种环境，如集成测试、预发布和生产环境，每个环境的配置文件都会有所区别（如数据源的配置、开关量、路由规则等），如果没有一个集中的配置管理中心，大规模更新每个环境的配置信息就会非常麻烦，工作量大的同时也会容易造成人为的失误。Spring Cloud Config 基于云端存储配置信息，它具有中心化、版本控制、支持动态更新、平台独立、语言独立等特性。Spring Cloud Config 比较容易上手使用，同时也能够满足大部分公司的需求。

　　Spring Cloud Config 的精妙之处在于它的配置存储于 Git，这就天然地把配置的修改、权限、版本等问题隔离在外。通过这个设计使得 Spring Cloud Config 整体很简单，不过也带来了一些不便之处。笔者认为，Spring Cloud Config 的功能可以进一步增加和丰富，如：

- ❑ 配置界面：一个界面管理不同环境、不同集群配置。
- ❑ 实例配置监控：可以方便地看到当前哪些客户端在使用哪些配置。

Chapter 10 第 10 章

消息驱动：Spring Cloud Stream

Spring Cloud Stream 是 Spring Cloud 微服务框架中构建消息驱动能力的组件。Stream 可以进行基于消息队列的消息通信，它使用 Spring Integration 连接消息中间件以实现消息事件驱动。基于 Stream，开发者可以实现与消息队列相关的消息驱动型应用，Spring Cloud 的消息总线 Bus 就是基于 Stream 实现的。

本章的第一小节主要讲解了消息队列的概念；第二小节主要讲解了 Stream 的基础应用，给出相关的代码示例；第三小节则主要讲解了 Stream 框架的实现原理和源码分析；第四小节是 Stream 的进阶应用教程。

10.1 消息队列

消息队列中间件是分布式系统中最为重要的组件之一，主要用于解决应用耦合、异步消息和流量削锋等问题，是大型分布式系统不可缺少的中间件。消息队列技术是分布式应用间交换信息的一种技术，消息可驻留在内存或磁盘上，以队列形式存储消息，直到它们被应用程序消费。通过消息队列，应用程序可以相对独立地运行，它们不需要知道彼此的位置，只需要给消息队列发送消息并且处理从消息队列发送来的消息。

消息队列的主要特点是异步处理和解耦：其主要的使用场景就是将比较耗时而且不需要同步返回结果的操作作为消息放入消息队列，从而实现异步处理；同时由于使用了消息队列，只要保证消息格式不变，消息的发送方和接受者并不需要彼此直接联系，也不会受到对方的影响，即解耦。

消息队列常用组件

本节以 RabbitMQ 为例，介绍消息队列的常用组件和相关概念。RabbitMQ 中将整个消息队列应用分为下列几个组件：消息生产者（Producer）、交换器（Exchange）、队列（Queue）和消息消费者（Consumer）。

消息生产者就是生成消息并将消息发送给消息队列的应用，而消息消费者则是接受消息队列发送的消息并消费的应用。

图 10-1 是 RabbitMQ 内部用于存储消息的队列，每个队列都有唯一的名称（queue_name），用来让消息消费者进行订阅。生产者生产的消息会存储到队列中，消息消费者则从队列中获取并消费消息，三者的关系如图 10-2 所示。

图 10-1　队列

多个消息消费者可以订阅同一个队列，这时队列中的消息会分配给多个消息消费者进行处理，而不是每一个消息消费者都收到所有的消息如图 10-3 所示。

虽然消息生产者可以直接将消息发送到队列中，但是在 RabbitMQ 中，生产者发送给消息队列的所有消息都要先经过交换器，再由交换器将消息路由到一个或者多个队列中，图 10-4 展示了交换器和队列之间的关系。

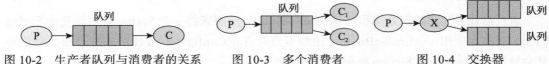

图 10-2　生产者队列与消费者的关系　　图 10-3　多个消费者　　图 10-4　交换器

在使用交换器时，需要将交换器和队列绑定（Binding）连接起来。在绑定的同时，一般会指定一个绑定键值（Binding key）；而生产者在将消息发送给交换器时，一般会指定一个路由键值（Routing key），用来指定这个消息的路由规则，而这个路由键值需要和交换器自身的类型（Type）以及绑定键值联合使用才能生效。

一般情况下当路由键值和绑定键值相匹配时，将消息路由到对应的队列中。但是绑定键值并不是所有情况都生效的，它依赖于交换器的类型。RabbitMQ 常用的交换器类型有 fanout、direct、topic 和 headers 四种。

- fanout 类型的 Exchange 的路由规则会把所有发送到该交换器的消息路由到所有与它绑定的队列中，可以将 fanout 类型简单地理解为广播类型。
- direct 类型的交换器路由规则会把消息路由到那些绑定键值和该消息的路由键值完全匹配的队列中。
- topic 类型的交换器在匹配规则上对 direct 进行了扩展，它与 direct 类型的交换器类似，也是将消息路由到绑定键值与该消息路由键值相匹配的队列中，但这里的匹配规则有些不同，它提供一些模糊匹配规则。
- headers 类型的交换器不依赖于路由键值与绑定键值的匹配规则来路由消息，而是根据发送的消息内容中的 headers 属性进行匹配。在绑定队列与交换器时指定一组键值对；当消息发送到交换器时，RabbitMQ 会取到该消息的 headers（也是一个键值对的形式），对比其中的键值对是否完全匹配队列与交换器绑定时指定的键值对；如果完全匹配则消息会路由到该队列，否则不会路由到该队列。

RabbitMQ 发送消息的流程为：首先获取网络连接，之后依次获取通道，定义交换器和队列。使用一个绑定键值将队列绑定到一个交换器上，通过指定交换器和路由键值将消息发送到对应的队列上，最后消费者在接收时也是先获取连接、通道。然后指定一个队列，从队列上获取消息，它对交换器、路由键值及如何绑定都不关心，从对应的队列上获取消息就可以。

10.2 基础应用

本小节主要讲述 Spring Cloud Stream 的编程模型。Spring Cloud Stream 提供了一系列预先定义的注解来声明输入型和输出型通道，业务系统基于这些通道与消息中间件进行通信，而不是直接与消息中间件进行通信。

10.2.1 声明和绑定通道

通过给业务应用的配置类添加 @EnableBinding 注解来将一个 Spring 应用转变成 Spring Cloud Stream 应用。@EnableBinding 注解本身拥有 @Configuration 注解来进行相关配置并且会触发 Spring Cloud Stream 框架的初始化机制。

@EnableBinding 注解可以设置声明输入型和输出型通道的接口类作为其 value 属性值。Spring Cloud Stream 提供了预先设置的三种接口来定义输入型通道和输出型通道，它们是 Source、Sink 和 Processor。Source 用来声明输出型通道，它的通道名称为 output；Sink 用来声明输入型通道，它的通道名称为 input；Processor 则用来声明输出输入型的通道。

10.2.2 自定义通道

使用 @Input 和 @Output 注解，编程人员可以给每个通道一个自定义的名称，使用这个自定义通道，可以与消息队列中相应的通道进行交互。

```
public interface MessageInput{
    @Input("input")
    SubscribableChannel input();
}
public interface MessageOutput{
    @Output("output")
    MessageChannel output();
}
```

如上述代码所示，MesageInput 和 MessageOutput 类分别声明了名称为 input 的输入型通道和名称为 output 的输出型通道。

10.2.3 接收消息

使用 Spring Cloud Stream 的 @StreamListener 注解可以进行消息的接收和消费。@StreamListener

注解基于 Spring Messaging 注解（比如说 @MessageMapping、@JmsListener、@RabbitListener），除此之外，该注解添加了内容（content）类型管理和类型强制等特性。

Spring Cloud Stream 提供了可扩展的消息转换（MessageConverter）机制来处理数据转换，并将转换后的数据分配给对应的被 @StreamListener 修饰的方法。下面这个例子展示了一个处理外部订单消息的应用：

```
@EnableBinding(Sink.class)
public class OrderHandler {
    @Autowired
    OrderService orderService;
    @StreamListener(Sink.INPUT)
    public void handle(Order order) {
        orderService.handle(order);
    }
}
```

假设，输入的 Message 对象有一个 String 类型的 payload 和一个值为 application/json 的 contentType。在使用 @StreamListener 时，MessageConverter 会使用消息的 contentType 来解析 String 类型的 payload 并赋值给 Order 对象。

就像其他的 Spring Messaging 方法一样，由 @StreamListener 注解的方法的参数可以使用 @Payload 和 @Headers 进行注解。对于返回数据的方法，必须使用 @SendTo 注解来指定该返回数据发送到哪个输出型通道，如下所示：

```
@EnableBinding(Processor.class)
public class TransformProcessor {
    @Autowired
    VotingService votingService;
    @StreamListener(Processor.INPUT)
    @SendTo(Processor.OUTPUT)
    public VoteResult handle(Vote vote) {
        return votingService.record(vote);
    }
}
```

Spring Cloud Stream 支持将消息分配到多个 @StreamListener 修饰的方法。为了能使用该分配机制，该方法不能有返回值，而且必须是单一类消息的处理方法。

使用注解的 condition 属性中的 SpEL 表达式可以设置 @StreamListener 接收消息的判断条件。所有匹配了该条件判断的方法都会在同一个线程中被调用，但是方法调用相对顺序不能保证。

下面就是一个 @StreamListener 根据头部属性分配消息的例子。在这个例子中，所有头部属性 type 对应值为 food 的消息都会分配给 receiveFoodOrder 方法，所有头部属性 type 对应值为 computer 的消息都会分配给 receiveComputeOrder 方法：

```
@EnableBinding(Sink.class)
@EnableAutoConfiguration
public static class TestPojoWithAnnotatedArguments {
```

```
@StreamListener(target = Sink.INPUT, condition = "headers['type']=='food'")
public void receiveFoodOrder(@Payload FoodOrder foodOrder) {
    // handle the message
}
@StreamListener(target = Sink.INPUT, condition = "headers['type']=='computer'")
public void receiveComputeOrder(@Payload ComputeOrder computeOrder) {
    // handle the message
}
```

10.2.4 配置

开发人员在 application.yml 中设置 Stream 的相关配置，可以 bindings 下面声明通道的绑定信息，并且指定进行通道绑定的绑定器，最后使用 binders 来声明使用的绑定器。下面的配置就声明了名为 output 和 input 的两个 binding 和一个名为 rabbit1 的绑定器：

```yaml
// application.yml
cloud:
  stream:
    bindings:
      output:
        content-type: application/x-java-object;type=com.example.demo.entity.Message
        destination: msg
        binder: rabbit1
      input:
        content-type: application/x-java-object;type=com.example.demo.entity.Message
        destination: msg
        binder: rabbit1
    binders:
      rabbit1:
        type: rabbit
        environment:
          spring:
            rabbitmq:
              host: 127.0.0.1
              port: 5672
              username: stream
              password: stream
              virtual-host: /sc
```

10.3 源码分析

Stream 首先会动态注册相关 BeanDefinition，并且处理 @StreamListener 注解；然后在 Bean 实例初始化之后，会调用 BindingService 的相关方法进行服务绑定；BindingService 在绑定服务时会首先获取特定的 Binder 绑定器，然后绑定生产者和消费者；最后 Stream 的相关实例就会进行发送和接受消息的处理，如图 10-5 所示。

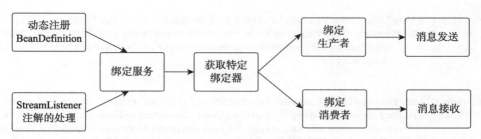

图 10-5　Stream 源码流程图

10.3.1　动态注册 BeanDefinition

@EnableBinding 注解是 Stream 框架开始生效运转的起点，它会将消息通道绑定到自己修饰的目标实例上，从而让这些实例具备与消息队列进行消息交互的能力。

@EnableBinding 使用 @Import 注解导入了三个类，分别是：BindingServiceConfiguration、BindingBeansRegistrar 和 BinderFactoryConfiguration。EnableBinding 注解的具体实现如下所示：

```
@Import({ BindingServiceConfiguration.class, BindingBeansRegistrar.class,
    BinderFactoryConfiguration.class,
        SpelExpressionConverterConfiguration.class })
@EnableIntegration
public @interface EnableBinding {
    Class<?>[] value() default {};
}
```

BindingBeansRegistrar 实现了 ImportBeanDefinitionRegistrar 接口，用于向 Spring 容器动态注册 BeanDefinition。而 BindingBeansRegistrar 的作用是注册声明通道的接口类的 BeanDefinition，从而能够获取这些接口类的实例，并使用这些实例进行消息的发送和接收。

registerBeanDefinitions 方法要动态注册 @EnableBinding 注解声明的接口类的 BeanDefinition，比如说基础应用中的 MessageOutput。它会遍历 @EnableBinding 的 value 属性值的类数组，然后依次调用 BindingBeanDefinitionRegistryUtils 的 registerBindingTargetBeanDefinitions 和 registerBindingTargetsQualifiedBeanDefinitions 方法进行 BeanDefinition 的动态注册，如下所示：

```
//BindingBeansRegistrar.java
public void registerBeanDefinitions(AnnotationMetadata metadata,
        BeanDefinitionRegistry registry) {
    ...
    //collectClass获取EnableBinding注解的value值，也就是Sink,Output或者自定义bound通道
      的接口的Class
    //比如说MessageOutput和MessageInput接口的Class对象
    //因为这些接口的实例需要自动装配，所以这里必须给出这些实例的定义（BeanDefinition）
    //这里的getClassName获得的类信息是使用@EnableBinding注解的类信息
    for (Class<?> type : collectClasses(attrs, metadata.getClassName())) {
        BindingBeanDefinitionRegistryUtils.
```

```
registerBindingTargetBeanDefinitions(type,type.getName(), registry);
        BindingBeanDefinitionRegistryUtils.registerBindingTargetsQualifiedBeanDefinitions(
            ClassUtils.resolveClassName(metadata.getClassName(), null), type,
            registry);
    }
}
private Class<?>[] collectClasses(AnnotationAttributes attrs, String className) {
    EnableBinding enableBinding = AnnotationUtils.synthesizeAnnotation(attrs,
        EnableBinding.class, ClassUtils.resolveClassName(className, null));
    // 通过 EnableBinding 注解获取其值，比如说 InputController 类使用了 @EnableBinding 注解，
      其 value 值
    // 为 MessageOutput，这里返回的 Class 数组就由 MessageOutput.class 组成。
    return enableBinding.value();
}
```

registerBindingTargetBeanDefinitions 方法会调用 ReflectionUtils 的 doWithMethods 方法来扫描接口类中的所有方法，然后为被 @Input 和 @Output 注解修饰的函数注册其返回值 SubscribableChannel 或者 MessageChannel 的 BeanDefinition，如下所示：

```
// BindingBeanDefinitionRegistryUtils.java
public static void registerBindingTargetBeanDefinitions(Class<?> type, final String
    bindingTargetInterfaceBeanName,
        final BeanDefinitionRegistry registry) {
    //该方法通过Reflection来处理MessageOutput和MessageInput类中的@Input和@Output注解的函数
    //因为这些函数需要返回MessageChannel或者SubscribableChannel实例
    ReflectionUtils.doWithMethods(type, new MethodCallback() {
        @Override
        public void doWith(Method method) throws IllegalArgumentException,
            IllegalAccessException {
            Input input = AnnotationUtils.findAnnotation(method, Input.class);
            if (input != null) {
                String name = getBindingTargetName(input, method);
                registerInputBindingTargetBeanDefinition(input.value(), name,
                    bindingTargetInterfaceBeanName,
                    method.getName(), registry);
            }
            Output output = AnnotationUtils.findAnnotation(method, Output.class);
            if (output != null) {
                String name = getBindingTargetName(output, method);
                registerOutputBindingTargetBeanDefinition(output.value(), name,
                    bindingTargetInterfaceBeanName,
                    method.getName(), registry);
            }
        }
    });
}
```

@Input 和 @Output 注解修饰的函数返回值 SubscribableChannel，或者 MessageChannel 的 BeanDefinition 的注册由 registerBindingTargetBeanDefinition 完成，使用 RootBeanDefinition 的 setFactoryBeanName 和 setUniqueFactoryMethodName 方法，将该 Bean 的实例化交给工厂类的对应函数完成。registerBindingTargetBeanDefinition 方法的具体实现如下所示：

```java
// BindingBeanDefinitionRegistryUtils.java
private static void registerBindingTargetBeanDefinition(Class<? extends Annotation>
    qualifier,
        String qualifierValue, String name, String bindingTargetInterfaceBeanName,
        String bindingTargetInterfaceMethodName, BeanDefinitionRegistry registry)
{
    RootBeanDefinition rootBeanDefinition = new RootBeanDefinition();
        //该Bean的实例由下面两行注册的FactoryBean和相应的factorymethod生成。
    rootBeanDefinition.setFactoryBeanName(bindingTargetInterfaceBeanName);
    rootBeanDefinition.setUniqueFactoryMethodName(bindingTargetInterfaceMethodName);
    rootBeanDefinition.addQualifier(new AutowireCandidateQualifier(qualifier, qualifierValue));
    registry.registerBeanDefinition(name, rootBeanDefinition);
}
```

如上述源码和图 10-6 所示，bindingTargetInterfaceBeanName 的值是声明通道的接口类的类名，而 bindingTargetInterfaceMethodName 的值对应函数的名称。Stream 会为每个声明通道的接口类注册一个对应的工厂类，并且每个接口类中的函数都会有一个对应的工厂函数，所以这些函数的返回对象由对应的工厂函数来生成。

```
▶ ⓟ parent = {Class@3125} "class com.example.demo.OutputController" … Navigate
▶ ⓟ type = {Class@3097} "interface com.example.demo.message.MessageOutput" … Navigate
▶ ⓟ registry = {DefaultListableBeanFactory@3036} "org.springframework.beans.factory.support.DefaultListableBeanFactory@56113384:
```

图 10-6　registerBindingTargetBeanDefinition 调试

registerBindingTargetsQualifiedBeanDefinitions 是在注册 registerBindingTargetBeanDefinition 时使用到的工厂类 BeanDefinition，这个工厂类用来生成 registerBindingTargetBeanDefinition 注册的 Bean 的实例，如下所示：

```java
// BindingBeanDefinitionRegistryUtils.java
public static void registerBindingTargetsQualifiedBeanDefinitions(Class<?> parent,
    Class<?> type,
        final BeanDefinitionRegistry registry) {
    ...
    RootBeanDefinition rootBeanDefinition = new RootBeanDefinition(BindableProxy
        Factory.class);
    rootBeanDefinition.addQualifier(new AutowireCandidateQualifier(Bindings.class,
        parent));
    rootBeanDefinition.getConstructorArgumentValues().addGenericArgumentValue(type);
    //type.getName和registerBindingTargetBeanDefinition中的bindingTargetInterfaceBeanName相同
    registry.registerBeanDefinition(type.getName(), rootBeanDefinition);
    ...
}
```

如图 10-7 所示，Stream 将 BindableProxyFactory 注册成了名称为 com.example.demo.message.MessageOutput 的 FactoryBean。通过 BindableProxyFactory，Spring 容器可以获取到 registerBindingTargetBeanDefinitions 方法中所注册的 Bean 实例。比如说 MessageOutput 中 outputMessagefunction 方法返回的 MessageChannel 实例。

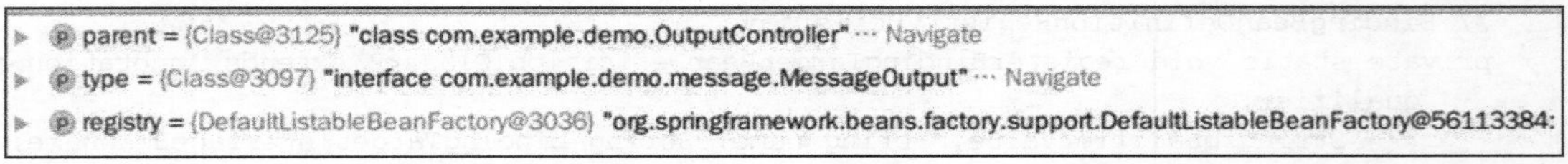

```
▶ ⓟ parent = {Class@3125} "class com.example.demo.OutputController" ... Navigate
▶ ⓟ type = {Class@3097} "interface com.example.demo.message.MessageOutput" ... Navigate
▶ ⓟ registry = {DefaultListableBeanFactory@3036} "org.springframework.beans.factory.support.DefaultListableBeanFactory@56113384:
```

图 10-7　registerBindingTargetsQualifiedBeanDefinitions 调试图

10.3.2　绑定服务

BindableProxyFactory 是 Spring Cloud Stream 比较核心的一个类，它最为重要的一项职责是声明通道接口类中相关类型的实例。

```
//BindableProxyFactory.java
public void afterPropertiesSet() throws Exception {
    //afterPropertiesSet是InitializingBean的接口，在实现的bean的所有属性都被设置之后会被
      BeanFactory调用
    ReflectionUtils.doWithMethods(this.type, new ReflectionUtils.MethodCallback() {
        @Override
        public void doWith(Method method) throws IllegalArgumentException {
            Output output = AnnotationUtils.findAnnotation(method, Output.class);
            //通过ReflectionUtils来处理所有被Output注解注释的method,比如说MessageOutput
              中的outputMessagefunction方法。
            String name = BindingBeanDefinitionRegistryUtils.
                getBindingTargetName(output, method);
            Class<?> returnType = method.getReturnType();
            //name就是Output注解的value值，returnType是该method的返回值类型
            BindableProxyFactory.this.outputHolders.put(name, new BoundTarget
                Holder(getBindingTargetFactory(returnType)
                .createOutput(name), true));
            ...
        }
    });
}
```

上述代码的作用是处理所有由 @Output 注解注释的函数，生成函数返回值类型的实例存储在 BoundTargetHolder 变量中供后续使用。getBindingTargetFactory 方法会返回 SubscribableChannelBindingTargetFactory 实例，它会在 createOutput 方法中返回一个 DirectChannel 实例。该实例会存储起来，供 BindableProxyFactory 之后使用。

在之前的代码中，名称为 output 的 BeanDefinition 将 BindableProxyFactory 设置成其实例工厂类（BeanFactory），并将 outputMessagefunction 方法设置成其实例的工厂函数（BeanFactoryMethod）。所以当 Spring 容器创建该实例时，会调用 BindableProxyFactory 的 outputMessagefunction 方法，由于 BindableProxyFactory 实现了 MethodInterceptor 接口，所以就调用到了其 invoke 方法。BindableProxyFactory 的 invoke 方法的具体实现如下所示：

```
//BindableProxyFactory.java
@Override
```

```java
public synchronized Object invoke(MethodInvocation invocation) throws Throwable {
    Method method = invocation.getMethod();
    //首先利用缓存返回
    Object boundTarget = targetCache.get(method);
    if (boundTarget != null) {
        return boundTarget;
    }
    ... //input相关的处理
    Output output = AnnotationUtils.findAnnotation(method, Output.class);
    if (output != null) {
        String name = BindingBeanDefinitionRegistryUtils.getBindingTargetName
            (output, method);
        //直接使用之前afterPropertiesSet生成的outputHolders返回相应的channel对象
        boundTarget = this.outputHolders.get(name).getBoundTarget();
        targetCache.put(method, boundTarget);
        return boundTarget;
    }
    return null;
}
```

BindingService 用于将通道通过绑定器 Binder 与相应的消息队列进行绑定，绑定器有多种实现，比如说 spring-cloud-stream-binder-rabbit 和 spring-cloud-stream-binder-kafka。在这个过程中涉及多个类的协作，包括 OutputBindingLifecycle 和 BindableProxyFactory 等。

SmartLifecycle 是 Spring 框架中处理 Bean 实例生命周期的接口。其 start 方法加载 BindingService 实例并且调用之前注册的 BindableProxyFactory 实例的 bindOutputs 方法。OutputBindingLifecycle 将调用名为 MessageOutput 和 MessageInput 这两个 BindableProxyFactory 实例的 bindOutput 方法。start 方法的具体实现如下所示：

```java
//OutputBindingLifecycle.java
public void start() {
    ...
    BindingService bindingService = this.applicationContext
            .getBean(BindingService.class);
    //BindableProxyFactory就是Bindable.class类型的Bean,所以这里就是将所有BindableProxyFactory
    //   的实例都获取到，然后分别调用bindOutputs
    Map<String, Bindable> bindables = this.applicationContext
            .getBeansOfType(Bindable.class);
    for (Bindable bindable : bindables.values()) {
        bindable.bindOutputs(bindingService);
    }
    ...
}
```

图 10-8 展示的是示例代码中运行到 start 方法的调试信息，bindables 数组中包含对应定义通道的接口类的 BindableProxyFactory 实例，比如说 MessageInput。bindOutputs 方法用于将之前 afterPropertiesSet 方法调用时生成的 SubscribableChannel 和 MessageChannel 实例交给 BindingService 进行绑定。

```
▶ ≡ this = {OutputBindingLifecycle@5608}
▶ ≡ bindingService = {BindingService@5611}
▼ ≡ bindables = {LinkedHashMap@5612} size = 3
   ▶ ≡ 0 = {LinkedHashMap$Entry@5617} "dynamicDestinationsBindable" ->
   ▶ ≡ 1 = {LinkedHashMap$Entry@5618} "&com.example.demo.message.MessageInput" ->
   ▼ ≡ 2 = {LinkedHashMap$Entry@5619} "&com.example.demo.message.MessageOutput" ->
      ▶ ≡ key = "&com.example.demo.message.MessageOutput"
      ▶ ≡ value = {BindableProxyFactory@5625}
  ⚐ this.running = false
  ⚐ this.applicationContext = {AnnotationConfigEmbeddedWebApplicationContext@3039} "org.springframework.boot.context.embedded.AnnotationConfigEmbeddedWe...
```

图 10-8 OutputLifecycle 的 start 方法调试图

10.3.3 获取绑定器

在探索 BindingService 的 bindProducer 方法之前，我们先来看一下 BindingService 的构造过程，因为其中涉及了 application.yml 文件配置信息的读取和 spring-cloud-stream-binder-rabbit 等绑定器（Binder）的加载，如下所示：

```java
//BindingService.java
public <T> Binding<T> bindProducer(T output, String outputName) {
    String bindingTarget = this.bindingServiceProperties
            .getBindingDestination(outputName);
    //根据outputName和output的class信息来获取绑定器
    Binder<T, ?, ProducerProperties> binder = (Binder<T, ?, ProducerProperties>)
            getBinder(
            outputName, output.getClass());
    ProducerProperties producerProperties = this.bindingServiceProperties
            .getProducerProperties(outputName);
    ...
    //通过获取的绑定器来绑定具体的通道和消息队列
    Binding<T> binding = binder.bindProducer(bindingTarget, output,
            producerProperties);
    this.producerBindings.put(outputName, binding);
    return binding;
}
private <T> Binder<T, ?, ?> getBinder(String channelName, Class<T> bindableType) {
    String binderConfigurationName = this.bindingServiceProperties.getBinder
        (channelName);
    return binderFactory.getBinder(binderConfigurationName, bindableType);
}
```

由上述代码我们可以看到，通过 DefaultBinderFactory 的 getBinder 方法获取具体的 Binder 来对相应的 MessageChannel 实例进行绑定。

BinderFactoryConfiguration 是一个配置类，它会初始化两个重要的 Bean 实例，分别是 BinderTypeRegistry 和 DefaultBinderFactory。BinderTypeRegistry 负责通过名称来管理 BinderType；而 DefaultBinderFactory 负责获取具体的 Binder。接下来，我们先来看一下 BinderTypeRegistry 的初始化，它和自定义消息队列 Binder 有密切关系。BinderFactoryConfiguration 的相关配置如下所示：

```java
//BinderFactoryConfiguration.java
@Bean
@ConditionalOnMissingBean(BinderTypeRegistry.class)
public BinderTypeRegistry binderTypeRegistry(ConfigurableApplicationContext
    configurableApplicationContext) {
    Map<String, BinderType> binderTypes = new HashMap<>();
    ClassLoader classLoader = configurableApplicationContext.getClassLoader();
    //先获取ClassLoader，通过ClassLoader来获取META-INF下的spring.binders配置文件
    if (classLoader == null) {
        classLoader = BinderFactoryConfiguration.class.getClassLoader();
    }
    //getResources方法会将所有jar包中的META-INF/spring.binders文件的URL都获取到，然后由
      parseBinderConfigurations方法来读取这些文件中的配置
    Enumeration<URL> resources = classLoader.getResources("META-INF/spring.
        binders");
    while (resources.hasMoreElements()) {
        URL url = resources.nextElement();
        UrlResource resource = new UrlResource(url);
        for (BinderType binderType : parseBinderConfigurations(classLoader, resource)) {
            binderTypes.put(binderType.getDefaultName(), binderType);
        }
    }
    //将所有获取的BinderType对象都存放在DefaultBinderTypeRegistry对象中
    return new DefaultBinderTypeRegistry(binderTypes);
}
```

parseBinderConfigurations 方法主要是解析 META-INF/spring.binders 文件中有关 Binder 的配置，生成 BinderType 对象。BinderType 代表着 Binder 接口的具体实现，如下所示：

```java
//BinderFactoryConfiguration.java
static Collection<BinderType> parseBinderConfigurations(ClassLoader classLoader,
    Resource resource)
        throws IOException, ClassNotFoundException {
    //从spring.binders文件中读取配置，一般都是对应消息队列Binder的配置类，比如Rabbit的绑定
      器依赖包就提供了org.springframework.cloud.stream.binder.rabbit.config.
      RabbitServiceAutoConfiguration类
    Properties properties = PropertiesLoaderUtils.loadProperties(resource);
    Collection<BinderType> parsedBinderConfigurations = new ArrayList<>();
    for (Map.Entry<?, ?> entry : properties.entrySet()) {
        String binderType = (String) entry.getKey();
        String[] binderConfigurationClassNames = StringUtils
            .commaDelimitedListToStringArray((String) entry.getValue());
        Class<?>[] binderConfigurationClasses = new Class[binderConfigurationCla
            ssNames.length];
        int i = 0;
        for (String binderConfigurationClassName : binderConfigurationClassNames) {
            //通过反射技术获取到对应BinderConfiguration的类，然后封装成BinderType对象
            binderConfigurationClasses[i++] = ClassUtils.forName(binderConfigura
                tionClassName, classLoader);
        }
        parsedBinderConfigurations.add(new BinderType(binderType, binderConfigurationClasses));
    }
    return parsedBinderConfigurations;
```

}

使用 RabbitMQ 的 Binder 会引入名为 spring-cloud-starter-stream-rabbitmq 的依赖，其 META-INF/spring.binders 文件的内容如下所示。BinderType 就是将 RabbitServiceAutoConfiguration 类信息封装起来，供 Stream 获取 Binder 实例时使用。

```
rabbit:\
org.springframework.cloud.stream.binder.rabbit.config.RabbitServiceAutoConfiguration
```

DefaultBinderFactory 的初始化依赖于 BinderTypeRegistry 获得的 BinderType 列表。DefaultBinderFactory 的初始化如下所示：

```java
//BinderFactoryConfiguration.java
@Bean
@ConditionalOnMissingBean(BinderFactory.class)
public DefaultBinderFactory binderFactory() {
    DefaultBinderFactory binderFactory = new DefaultBinderFactory(getBinderConfi
        gurations());
    binderFactory.setDefaultBinder(bindingServiceProperties.getDefaultBinder());
    binderFactory.setListeners(binderFactoryListeners);
    return binderFactory;
}
```

getBinderConfigurations 方法将是将 bindingServiceProperties 变量中的 BinderProperties 与 BinderTypeRegistry 中的 BinderType 进行结合，封装成 BinderConfiguration 对象。BinderProperties 封装了 Stream 从 application.yml 文件中读取的关于 Binder 的配置信息，而 BinderType 则是具体 Binder 的实现类信息。getBinderConfigurations 方法的具体实现如下所示：

```java
//BinderFactoryConfiguration.java
public Map<String, BinderConfiguration> getBinderConfigurations() {
    Map<String, BinderConfiguration> binderConfigurations = new HashMap<>();
    Map<String, BinderProperties> declaredBinders = this.bindingServiceProperties.
        getBinders();
    boolean defaultCandidatesExist = false;
    Iterator<Map.Entry<String, BinderProperties>> binderPropertiesIterator =
        declaredBinders.entrySet().iterator();
    ...
    List<String> existingBinderConfigurations = new ArrayList<>();
    for (Map.Entry<String, BinderProperties> binderEntry : declaredBinders.entrySet()) {
        BinderProperties binderProperties = binderEntry.getValue();
        if (this.binderTypeRegistry.get(binderEntry.getKey()) != null) {
            //关键点是通过binderTypeRegistry获取BinderType
            binderConfigurations.put(binderEntry.getKey(),
                new BinderConfiguration(this.binderTypeRegistry.get(binderEntry.
                    getKey()),
                        binderProperties.getEnvironment(), binderProperties.
                            isInheritEnvironment(),
                        binderProperties.isDefaultCandidate()));
            existingBinderConfigurations.add(binderEntry.getKey());
```

```
            }
            ...
        }
        if (!defaultCandidatesExist) {
            for (Map.Entry<String, BinderType> binderEntry : this.binderTypeRegistry.
                getAll().entrySet()) {
                if (!existingBinderConfigurations.contains(binderEntry.getKey())) {
                    binderConfigurations.put(binderEntry.getKey(), new BinderConfiguration
                        (binderEntry.getValue(),
                            new Properties(), true, true));
                }
            }
        }
        return binderConfigurations;
    }
```

通过上述的代码的执行，可以获得如图 10-9 所示的 binderConfigurations。这些信息都是从示例代码的 application.yml 文件中获取的。需要注意的是 BinderConfiguration 的 configurationClass 属性值是 RabbitServiceAutoConfiguration 类，该类正是上述 META-INF/spring.binders 文件中的配置类。

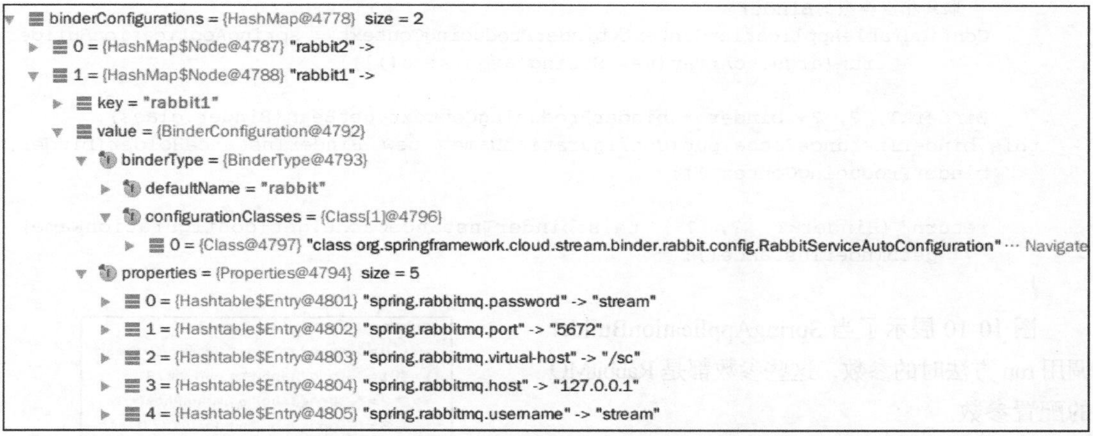

```
▼ ▤ binderConfigurations = {HashMap@4778} size = 2
  ▶ ▤ 0 = {HashMap$Node@4787} "rabbit2" ->
  ▼ ▤ 1 = {HashMap$Node@4788} "rabbit1" ->
    ▶ ▤ key = "rabbit1"
    ▼ ▤ value = {BinderConfiguration@4792}
      ▼ ▤ binderType = {BinderType@4793}
        ▶ ▤ defaultName = "rabbit"
        ▼ ▤ configurationClasses = {Class[1]@4796}
          ▶ ▤ 0 = {Class@4797} "class org.springframework.cloud.stream.binder.rabbit.config.RabbitServiceAutoConfiguration" ... Navigate
      ▼ ▤ properties = {Properties@4794} size = 5
        ▶ ▤ 0 = {Hashtable$Entry@4801} "spring.rabbitmq.password" -> "stream"
        ▶ ▤ 1 = {Hashtable$Entry@4802} "spring.rabbitmq.port" -> "5672"
        ▶ ▤ 2 = {Hashtable$Entry@4803} "spring.rabbitmq.virtual-host" -> "/sc"
        ▶ ▤ 3 = {Hashtable$Entry@4804} "spring.rabbitmq.host" -> "127.0.0.1"
        ▶ ▤ 4 = {Hashtable$Entry@4805} "spring.rabbitmq.username" -> "stream"
```

图 10-9　BinderConfigurations 中的数据

到了这里，再来回过头来看一下 DefaultBinderFactory 的 getBinderInstance 方法。我们将看到 Stream 是如何使用 binderConfiguration 中的信息来获取 Binder 实例的。这其中涉及的原理对大家了解 Spring Bean 的初始化机制很有帮助。

getBinderInstance 方法中会生成一个 ConfigurableApplicationContext 上下文实例来创建 Binder 实例。在创建 ConfigurableApplicationContext 实例时，会将 BinderConfiguration 设置到 SpringApplicationBuilder 中，从而可以在 ConfigurableApplicationContext 调用 getBinder 方法时，使用到 BinderConfiguration 的属性和配置，生成 BinderConfiguration 中设置的具体类型的 Binder 实现，具体实现如下所示：

```java
//DefaultBinderFactory.java
private <T> Binder<T, ?, ?> getBinderInstance(String configurationName) {
    BinderConfiguration binderConfiguration = this.binderConfigurations.get
        (configurationName);
    Properties binderProperties = binderConfiguration.getProperties();
    ArrayList<String> args = new ArrayList<>();
    for (Map.Entry<Object, Object> property : binderProperties.entrySet()) {
        args.add(String.format("--%s=%s", property.getKey(), property.getValue()));
    }
    ...
    args.add("--spring.jmx.default-domain=" + defaultDomain + "binder." +
        configurationName);
      List<Class<?>> configurationClasses = new ArrayList<Class<?>>(
          Arrays.asList(binderConfiguration.getBinderType()
.getConfigurationClasses()));
        //这里将configurationClass设置到SpringApplicationBuilder,即RabbitServiceAutoConfiguration类
      SpringApplicationBuilder springApplicationBuilder = new SpringApplicationBuilder()
.sources(configurationClasses
.toArray(new Class<?>[]
{})).bannerMode(Mode.OFF).web(false);
      springApplicationBuilder.parent(this.context);
        //上边的这些代码都是为了初始化一个ConfigurableApplicationContext对象,这个对象可以获
          取到相应类型的Binder
      ConfigurableApplicationContext binderProducingContext = springApplicationBuilder
            .run(args.toArray(new String[args.size()]));

    Binder<T, ?, ?> binder = binderProducingContext.getBean(Binder.class);
this.binderInstanceCache.put(configurationName, new BinderInstanceHolder(binder,
binderProducingContext));

    return (Binder<T, ?, ?>) this.binderInstanceCache.get(configurationName).
        getBinderInstance();
}
```

图 10-10 展示了当 SpringApplicationBuilder 调用 run 方法时的参数,这些参数都是 RabbitMQ 的配置参数。

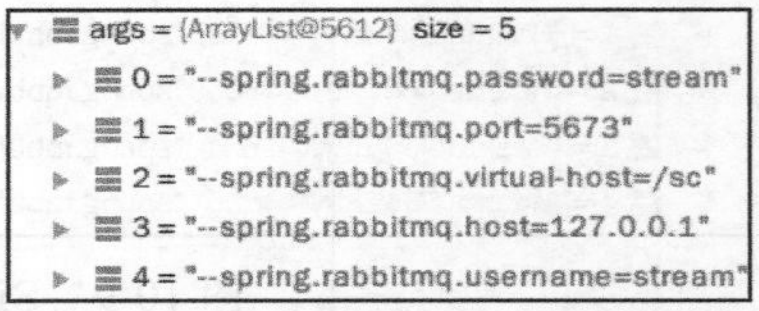

RabbitServiceAutoConfiguration 是 一个配置类,并且引入了 RabbitMessageChannelBinderConfiguration 这个配置类,在该配置类中会初始化 RabbitMessageChannelBinder 的实例。下述代码中生成的 RabbitMessageChannelBinder 就是上述代码中 getBean 方法获取的 Binder 对象:

图 10-10 SpringApplicationBuilder run 时的参数

```java
//RabbitMessageChannelBinderConfiguration.java
@Bean
RabbitMessageChannelBinder rabbitMessageChannelBinder() {
    RabbitMessageChannelBinder binder = new RabbitMessageChannelBinder(rabbitCon
        nectionFactory, rabbitProperties,
            provisioningProvider());
    binder.setCodec(codec);
```

```
        binder.setAdminAddresses(rabbitBinderConfigurationProperties.getAdminAddresses());
        binder.setCompressingPostProcessor(gZipPostProcessor());
        binder.setDecompressingPostProcessor(deCompressingPostProcessor());
        binder.setNodes(rabbitBinderConfigurationProperties.getNodes());
        binder.setExtendedBindingProperties(rabbitExtendedBindingProperties);
        return binder;
    }
```

10.3.4 绑定生产者

了解了 Binder 实例的初始化之后，再回到 BindingService 的 bindProducer 方法，该函数会调用 Binder 的 bindProducer 方法，而 bindProducer 方法会先调用 AbstractMessageChannelBinder 的 doBindProducer 方法。

doBindProducer 做了三件事情，首先调用 ProvisioningProvider 的 provisionProducerDestination 方法来创建 ProducerDestination；然后调用 createProducerMessageHandler 方法来创建 MessageHandler；最后调用 SubscribableChannel 的 subscribe 方法来绑定 MessageHandler。doBindProducer 方法的实现如下所示：

```java
//AbstractMessageChannelBinder.java
public final Binding<MessageChannel> doBindProducer(final String destination,
    MessageChannel outputChannel,
        final P producerProperties) throws BinderException {
    Assert.isInstanceOf(SubscribableChannel.class, outputChannel,
            "Binding is supported only for SubscribableChannel instances");
    final MessageHandler producerMessageHandler;
    final ProducerDestination producerDestination;
    //调用provisioningProvider来生成ProducerDestination，
    //比如RabbitMQ的就是RabbitExchangeQueueProvisioner
    producerDestination = this.provisioningProvider.provisionProducerDestination
        (destination, producerProperties);
    //createProducerMessageHandler由每个子类实现
    producerMessageHandler = createProducerMessageHandler(producerDestination,
        producerProperties);
    //将outputChannel和Messagehandler绑定起来，而MessageHandler又和producerDestination
        有联系，producerDestination和
    //具体的RabiitMQ的交换器或者队列有联系
    ((SubscribableChannel) outputChannel).subscribe(
            new SendingHandler(producerMessageHandler, !this.supportsHeadersNatively
                && HeaderMode.embeddedHeaders
                    .equals(producerProperties.getHeaderMode()), this.headersToEmbed,
                    producerProperties.isUseNativeEncoding())));
    return new DefaultBinding<MessageChannel>();
}
```

ProvisioningProvider 的 provisionProducerDestination 方法会创建一个 ProducerDestination 实例。下面的代码展示了 RabbitExchangeQueueProvisioner 的具体实现，在这个方法中就涉及了 RabbitMQ 的 API 的一些操作，例如初始化交换器和队列，代码如下所示：

```
//ProvisioningProvider.java
public ProducerDestination provisionProducerDestination(String name, ExtendedProducerProperties
    <RabbitProducerProperties> producerProperties) {
    final String exchangeName = applyPrefix(producerProperties.getExtension().
        getPrefix(), name);
    Exchange exchange = buildExchange(producerProperties.getExtension(),
        exchangeName);
    if (producerProperties.getExtension().isDeclareExchange()) {
        //调用rabbitadmin来声明exchange
        declareExchange(exchangeName, exchange);
    }
    Binding binding = null;
    for (String requiredGroupName : producerProperties.getRequiredGroups()) {
        String baseQueueName = producerProperties.getExtension().isQueueNameGroupOnly()
                ? requiredGroupName : (exchangeName + "." + requiredGroupName);
        if (!producerProperties.isPartitioned()) {
            //创建队列,declareQueue和declareExchange一样,会调用RabbitMQ的原生API
            Queue queue = new Queue(baseQueueName, true, false, false,
                queueArgs(baseQueueName, producerProperties.getExtension(), false));
            declareQueue(baseQueueName, queue);
            autoBindDLQ(baseQueueName, baseQueueName, producerProperties.
                getExtension());
            if (producerProperties.getExtension().isBindQueue()) {
                binding = notPartitionedBinding(exchange, queue, producerProperties.
                    getExtension());
            }
        }
        ...
    }
    //生成ProducerDestination
    return new RabbitProducerDestination(exchange, binding);
}
```

declareExchange 方法最终会调用 RabbitAdmin 的 declareExchanges 方法来声明 Exchange。RabbitAdmin 调用的是 RabbitMQ Client 的原生 API 接口⊖，代码如下所示：

```
//ProvisioningProvider.java
private void declareExchanges(final Channel channel, final Exchange... exchanges)
throws IOException {
    for (final Exchange exchange : exchanges) {
        if (!isDeclaringDefaultExchange(exchange)) {
            try {
                Map<String, Object> arguments = exchange.getArguments();
                ...
                channel.exchangeDeclare(exchange.getName(), DELAYED_MESSAGE_
                    EXCHANGE, exchange.isDurable(),
                        exchange.isAutoDelete(), exchange.isInternal(), arguments);
            }
            catch (IOException e) {
                logOrRethrowDeclarationException(exchange, "exchange", e);
```

⊖ RabbitMQ API 的文档请查阅 https://www.rabbitmq.com/getstarted.html

```
            }
          }
        }
}
```

doBindProducer 方法做的第二件事情是调用 createProducerMessageHandler 方法创建 MessageHandler 实例。AmqpOutboundEndpoint 是 MessageHandler 的子类，用来实现与 RabbitMQ 的交互。createProducerMessageHandler 方法的具体实现如下所示：

```
//AbstractMessageChannelBinder.java
protected MessageHandler createProducerMessageHandler(final ProducerDestination
    producerDestination,
        ExtendedProducerProperties<RabbitProducerProperties> producerProperties,
            MessageChannel errorChannel) {
    String prefix = producerProperties.getExtension().getPrefix();
    String exchangeName = producerDestination.getName();
    String destination = StringUtils.isEmpty(prefix) ? exchangeName : exchangeName.
        substring(prefix.length());
    final AmqpOutboundEndpoint endpoint = new AmqpOutboundEndpoint(
            buildRabbitTemplate(producerProperties.getExtension(), errorChannel
                != null));
    endpoint.setExchangeName(producerDestination.getName());
    RabbitProducerProperties extendedProperties = producerProperties.getExtension();
    boolean expressionInterceptorNeeded = expressionInterceptorNeeded(extendedPr
        operties);

    endpoint.setDefaultDeliveryMode(extendedProperties.getDeliveryMode());
    endpoint.setBeanFactory(this.getBeanFactory());
    endpoint.afterPropertiesSet();
    return endpoint;
}
```

第三件事情就是最后调用 SubscribableChannel 的 subscribe 方法来绑定 MessageHandler，到此绑定生产者的流程结束。

10.3.5　消息发送的流程

Stream 框架通过 MessageChannel 发送消息，MessageChannel 首先将消息发送给通过 doBindProducer 创建（参见 10.3.4 节）并绑定的 SendingHandler 对象，然后交由 MessageHandler 做发送前的准备工作，最后由 RabbitTemplate 调用 RabbitMQ 客户端的 Channel 实例将消息发送给 RabbitMQ 的服务端。如图 10-11 所示。

图 10-11　消息发送流程图

SendingHandler 是 AbstractMessageHandler 的子类，MessageChannel 有消息时，会将消息分发给 SendingHandeer 的 handleMessageInternal 方法，然后再由 SendingHandler 转发给对应的 MessageHandler。handleMessageInternal 方法的具体实现如下所示：

```java
//SendingHandler.java
protected void handleMessageInternal(Message<?> message) throws Exception {
    Message<?> messageToSend = this.useNativeEncoding ? message : this.serialize
        AndEmbedHeadersIfApplicable(message);
    this.delegate.handleMessage(messageToSend);
}
```

delegate 就是之前 createProducerMessageHandler 方法生成的 MessageHandler 实例，也就是 AmqpOutboundEndpoint 对象。其 handleMessage 方法最终会调用该类的 handleRequestMessage 方法。

handleRequestMessage 方法会生成 exchangeName 和 routingKey，然后根据是否需要返回值，来调用 sendAndReceive 方法或则 send 方法，如下所示：

```java
//AmqpOutboundEndpoint.java
protected Object handleRequestMessage(Message<?> requestMessage) {
    CorrelationData correlationData = this.generateCorrelationData(requestMessage);
    String exchangeName = this.generateExchangeName(requestMessage);
    String routingKey = this.generateRoutingKey(requestMessage);
    if (this.expectReply) {
        return this.sendAndReceive(exchangeName, routingKey, requestMessage,
            correlationData);
    } else {
        this.send(exchangeName, routingKey, requestMessage, correlationData);
        return null;
    }
}
```

send 方法则是首先使用 MessageConverter 来转换消息对象，然后调用 RabbitTemplate 对象的 send 方法来发送请求，如下所示：

```java
//AmqpOutboundEndpoint.java
private void send(String exchangeName, String routingKey, Message<?> requestMessage,
    CorrelationData correlationData) {
    MessageConverter converter = ((RabbitTemplate)this.amqpTemplate).getMessageConverter();
    org.springframework.amqp.core.Message amqpMessage = MappingUtils.
        mapMessage(requestMessage, converter, this.getHeaderMapper(), this.
        getDefaultDeliveryMode(), this.isHeadersMappedLast());
    this.addDelayProperty(requestMessage, amqpMessage);
    ((RabbitTemplate)this.amqpTemplate).send(exchangeName, routingKey, amqpMessage,
        correlationData);
}
```

RabbitTemplate 是简化与 RabbitMQ 消息交互的工具类，它封装了 RabbitmMQ 的 Client 端原生 API 接口。其 send 方法最终会调用 doSend 方法。RabbitTemplate 的 doSend 方法则会调用 sendToRabbit 方法，进行真正的消息发送。sendToRabbit 调用了 RabbitMQ

中 Channel 对象的原生 API 接口 basicPublish 方法，将消息发送给 RabbitMQ 消息队列，如下所示：

```java
//RabbitTemplate.java
protected void sendToRabbit(Channel channel, String exchange, String routingKey,
    boolean mandatory,
        Message message) throws IOException {
    BasicProperties convertedMessageProperties = this.messagePropertiesConverter
            .fromMessageProperties(message.getMessageProperties(), this.encoding);
    //调用了RabbitMQ的原生API，使用channel来发布消息
    channel.basicPublish(exchange, routingKey, mandatory, convertedMessageProperties,
        message.getBody());
}
```

10.3.6　StreamListener 注解的处理

前面的小节主要介绍了输出型通道初始化和消息发送的过程，下面我们就来看一看输入型通道和消息接收的过程。

如果使用者想要使用 Stream 来接收并处理一个消息队列发送过来的信息，可以通过下列的代码实现：

```java
@EnableBinding(Sink.class)
@EnableAutoConfiguration
public static class TestController{
    @StreamListener(target = Sink.INPUT, condition = "headers['type']=='food'")
    public void receiveFoodOrder(@Payload FoodOrder foodOrder) {
        // handle the message
    }
    @StreamListener(target = Sink.INPUT, condition = "headers['type']=='compute'")
    public void receiveComputeOrder(@Payload ComputeOrder computeOrder) {
        // handle the message
    }
}
```

@StreamListener 是一个可以修饰函数的注解，被它修饰的函数会用来接收输入型通道的消息，下面就来探究一下 Spring Cloud Stream 是如何将消息队列 MessageInput.INPUT_MESSAGE 这个通道传来的消息分配到对应的函数。

Stream 定义了 StreamListenerAnnotationBeanPostProcessor 类用来处理项目中的注解 @StreamListener，StreamListenerAnnotationBeanPostProcessor 实现了 BeanPostProcessor 接口，用来在 Bean 初始化之前和之后两个时间点对 Bean 实例进行处理。postProcessAfterInitialization 是在 Bean 实例初始化之后被调用的方法，它会遍历 Bean 实例中的所有函数，处理那些被 @StreamListener 注解修饰的函数，如下所示：

```java
//StreamListenerAnnotationBeanPostProcessor.java
public final Object postProcessAfterInitialization(final Object bean, final
    String beanName) throws BeansException {
    Class<?> targetClass = AopUtils.isAopProxy(bean) ? AopUtils.getTargetClass
        (bean) : bean.getClass();
```

```java
//获取到该类对象所有方法
Method[] uniqueDeclaredMethods = ReflectionUtils.getUniqueDeclaredMethods(ta
    rgetClass);
for (Method method : uniqueDeclaredMethods) {
    StreamListener streamListener = AnnotatedElementUtils.findMergedAnnotation(method,
            StreamListener.class);
    //判断该method是否被@StreamListener修饰
    if (streamListener != null && !method.isBridge()) {
        //获取streamListener的基本信息
        streamListener = postProcessAnnotation(streamListener, method);
        String methodAnnotatedInboundName = streamListener.value();
        String methodAnnotatedOutboundName = StreamListenerMethodUtils.getOu
            tboundBindingTargetName(method);
        int inputAnnotationCount = StreamListenerMethodUtils.inputAnnotationCount
            (method);
        int outputAnnotationCount = StreamListenerMethodUtils.outputAnnotationCount
            (method);
        boolean isDeclarative = checkDeclarativeMethod(method, methodAnnotatedInboundName,
                methodAnnotatedOutboundName);
        StreamListenerMethodUtils.validateStreamListenerMethod(method,
            inputAnnotationCount,
                outputAnnotationCount, methodAnnotatedInboundName,
                methodAnnotatedOutboundName,
                isDeclarative, streamListener.condition());
        //将method,streamListener的信息暂时存储
        registerHandlerMethodOnListenedChannel(method, streamListener, bean);
    }
}
return bean;
}
```

如下代码所示，registerHandlerMethodOnListenedChannel 方法会将 Method 和 @StreamListener 注解的相关信息组装成 StreamListenerHandlerMethodMapping 对象存储到 StreamListenerAnnotationBeanPostProcessor 的 mappedListenerMethods 对象中，供接下来统一处理时使用：

```java
//StreamListenerAnnotationBeanPostProcessor.java
protected final void registerHandlerMethodOnListenedChannel(Method method,
    StreamListener streamListener,
        Object bean) {
    final String defaultOutputChannel = StreamListenerMethodUtils.getOutboundBin
        dingTargetName(method);
    StreamListenerMethodUtils.validateStreamListenerMessageHandler(method);
    mappedListenerMethods.add(streamListener.value(),
            new StreamListenerHandlerMethodMapping(bean, method, streamListener.
                condition(), defaultOutputChannel,
                streamListener.copyHeaders()));
}
```

StreamListenerAnnotationBeanPostProcessor 还实现了 SmartInitializingSingleton 接口，Spring 容器会在单例类型的实例创建结束时调用所有 SmartInitializingSingleton 接口

实现类的 afterSingletonsInstantiated 方法。StreamListenerAnnotationBeanPostProcessor 在 afterSingletonsInstantiated 方法中，将之前的 postProcessAfterInitialization 方法收集到的有关 @StreamListener 的信息进行处理。

afterSingletonsInstantiated 方法会遍历 mappedListenerMethods 对象的所有 entry，为每一个 StreamListenerHandlerMethodMapping 创建一个 StreamListenerMessageHandler 实例，然后根据是否需要条件处理，生成 DispatchingStreamListenerMessageHandler 的 ConditionalStreamListenerHandler 实例。接着生成一个 DispatchingStreamListenerMessageHandler 实例，将之前生成的所有 ConditionalStreamListenerHandler 实例传给它，最后根据 mappedBindingEntry 的 key 值，也就是之前 @StreamListener 的 value 值来获取 SubscribableChannel 实例，并调用其 subscribe 方法，将 DispatchingStreamListenerMessageHandler 实例注册给 SubscribableChannel。afterSingletonsInstantiated 方法的具体实现如下所示：

```java
//StreamListenerAnnotationBeanPostProcessor.java
public final void afterSingletonsInstantiated() {
    for (Map.Entry<String, List<StreamListenerHandlerMethodMapping>> mappedBindingEntry:
        mappedListenerMethods
            .entrySet()) {
        Collection<DispatchingStreamListenerMessageHandler.ConditionalStreamList
            enerHandler> handlers = new ArrayList<>();
        for (StreamListenerHandlerMethodMapping mapping : mappedBindingEntry.
            getValue()) {
            //使用函数AOP，为了将Message对象转化成函数输入参数的类型。按照上面例子的情况，也
                就是将Message转化为FooPojo或者BarPojo
            final InvocableHandlerMethod invocableHandlerMethod = this.
                messageHandlerMethodFactory
                    .createInvocableHandlerMethod(mapping.getTargetBean(),
                        checkProxy(mapping.getMethod(), mapping.getTargetBean()));
            StreamListenerMessageHandler streamListenerMessageHandler = new
                StreamListenerMessageHandler(
                invocableHandlerMethod, resolveExpressionAsBoolean(mapping.
                    getCopyHeaders(), "copyHeaders"),
                    springIntegrationProperties.getMessageHandlerNotPropagatedHe
                        aders());
            //处理streamlistener的condition参数问题，通过SPEL解释器来解析condition的
                value值，然后新建一个ConditionalStreamListenerHandler
            if (StringUtils.hasText(mapping.getCondition())) {
                String conditionAsString = resolveExpressionAsString(mapping.
                    getCondition(), "condition");
                Expression condition = SPEL_EXPRESSION_PARSER.parseExpression
                    (conditionAsString);
                handlers.add(new DispatchingStreamListenerMessageHandler.Conditi
                    onalStreamListenerHandler(
                    condition, streamListenerMessageHandler));
            }
        }
        DispatchingStreamListenerMessageHandler handler = new DispatchingStreamL
            istenerMessageHandler(
            handlers, this.evaluationContext);
```

```
        //把之前生成的DispatchingStreamListenerMessageHandler对象注册到SubscribableChannel
            对象上。
        applicationContext.getBean(mappedBindingEntry.getKey(), SubscribableChannel.
            class).subscribe(handler);
    }
    this.mappedListenerMethods.clear();
}
```

ApplicationContext 的 getBean 方法获取 SubscribableChannel 实例时, 就会用到 BindingBeansRegistrar 中 registerBindingTargetBeanDefinitions 方法注册的 BeanDefinition, 最后由 BindableProxyFactory 来生成 SubscribableChannel 实例。

当 SubscribableChannel 接收到消息时, 会调用 DispatchingStreamListenerMessageHandler 的 handleRequestMessage 方法。该方法会调用 findMatchingHandlers 方法获取所有适合处理这个 Message 的 ConditionalStreamListenerHandler 实例, 然后调用每个 ConditionalStreamListenerHandler 的 handleMessage 方法。handleRequestMessage 方法的实现如下所示:

```
//DispatchingStreamListenerMessageHandler.java
protected Object handleRequestMessage(Message<?> requestMessage) {
    Collection<ConditionalStreamListenerHandler> matchingHandlers = findMatching
        Handlers(requestMessage);
    if (matchingHandlers.size() == 0) {
        return null;
    }
    else if (matchingHandlers.size() > 1) {
        for (ConditionalStreamListenerHandler matchingMethod : matchingHandlers) {
            matchingMethod.handleMessage(requestMessage);
        }
        return null;
    }
    else {
        final ConditionalStreamListenerHandler singleMatchingHandler = matchingHandlers.
            iterator().next();
        singleMatchingHandler.handleMessage(requestMessage);
        return null;
    }
}
```

如下代码所示, findMatchingHandlers 方法主要是根据 ConditionalStreamListenerHandler 的条件判断的 Expression 实例来判断 ConditionalStreamListenerHandler 是否适合处理当前这个 Message。如果 ConditionalStreamListenerHandler 的 getCondition 方法返回的 Expression 实例为 null, 那么认为该 ConditionalStreamListenerHandler 适合处理该 Message; 如果 getCondition 方法返回的 Expression 实例不为 null, 则调用其 getValue 方法来判断是否适合处理该 Message。

```
//DispatchingStreamListenerMessageHandler.java
private Collection<ConditionalStreamListenerHandler> findMatchingHandlers(Message<?>
    message) {
    ArrayList<ConditionalStreamListenerHandler> matchingMethods = new ArrayList<>();
```

```
        for (ConditionalStreamListenerHandler conditionalStreamListenerHandlerMethod:
            this.handlerMethods) {
          if (conditionalStreamListenerHandlerMethod.getCondition() == null) {
             matchingMethods.add(conditionalStreamListenerHandlerMethod);
          }
          else {
             boolean conditionMetOnMessage = conditionalStreamListenerHandlerMethod.
                getCondition().getValue(
                    this.evaluationContext, message, Boolean.class);
             if (conditionMetOnMessage) {
                matchingMethods.add(conditionalStreamListenerHandlerMethod);
             }
          }
        }
        return matchingMethods;
    }
```

其中，ConditionalStreamListenerHandler 的 handleMessage 方法就是直接调用它的 StreamListenerMessageHandler 成员变量的 handleMessage 方法，如下所示：

```
//ConditionalStreamListenerHandler.java
@Override
public void handleMessage(Message<?> message) throws MessagingException {
    this.streamListenerMessageHandler.handleMessage(message);
}
```

StreamListenerMessageHandler 是 AbstractReplyProducingMessageHandler 的子类，其 handleMessage 方法最终会调用自己声明的 handleRequestMessage 方法。

handleRequestMessage 方法调用了 InvocableHandlerMethod 的 invoke 方法来调用对应的由 @StreamListener 注解修饰的方法，如下所示：

```
//StreamListenerMessageHandler.java
@Override
protected Object handleRequestMessage(Message<?> requestMessage) {
    try {
        return this.invocableHandlerMethod.invoke(requestMessage);
    }
    catch (Exception e) {
        if (e instanceof MessagingException) {
            throw (MessagingException) e;
        }
        else {
            throw new MessagingException(requestMessage,
                "Exception thrown while invoking " + this.invocableHandlerMethod.
                    getShortLogMessage(), e);
        }
    }
}
```

其中，InvocableHandlerMethod 对象是在 StreamListenerAnnotationBeanPostProcessor 的 afterSingletonsInstantiated 方法中创建并赋予给 StreamListenerMessageHandler 的。

InvocableHandlerMethod 对象包含了由 @StreamListener 修饰的函数和对应的 Bean 实例，如下所示：

```
//StreamListenerMessageHandler.java
final InvocableHandlerMethod invocableHandlerMethod = this.messageHandlerMethodFactory
    .createInvocableHandlerMethod(mapping.getTargetBean(), checkProxy(mapping.
        getMethod(), mapping.getTargetBean())); 
```

图 10-12 展示的是 Message 对象传递的流程，从 SubscribableChannel 接收到 RabbitMQ 传递过来的 Message 对象后，会调用 DispatchingStreamListenerMessageHandler 进行消息分发，这个处理器会根据 ContitionStreamListenerHandler 的条件判断来传递消息，最终消息经过 InvocableHandlerMethod 传递给 TestController 的对应函数。

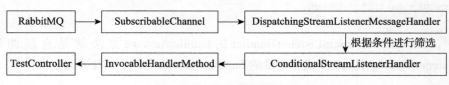

图 10-12　消息接收流程图

下一小节会讲解 RabbitMQ 和 SubscribableChannel 之间的消息传递原理，也就是 RabbitMessageChannelBinder 的 doBindConsumer 方法的原理。

10.3.7　绑定消费者

RabbitMessageChannelBinder 的 doBindConsumer 方法会将输入型通道所对应的 SubscribableChannel 和具体的消息队列实体相连。doBindConsumer 方法主要做了三件事情，一是调用 ProvisioningProvider 的 provisionConsumerDestination 方法来创建 ConsumerDestination，二是调用 createConsumerEndpoint 方法创建 MessageProducer 实例，三是生成 DefaultBinding 实例，如下所示：

```
//RabbitMessageChannelBinder.java
@Override
public final Binding<MessageChannel> doBindConsumer(String name, String group,
    MessageChannel inputChannel,
        final C properties) throws BinderException {
    MessageProducer consumerEndpoint = null;
    try {
        ConsumerDestination destination = this.provisioningProvider.provisionCon
            sumerDestination(name, group, properties);
        consumerEndpoint = createConsumerEndpoint(destination, group, properties);
        if (consumerEndpoint instanceof Lifecycle) {
            ((Lifecycle) consumerEndpoint).start();
        }
        return new DefaultBinding<MessageChannel>(name, group, inputChannel,
            consumerEndpoint instanceof Lifecycle ? (Lifecycle) consumerEndpoint :
                null) {
            @Override
```

```
            protected void afterUnbind() {
                ...
            }
        };
    }
    catch (Exception e) {
        ...
    }
}
```

RabbitExchangeQueueProvisioner 的 provisionConsumerDestination 方法主要是创建并声明消息队列的 Exchange 和 Queue 实例，然后将二者进行绑定。provisionConsumerDestination 方法的具体实现如下所示：

```
//RabbitExchangeQueueProvisioner.java
@Override
public ConsumerDestination provisionConsumerDestination(String name, String group,
        ExtendedConsumerProperties<RabbitConsumerProperties> properties) {
    boolean anonymous = !StringUtils.hasText(group);

    String  baseQueueName = anonymous ? groupedName(name, ANONYMOUS_GROUP_NAME_
        GENERATOR.generateName())
            : properties.getExtension().isQueueNameGroupOnly() ? group : groupedName
                (name, group);

    String prefix = properties.getExtension().getPrefix();
    //创建并声明Exchange
    final String exchangeName = applyPrefix(prefix, name);
    Exchange exchange = buildExchange(properties.getExtension(), exchangeName);
    if (properties.getExtension().isDeclareExchange()) {
        declareExchange(exchangeName, exchange);
    }
    ...
    String queueName = applyPrefix(prefix, baseQueueName);
    //创建被声明Queue
    Queue queue = new Queue(queueName, false, true, true, queueArgs(queueName,
        properties.getExtension(), false));
    declareQueue(queueName, queue);

    //绑定Exchange和Queue
    Binding binding = declareConsumerBindings(name, properties, exchange, partitioned,
        queue);
    return new RabbitConsumerDestination(queue, binding);
}
```

buildExchange 方法使用 ExchangeBuilder 来构建 Exchange 实例。Exchange 的全限定名称为 org.springframework.amqp.core.Exchange，它是 Spring 框架对于 AMQP 协议中 Exchange 组件的接口，是对所有基于 AMQP 协议的消息队列的 Exchange 对象的抽象。所以 Stream 在创建出 Exchange 实例之后，还需要调用 declareExchange 方法在消息队列服务器端声明该 Exchange，代码如下所示：

```
//RabbitExchangeQueueProvisioner.java
private Exchange buildExchange(RabbitCommonProperties properties, String exchangeName) {
    try {
        ExchangeBuilder builder = new ExchangeBuilder(exchangeName, properties.
            getExchangeType());
        builder.durable(properties.isExchangeDurable());
        if (properties.isExchangeAutoDelete()) {
            builder.autoDelete();
        }
        if (properties.isDelayedExchange()) {
            builder.delayed();
        }
        return builder.build();
    }
    catch (Exception e) {
        throw new ProvisioningException("Failed to create exchange object", e);
    }
}
```

declareExchange 方法会调用 RabbitAdmin 与消息队列服务端进行交互，在消息队列中创建 Exchange 组件，然后将 Exchange 实例添加到 GenericApplicationContext 上下文的实例列表中，如下所示：

```
//RabbitExchangeQueueProvisioner.java
private void declareExchange(final String rootName, final Exchange exchange) {
    this.rabbitAdmin.declareExchange(exchange);
    addToAutoDeclareContext(rootName + ".exchange", exchange);
}
```

RabbitAdmin 的 declareExchange 方法调用了 RabbitTemplate 实例的 execute 方法，并在回调方法中调用了自身的 declareExchanges 方法，如下所示：

```
//RabbitAdmin.java
public void declareExchange(final Exchange exchange) {
    try {
        this.rabbitTemplate.execute(new ChannelCallback<Object>() {
            @Override
            public Object doInRabbit(Channel channel) throws Exception {
                declareExchanges(channel, exchange);
                return null;
            }
        });
    }
    catch (AmqpException e) {
        logOrRethrowDeclarationException(exchange, "exchange", e);
    }
}
```

RabbitAdmin 的 declareExchanges 方法直接调用 RabbitMQ 客户端提供的 API，向 RabbitMQ 服务端注册一个 Exchange 组件，其具体实现如下所示：

```
//RabbitAdmin.java
```

```java
private void declareExchanges(final Channel channel, final Exchange... exchanges)
    throws IOException {
    for (final Exchange exchange : exchanges) {
        try {
            ...
            channel.exchangeDeclare(exchange.getName(), exchange.getType(),
                exchange.isDurable(),
                exchange.isAutoDelete(), exchange.isInternal(), exchange.getArguments());
        }
        catch (IOException e) {
            logOrRethrowDeclarationException(exchange, "exchange", e);
        }
    }
}
```

provisionConsumerDestination 方法中对队列和绑定的操作类似于 Exchange，最终都是通过调用 RabbitTemplate 的相应方法，使用 RabbitMQ 的 API 来实现的。

doBindConsumer 做的第二件事情就是调用 createConsumerEndpoint 方法来创建 MessageProducer 实例。createConsumerEndpoint 方法会首先创建 SimpleMessageListenerContainer 实例，然后在其基础上再创建 AmqpInboundChannelAdapter 实例并返回，如下所示：

```java
//RabbitMessageChannelBinder.java
@Override
protected MessageProducer createConsumerEndpoint(ConsumerDestination consumerDestination,
    String group,
        ExtendedConsumerProperties<RabbitConsumerProperties> properties) {
    String destination = consumerDestination.getName();
    SimpleMessageListenerContainer listenerContainer = new SimpleMessageListenerContainer(
            this.connectionFactory);
    ...
    listenerContainer.setQueueNames(consumerDestination.getName());
    listenerContainer.afterPropertiesSet();
    AmqpInboundChannelAdapter adapter = new AmqpInboundChannelAdapter(listenerCo
        ntainer);
    ...
    return adapter;
}
```

doBindConsumer 方法会判断 AmqpInboundChannelAdapter 是否实现了 Lifecycle 接口，如果是，则调用其 start 方法，开始启动监听消息的流程。Lifecycle 是定义 start 和 stop 生命周期管理方法的通用接口，它的典型使用场景是异步处理。AmqpInboundChannelAdapter 的 doStart 方法如下所示。

```java
//AmqpInboundChannelAdapter.java
protected void doStart() {
    //messageListenerContainer 是 AbstractMessageListenerContainer 类型的,
    'SimpleMessageListenerContainer' 是 'AbstractMessageListenerContainer' 的一个子类,
        最终会调用到其 doStart() 方法。
    this.messageListenerContainer.start();
}
```

如上所示，AmqpInboundChannelAdapter 的 doStart 方法十分简单，直接调用了 SimpleMessageListenerContainer 的 start 方法。

SimpleMessageListenerContainer 的 doStart 方法通常会调用 initializeConsumers 方法来初始化 BlockingQueueConsumer 实例，然后，在 BlockingQueueConsumer 的基础上创建 AsyncMessageProcessingConsumer 实例，最后使用 Java 线程池的 Executor 实例执行 AsyncMessageProcessingConsumer，如下所示：

```java
//SimpleMessageListenerContainer.java
@Override
protected void doStart() throws Exception {
    ...
    super.doStart();
    synchronized (this.consumersMonitor) {
        //初始化BlockingQueueConsumer对象，该对象调用RabbitMQ原生接口，与RabbitMQ直接交互
        int newConsumers = initializeConsumers();
        Set<AsyncMessageProcessingConsumer> processors = new HashSet<AsyncMessag
            eProcessingConsumer>();
        for (BlockingQueueConsumer consumer : this.consumers) {
            AsyncMessageProcessingConsumer processor = new AsyncMessageProcessin
                gConsumer(consumer);
            processors.add(processor);
            getTaskExecutor().execute(processor);
        }
        ...
    }
}
```

initializeConsumers 方法会依据成员变量 concurrentConsumers 的值来创建相应个数的 BlockingQueueConsumer 实例。concurrentConsumers 的值默认为 1，Stream 推荐当队列中的消息过多需要对消费端进行扩容时将该值增大，从而创建更多的 BlockingQueueConsumer 实例来接受消息，但是这样会影响到有序消息的处理，initializeConsumers 的代码如下所示：

```java
//SimpleMessageListenerContainer.java
protected int initializeConsumers() {
    int count = 0;
    synchronized (this.consumersMonitor) {
        if (this.consumers == null) {
            this.cancellationLock.reset();
            this.consumers = new HashSet<BlockingQueueConsumer>(this.concurrentConsumers);
            //根据concurrentConsumers成员变量的值来生成相应数量的BlockingQueueConsumer实例
            for (int i = 0; i < this.concurrentConsumers; i++) {
                BlockingQueueConsumer consumer = createBlockingQueueConsumer();
                this.consumers.add(consumer);
                count++;
            }
        }
    }
    return count;
}
```

AsyncMessageProcessingConsumer 是 SimpleMessageListenerContainer 的内部类，实现了 Runnable 接口。在它的 run 方法中会调用 BlockingQueueConsumer 的 start 方法。

BlockingQueueConsumer 封装了 RabbitMQ 的 Client API，它在 start 方法中首先通过 RabbitResourceHolder 实例获取 RabbitMQ 的 Channel 实例，然后为该通道初始化 InternalConsumer 实例，最后调用 consumeFromQueue 方法使用 InternalConsumer 订阅消息队列中的 Queue。BlockingQueueConsumer 的 start 方法如下所示：

```java
//BlockingQueueConsumer.java
public void start() throws AmqpException {
    this.thread = Thread.currentThread();
    this.resourceHolder = ConnectionFactoryUtils.getTransactionalResourceHolder(
        this.connectionFactory,
            this.transactional);
    this.channel = this.resourceHolder.getChannel();
    this.consumer = new InternalConsumer(this.channel);
    ...
    try {
        if (!cancelled()) {
            for (String queueName : this.queues) {
                if (!this.missingQueues.contains(queueName)) {
                    consumeFromQueue(queueName);
                }
            }
        }
    }
    catch (IOException e) {
        throw RabbitExceptionTranslator.convertRabbitAccessException(e);
    }
}
```

consumeFromQueue 方法将调用 Channel 的 basicConsume 方法来让 InternalConsumer 订阅对应的 Queue，如下所示：

```java
//BlockingQueueConsumer.java
private void consumeFromQueue(String queue) throws IOException {
    //RabbitMQ的原生API，监听quque队列，如果有消息就回调InternalConsumer的对应方法。
    String consumerTag = this.channel.basicConsume(queue, this.acknowledgeMode.
        isAutoAck(),
    (this.tagStrategy != null ? this.tagStrategy.createConsumerTag(queue) : ""), this.
        noLocal, this.exclusive,
            this.consumerArgs, this.consumer);
    if (consumerTag != null) {
        this.consumerTags.put(consumerTag, queue);
    }
}
```

至此，从 RabbitMQ 服务端到 SubscribableChannel 的整个链路便已经绑定成功，就等着消息队列将接收到的消息发送过来。

10.3.8 消息的接收

图 10-13 展示了消息从 RabbitMQ 服务端到 SubscribableChannel 的流转过程，首先接收到消息的是 InternalConsumer。

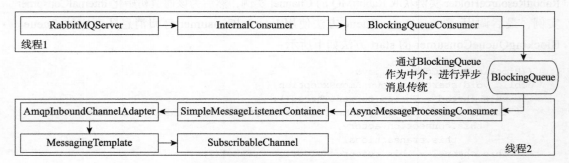

图 10-13　从 RabbitMQ 到 SubscribableChannel 的消息传递图

InternalConsumer 继承了 RabbitMQ Client API 中的 DefaultConsumer，当 RabbitMQ 对应的队列接收到消息之后，就会调用它的 handleDelivery 方法进行处理。

因为 InternalConsumer 是 BlockingQueueConsumer 的内部类，所以在 handleDelivery 方法中将 RabbitMQ 的消息封装成 Delivery 对象，并将其添加到 BlockingQueueConsumer 的成员变量 queue 队列中。handleDelivery 方法的定义如下所示：

```java
//InternalConsumer.java
public void handleDelivery(String consumerTag, Envelope envelope, AMQP.
    BasicProperties properties, byte[] body)
        throws IOException {
    try {
        ...
        BlockingQueueConsumer.this.queue.put(new Delivery(consumerTag, envelope,
            properties, body));
    }
    catch (InterruptedException e) {
        Thread.currentThread().interrupt();
    }
}
```

BlockingQueueConsumer 的成员变量 queue 是一个 BlockingQueue 类型的阻塞队列，用来暂时存储消息队列发送过来的消息。SimpleMessageListenerContainer 会启动一个 Runnable 来轮询调用 BlockingQueueConsumer 的 nextMessage 方法，处理阻塞队列中的消息。该 Runnable 就是 SimpleMessageListenerContainer 的内部类 AsyncMessageProcessingConsumer。AsyncMessageProcessingConsumer 的具体定义如下：

```java
//SimpleMessageListenerContainer.java
private final class AsyncMessageProcessingConsumer implements Runnable {
    ...
    @Override
```

```
public void run() {
    ...
    try {
        this.consumer.start();
        while (isActive(this.consumer) || this.consumer.hasDelivery() ||
            !this.consumer.cancelled()) {
                boolean receivedOk = receiveAndExecute(this.consumer);  // At
                    least one message received
        }
    }
    catch (InterruptedException e) {
        ...
    }
    ...
}
```

如下代码所示,doReceiveAndExecute 方法会调用 BlockingQueueConsumer 的 nextMessage 方法来获取一个 Message 实例,并且调用 executeListener 方法将该消息传递下去:

```
//SimpleMessageListenerContainer.java AsyncMessageProcessingConsumer
private boolean doReceiveAndExecute(BlockingQueueConsumer consumer) throws
    Throwable { //NOSONAR
    Channel channel = consumer.getChannel();
    for (int i = 0; i < this.txSize; i++) {
        //调用BlockingQueueConsumer的nextMessage方法,从其内部的阻塞队列queue中取出一个
            消息对象
        Message message = consumer.nextMessage(this.receiveTimeout);
        if (message == null) {
            break;
        }
        try {
            executeListener(channel, message);
        }
        ...
    }
    return consumer.commitIfNecessary(isChannelLocallyTransacted());
}
```

而 executeListener 方法最终会将消息传递给 AmqpInboundChannelAdapter。它的 sendMessage 方法会调用 MessagingTemplate 对象的 send 方法将消息发送给对应的 SubscribableChannel,其具体实现如下所示:

```
//AmqpInboundChannelAdapter.java
protected void sendMessage(Message<?> message) {
    if (message == null) {
        throw new MessagingException("cannot send a null message");
    } else {
        this.messagingTemplate.send(this.getOutputChannel(), message);
    }
    //....有代码删除
}
```

MessagingTemplate 会将对应的消息发送给 SubscribableChannel,然后 SubscribableChannel

会调用 DispatchingStreamListenerMessageHandler 将消息进行分配，最终将消息传递给对应的由 @StreamListener 修饰的方法，如下所示：

```java
//DispatchingStreamListenerMessageHandler.java
protected final void doSend(MessageChannel channel, Message<?> message, long timeout) {
    Message<?> messageToSend = message;
    MessageHeaderAccessor accessor = MessageHeaderAccessor.getAccessor(message,
        MessageHeaderAccessor.class);
    if (accessor != null && accessor.isMutable()) {
        accessor.removeHeader(this.sendTimeoutHeader);
        accessor.removeHeader(this.receiveTimeoutHeader);
        accessor.setImmutable();
    } else if (message.getHeaders().containsKey(this.sendTimeoutHeader) ||
        message.getHeaders().containsKey(this.receiveTimeoutHeader)) {
        messageToSend = MessageBuilder.fromMessage(message).setHeader(this.
            sendTimeoutHeader, (Object)null).setHeader(this.receiveTimeoutHeader,
            (Object)null).build();
    }

    boolean sent = timeout >= 0L ? channel.send(messageToSend, timeout) :
        channel.send(messageToSend);
}
```

如上述代码所示，doSend 方法会先使用 MessageHeaderAccessor 设置 Message 的头部信息，然后再调用 MessageChannel 的 send 方法将消息发送给该通道。

10.4 进阶应用

10.4.1 Binder For RocketMQ

Spring Cloud Stream 为接入不同的消息队列提供了一整套的自定义机制，通过为每个消息队列开发一个 Binder 来接入该消息队列。目前官方认定的 Binder 为 RabbitMQ binder 和 Kafka binder。但是开发人员可以基于 Stream Binder 的机制来制定自己的 Binder。下面我们就构建一个简单的 RocketMQ 的 Binder。

按照源码分析中的步骤，首先在 resources/META-INF/spring.binders 文件中配置有关 RocketMQ 的 Configuration 类，该配置类会使用 @Import 来导入为 RocketMQ 定制的 RocketMessageChannelBinderConfiguration，如下所示：

```
rocket:\
org.springframework.cloud.stream.binder.rocket.config.RocketServiceAutoConfiguration
```

RocketMessageChannelBinderConfiguration 将会提供两个极其重要的 Bean 实例，分别为 RocketMessageChannelBinder 和 RocketExchangeQueueProvisioner。RocketMessageChannelBinder 主要用于通道和消息队列的绑定，而 RocketExchangeQueueProvisioner 封装了 RocketMQ 的相关 API，可以用于创建消息队列的基础组件。

RocketMessageChannelBinderConfiguration 的代码如下所示：

```
@Configuration
public class RocketMessageChannelBinderConfiguration {
    @Autowired
    private ConnectionFactory rocketConnectionFactory;
    @Autowired
    private RocketProperties  rocketProperties;
    @Bean
    RocketMessageChannelBinder rocketMessageChannelBinder() throws Exception {
        RocketMessageChannelBinder binder = new RocketMessageChannelBinder(this.
            rocketConnectionFactory,
                this.rocketProperties, provisioningProvider());
        return binder;
    }
    @Bean
    RocketExchangeQueueProvisioner provisioningProvider() {
        return new RocketExchangeQueueProvisioner(this.rocketConnectionFactory);
    }
}
```

RocketMessageChannelBinder 继承了抽象类 AbstractMessageChannelBinder，并实现了 producerMessageHandler 和 createConsumerEndpoint 方法。

MessageHandler 能向消息队列发送消息，createProducerMessageHandler 方法为了创建 MessageHandler 对象，将输出型通道的消息发送到消息队列上。createProducerMessageHandler 方法的实现如下所示：

```
protected MessageHandler createProducerMessageHandler(
        ProducerDestination destination,
        ExtendedProducerProperties<RocketProducerProperties> producerProperties,
        MessageChannel errorChannel)
        throws Exception {
    final AmqpOutboundEndpoint endpoint = new AmqpOutboundEndpoint(
            buildRocketTemplate(producerProperties.getExtension(), errorChannel
                != null));
    return endpoint;
}
```

MessageProducer 能够从消息队列接收消息，并将该消息发送给输入型通道，其具体实现如下所示：

```
@Override
protected MessageProducer createConsumerEndpoint(ConsumerDestination consumerDestination,
    String group,
ExtendedConsumerProperties<RocketConsumerProperties> properties) throws Exception {
    SimpleRocketMessageListenerContainer listenerContainer = new SimpleRocketMes
        sageListenerContainer();
    RocketInboundChannelAdapter rocketInboundChannelAdapter = new RocketInboundC
        hannelAdapter(listenerContainer);
    return rocketInboundChannelAdapter;
}
```

createConsumerEndpoint 方法首先创建了 SimpleRocketMessageListenerContainer 实例，然后将其作为参数创建 RocketInboundChannelAdapter 实例。RocketInboundChannelAdapter 是 MessageProducer 的实现类之一。

1. SimpleRocketMessageListenerContainer

SimpleRocketMessageListenerContainer 的 doStart 函数会对 RocketBlockingQueueConsumer 实例初始化，并且启动 SimpleRocketMessageListenerContainer 的 AsyncMessageProcessingConsumer。AsyncMessageProcessingConsumer 会无限循环地从 RocketBlockingQueueConsumer 中获取 RocketMQ 传递过来的消息，代码如下所示：

```
private void doStart() {
    synchronized (this.lifecycleMonitor) {
        this.active = true;
        this.running = true;
        this.lifecycleMonitor.notifyAll();
    }
    synchronized (this.consumersMonitor) {
        //初始化Consumer
        int newConsumers = initializeConsumers();          Set<SimpleRocketMessa
            geListenerContainer.AsyncMessageProcessingConsumer>
processors = new HashSet<>();
        //对于每个RocketBlockingQueueConsumer启动一个
        //AsyncMessageProcessingConsumer来执行任务
        for (RocketBlockingQueueConsumer consumer : this.consumers) {
                    SimpleRocketMessageListenerContainer.
AsyncMessageProcessingConsumer processor = new
SimpleRocketMessageListenerContainer.
AsyncMessageProcessingConsumer(consumer);
            processors.add(processor);
            getTaskExecutor().execute(processor);
        }
    }
}
```

SimpleRocketMessageListenerContainer.AsyncMessageProcessingConsumer 实现了 Runnable 接口，在 run 方法中会无限循环地调用 SimpleRocketMessageListenerContainer 本身的 receiveAndExecute 方法，其具体实现如下所示：

```
@Override
public void run() {
    if (!isActive()) {
        return;
    }
    try {
        //只要consumer的状态正常，就会一直循环
        while (isActive(this.consumer) || this.consumer.hasDelivery() || !this.
            consumer.cancelled()) {
            boolean receivedOk = receiveAndExecute(this.consumer);
        }
```

```
        } catch (Exception e) {
        }
//代码有删减
}
```

receiveAndExecute 方法最终的作用就是调用 RocketBlockingQueueConsumer 的 nextMessage 方法，然后再将消息通过 messageListener 的 onMessage 方法传递出去，代码如下所示：

```
private boolean doReceiveAndExecute(RocketBlockingQueueConsumer consumer) throws
    Throwable {
    for (int i = 0; i < this.txSize; i++) {
        Message message = consumer.nextMessage(this.receiveTimeout);
        if (message == null) {
            break;
        }
        try {
            //通过MessageListener的方法将消息传递出去
            this.messageListener.onMessage(message, null);
        }
        catch (ImmediateAcknowledgeAmqpException e) {
            break;
        }
        catch (Throwable ex) {
        }
    }
    return consumer.commitIfNecessary(true);
}
```

2. RocketBlockingQueueConsumer

RocketBlockingQueueConsumer 有一个阻塞队列，用来存储 RocketMQ 传递给 RocketBlockingQueueConsumer.InnerConsumer 的消息。如下代码所示，nextMessage 方法就是从阻塞队列中拉取一个消息并返回：

```
public Message nextMessage(long timeout) throws InterruptedException, ShutdownSignalException {
    Message message = handle(this.queue.poll(timeout, TimeUnit.MILLISECONDS));
    if (message == null && this.cancelled.get()) {
        throw new ConsumerCancelledException();
    }
    return message;
}
```

InnerConsumer 实现的 MessageListenerConcurrently 接口是 RocketMQ 中用于并发接受异步消息的接口，该接口可以接收 RocketMQ 发送过来的异步消息。而 InnerConsumer 在接收到消息之后，会将消息封装成 RocketDelivery 加入到阻塞队列中，如下所示：

```
private final class InnerConsumer implements MessageListenerConcurrently {
    public InnerConsumer(DefaultMQPushConsumer mqPushConsumer) {
        super();
        mqPushConsumer.registerMessageListener(this);
    }
```

```
    @Override
    public ConsumeConcurrentlyStatus consumeMessage(List<MessageExt> list,
        ConsumeConcurrentlyContext consumeConcurrentlyContext) {
        if (list == null || list.isEmpty()) {
            return ConsumeConcurrentlyStatus.CONSUME_SUCCESS;
        }
        try {
            for (MessageExt ext : list) {
                RocketBlockingQueueConsumer.this.queue.put(new RocketDelivery(ext));
            }
            return ConsumeConcurrentlyStatus.CONSUME_SUCCESS;
        } catch (InterruptedException e) {
            return ConsumeConcurrentlyStatus.RECONSUME_LATER;
        }
    }
}
```

RocketInboundChannelAdapter 实现了 MessageProducer 接口。它主要将 SimpleRocket MessageListenerContainer 传递过来的消息经过 MessageTemplate 传递给 MessageChannel。

3. RocketInboundChannelAdapter

接下来，我们看一下 RocketInboundChannelAdapter.Listener 的实现，这个实现是 RocketBlockingQueueConsumer 的 nextMessage 方法中的 messageListener：

```
public class Listener implements ChannelAwareMessageListener, RetryListener {
    public void onMessage(Message message, Channel channel) throws Exception {
        try {
            this.createAndSend(message, channel);
        } catch (RuntimeException var7) {
            ...
        }
    }
    private void createAndSend(Message message, Channel channel) {
        org.springframework.messaging.Message<Object> messagingMessage = this.
            createMessage(message, channel);
        RocketInboundChannelAdapter.this.sendMessage(messagingMessage);
    }
    private org.springframework.messaging.Message<Object> createMessage(Message
        message, Channel channel) {
        Object payload = RocketInboundChannelAdapter.this.messageConverter.
            fromMessage(message);
        org.springframework.messaging.Message<Object> messagingMessage =
            RocketInboundChannelAdapter.this.getMessageBuilderFactory().withPayload
            (payload).build();
        return messagingMessage;
    }
}
```

sendMessage 方法如同源码分析章节中讲解的一样，就是调用 MessageTemplate 来将消息发送给对应的 Channel 实例，如下所示：

```
protected void sendMessage(Message<?> message) {
```

```
    if (message == null) {
        throw new MessagingException("cannot send a null message");
    }
    if (this.shouldTrack) {
        message = MessageHistory.write(message, this, this.getMessageBuilderFactory());
    }
    try {
        this.messagingTemplate.send(getOutputChannel(), message);
    }
    catch (RuntimeException e) {
        if (!sendErrorMessageIfNecessary(message, e)) {
            throw e;
        }
    }
}
```

RocketProvisioningProvider 实现了 ProvisioningProvider 接口，它有两个方法，分别是 provisionProducerDestination 和 provisionConsumerDestination，用于创建 ProducerDestination 和 ConsumerDestination。RocketProvisioningProvider 的实现类似于 RabbitProvisioningProvider，只不过使用了 RocketAdmin 所实现的 RocketMQ 的相关 API。

10.4.2 多实例

当我们横向扩展 Spring Cloud Stream 应用时，每个应用实例都会知道还存在多少和自己相同的应用实例以及它自己的实例索引（instance index）。Spring Cloud Stream 通过 spring.cloud.stream.instanceCount 和 spring.cloud.stream.instanceIndex 属性表示应用实例的数量和索引。比如说，当前有三个订单消息应用实例，所有的三个应用实例都会将 spring.cloud.stream.instanceCount 属性设置为 3，并且三个应用分别把自己的 spring.cloud.stream.instanceIndex 属性设置为 0、1、2。

当 Spring Cloud Stream 应用通过 Spring Cloud Cloud Data Flow 部署时，上述属性会自动配置；当 Spring Cloud Stream 应用独立部署时，运维人员必须对上述属性进行正确设置。

在横向扩容场景，上述两个属性的正确配置对于分区（partitioning）行为十分重要，而且特定的 Binder 也依赖于上述两个属性来将消息分配到不同的应用实例（比如说 Kafka Binder）。

10.4.3 分区

通过设置 partitionKeyExpression，partitionKeyExtractorClass 和 partitionCount 属性可以将输出型的通道配置成分区发送的状态，分区发送的配置示例如下所示：

```
spring.cloud.stream.bindings.output.producer.partitionKeyExpression=payload.id
spring.cloud.stream.bindings.output.producer.partitionCount=5
```

基于上述的例子配置，Stream 会基于消息的 payload.id 来将消息分发到不同的分区中。对于每个发送到输出型通道的消息都会基于 partitionKeyExpression 属性来计算一个分区键值。partitionKeyExpression 是一个 SpringEL 表达式，它可以用来计算消息的分区键。

如果 SpringEL 表达式不能满足你的需求，可以通过设置 partitionKeyExtractorClass 属性来计算分区键值。partitionKeyExtractorClass 属性值为实现了 org.springframework.cloud.stream.binder.PartitionKeyExtractorStrategy 接口的类。虽然 SpEL 表达式通常可以满足一般性的需求，但是较为复杂的分区逻辑仍然需要使用 PartitionKeyExtractorStrategy 实现。partitionKeyExtractorClass 属性配置示例如下所示：

```
spring.cloud.stream.bindings.output.producer.partitionKeyExtractorClass=com.example.
    MyKeyExtractor
spring.cloud.stream.bindings.output.producer.partitionCount=5
```

一旦消息的分区键计算出来，分区选择进程就会计算出一个目标分区 ID，一般为 0 到 partitionCount - 1 之间的一个值。默认的计算表达式为 key.hashCode() % partitionCount，它适用于绝大多数场景。该表达式可以通过设置 partitionSelectorExpression 和 partitionSelectorClass 属性值来自定义。MyKeyExtractor 类的实例需要在配置时给出，并且指明它的全限定名称，具体实现如下代码所示：

```
@Bean(name="com.example.MyKeyExtractor")
public MyKeyExtractor extractor() {
    return new MyKeyExtractor();
}
```

PartitionKeyExtractorStrategy 接口只有一个 extractKey 方法，从参数 Message<?> 实例上获取分区所需要的分区键。而 partitionSelectorClass 所需的 PartitionSelectorStrategy 接口也只有一个 selectPartition 方法，它有两个参数，分别是分区键和分区应用实例数量。PartitionTestSupport 实现了上述两个接口，它将 Message 的 Payload 作为分区键，然后取出 Payload 字节数组的第一个字节值与应用实例数相除计算出目标分区。PartitionTestSupport 的具体实现如下所示：

```
public class PartitionTestSupport implements PartitionKeyExtractorStrategy,
    PartitionSelectorStrategy {
    @Override
    public int selectPartition(Object key, int divisor) {
        return ((byte[]) key)[0] % divisor;
    }
    @Override
    public Object extractKey(Message<?> message) {
        return message.getPayload();
    }
}
```

10.5 本章小结

Spring Cloud Stream 是一个用来为微服务应用构建消息驱动能力的框架。它可以基于 Spring Boot 来创建独立的、可用于生产的 Spring 应用程序。它通过使用 Spring Integration 来连接消息代理中间件以实现消息事件驱动。Spring Cloud Stream 为一些供应商的消息中间件产品提供了个性化的自动化配置实现，引入了发布订阅、消费组、分区的三个核心概念。通过自定义 Spring Cloud Stream 的 Binder 组件，开发者可以使用 Stream 和市面上的任何消息队列进行连接。

第 11 章 消息总线：Spring Cloud Bus

ESB（企业服务总线）是 SOA（面向服务架构）的一种常见的设计实践。基于总线的设计，借鉴了计算机内部硬件组成的设计思想：通过总线传输数据。在分布式系统中，不同子系统之间需要实现相互通信和远程调用，比较直接的方式就是"点对点"的通信方式，但是这样会暴露出一些很明显的问题：系统之间紧密耦合、配置和引用混乱、服务调用关系错综复杂、难以统一管理、异构系统之间存在不兼容等。基于总线的设计，可以解决上述问题。总线作为中枢系统，提供统一的服务入口，实现了服务统一管理、服务路由、协议转换、数据格式转换等功能。这样能够将不同系统有效地连接起来，并大大降低了连接数（每个子系统只需要和总线建立连接）和系统间连接拓扑的复杂度。

Spring Cloud 作为微服务架构综合性的解决方案，也提供了一套实现消息总线的组件：Spring Cloud Bus。Spring Cloud Bus 基于 Spring Cloud Stream，Spring Cloud Stream 屏蔽了底层消息中间件的差异性，在其之上封装成各种 Binder，在上一章已经具体讲解过。通过 Spring Cloud Bus，可以非常容易地搭建消息总线，同时实现了一些消息总线中的常用功能，比如与 Spring Cloud Config 一起实现微服务应用配置信息的动态更新等。

本章第一小节由配置中心的动态更新作为基础应用的案例，引入 Spring Cloud Bus 的应用，示例如何改造第 9 章中的配置服务中心；第二小节将会对 Spring Cloud Bus 的主要功能：事件的订阅与发布、事件监听的具体实现，结合源码进行分析；第三小节是应用进阶部分，定制一个注册事件，并完成事件的监听和发起。

11.1 基础应用

在第 9 章中，我们讲了微服务架构中配置中心的用法与实现。配置仓库中的配置信息变更，如何及时通知到应用服务，并刷新上下文信息？在前面章节，我们的做法是引入 Spring Boot 监控的依赖 spring-boot-starter-actuator。通过其提供的 /refresh 端点，配置到 Git 的 webhook，当提交代码到配置仓库时，将会触发钩子，使得应用服务能够刷新。但这样做显然是不够的。微服务的数量极多，手工维护极为复杂。消息总线能利用消息代理将各个服务连接起来，将消息路由到目标服务实例。图 11-1 是我们改进的架构图。

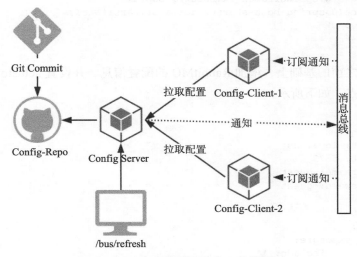

图 11-1　改进后的配置中心架构图

与之前的配置中心架构不一样的地方在于增加了消息总线。消息总线连接了 Config-Server 和 Config-Client。当配置提交到配置仓库时，利用 webhook 或者手动访问 /actuator/bus-refresh 端点，Config-Server 将会将变更消息通知到各个客户端。

主要涉及的服务包括：

- 配置服务器 Config-Server，从配置仓库中拉取配置，注册到服务发现组件 consul，并为配置客户端提供配置信息。
- 配置客户端 Config-Client-1，注册到 Consul，在启动时从配置服务器拉取配置。
- 配置客户端 Config-Client-2，同上。
- 配置仓库 Config-Repo，使用 Git 存储配置信息。

根据前面章节中配置中心的实现，我们进行相应的改写，以实施本章的设计方案。

11.1.1　配置服务器

消息中间件使用的是基于 AMQP（Advanved Message Queue）协议实现的 RabbitMQ。

1. 引入依赖

在之前的基础上，需要引入依赖 spring-cloud-starter-bus-amqp，这是 Spring Cloud Bus 基于 RabbitMQ 的实现。另外，还需要加上 spring-boot-starter-actuator，以使用其提供的上下文

刷新功能。代码如下所示：

```xml
<dependency>
    <groupId>org.springframework.cloud</groupId>
    <artifactId>spring-cloud-starter-bus-amqp</artifactId>
</dependency>
<dependency>
    <groupId>org.springframework.boot</groupId>
    <artifactId>spring-boot-starter-actuator</artifactId>
</dependency>
```

2. 配置文件

配置文件在之前的基础上，增加 RabbitMQ 的配置信息，并设置 actuator 在 HTTP 上暴露所有的控制端点，如下所示：

```yaml
spring:
    rabbitmq:
        host: localhost
        port: 5672
        username: guest
        password: guest
management:
    endpoints:
        web:
            exposure:
                include: "*"
```

11.1.2 配置客户端

配置客户端在之前的基础上，需要引入依赖 spring-cloud-starter-bus-amqp：

```xml
<dependency>
    <groupId>org.springframework.cloud</groupId>
    <artifactId>spring-cloud-starter-bus-amqp</artifactId>
</dependency>
```

配置文件中增加 rabbitmq 的配置信息：

```yaml
spring:
    rabbitmq:
        host: localhost
        port: 5672
        username: guest
        password: guest
```

11.1.3 结果验证

1）依次启动 Config Server 和两个 Config Client。在控制台的日志中，新增了两个端点 /actuator/bus-refresh 和 /actuator/bus-env 的端点信息，由 Spring Cloud Bus 提供。前一个端点 /actuator/bus-refresh 用于重新加载每个应用服务的配置；后一个端点 /actuator/bus-env 发送键值对来更新每个节点的 Spring 上下文环境，如下所示：

```
Mapped "{[/actuator/bus-env/{destination}],methods=[POST],consumes=[applicati
on/vnd.spring-boot.actuator.v2+json || application/json]}" onto public java.
lang.Object org.springframework.boot.actuate.endpoint.web.servlet.Abstrac
tWebMvcEndpointHandlerMapping$OperationHandler.handle(javax.servlet.http.
HttpServletRequest,java.util.Map<java.lang.String, java.lang.String>)
Mapped "{[/actuator/bus-refresh],methods=[POST]}" onto public java.lang.Object
org.springframework.boot.actuate.endpoint.web.servlet.AbstractWebMvcEndpointHan
dlerMapping$OperationHandler.handle(javax.servlet.http.HttpServletRequest,java.
util.Map<java.lang.String, java.lang.String>)
```

2）打开 rabbitmq 的管理界面，如图 11-2 所示，在"Exchanges"中，可以看到增加了 SpringCloudBus。这是 Spring Cloud Bus 自动创建的话题，所有增加了 Bus 依赖的服务，都会订阅该话题。

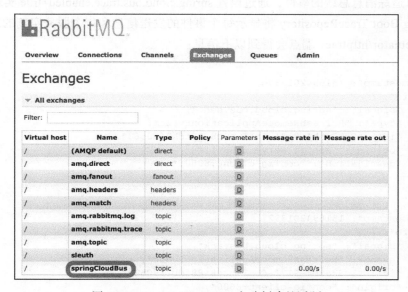

图 11-2　Spring Cloud Bus 自动创建的话题

3）Config Client 访问 /cloud/version 接口。Config-Client-1 和 Config-Client-2 返回的都是相同的结果：

```
Edgware.RELEASE
```

4）提交配置到配置仓库。将 config-client-dev.yml 中的 cloud.version 改为 Camden SR7。同时需要发送 POST 请求 /actuator /bus-refresh 到 Config Server。

从 Config Server 的控制台日志中可以看出，配置服务器重新获取配置仓库中的拉取配置，并显示了一处更新。

```
INFO 32051 --- [nio-8888-exec-9] .c.s.e.MultipleJGitEnvironmentRepository : Fetched
    for remote master and found 1 updates
```

同时，配置客户端的日志也相应地变化如下：

```
INFO 32054 --- [2q3v4kn4xg5_w-1] o.s.cloud.bus.event.RefreshListener     : Received 
    remote refresh request. Keys refreshed [config.client.version, cloud.version]
```

通过消息总线接收到配置更新，并更新到相应的变更字段 cloud.version。

5）配置客户端再次访问 /cloud/version 接口，可以看到返回的是更新之后的结果 Camden SR7。

6）通过查阅 /bus-refresh 端点信息，我们发现其支持更细粒度的更新推送。/actuator/bus-refresh /{ destination }，destination 对应服务名和端口，也可以手动设置。如只推送通知到 Config-Client-1，请求命令为 curl -v -X POST http://localhost:8888/actuator/bus-refresh/config-client:8000。

7）可以追踪消息总线的事件，通过设置 spring.cloud.bus.trace.enabled=true 实现。设置成功后，Spring Boot TraceRepository 将显示每个事件的发送信息和每个服务实例的所有 ACK。我们访问 /actuator/httptrace 端点会得到以下信息：

```
{
    "timestamp": 1516942013396,
    "info": {
        "signal": "spring.cloud.bus.sent",
        "type": "RefreshRemoteApplicationEvent",
        "id": "5c83a5b4-917f-4b51-b2c8-88fdb2f90563",
        "origin": "config-server:8888",
        "destination": "config-client:8000:**"
    }
},
{
    "timestamp": 1516942013927,
    "info": {
        "signal": "spring.cloud.bus.ack",
        "event": "RefreshRemoteApplicationEvent",
        "id": "5c83a5b4-917f-4b51-b2c8-88fdb2f90563",
        "origin": "config-client:8000",
        "destination": "config-client:8000:**"
    }
}
```

上面的追踪表示，RefreshRemoteApplicationEvent 从 config-server:8888 发出，并到达了 config-client:8000。

11.2 源码解析

Spring Cloud Bus 是在 Spring Cloud Stream 的基础上进行的封装，对于指定主题的消息的发布与订阅是通过 Spring Cloud Stream 的具体 Binder 实现。因此引入的依赖可以是 spring-cloud-starter-bus-amqp 或者是 spring-cloud-starter-bus-kafka，分别对应于 RabbitMQ 和 Kafka 绑定器的实现。如图 11-3 为消息总线的工作原理图，发送 / 接收端通过消息中间件连接，事件与消息相互转换，我们总结出 Spring Cloud Bus 的主要功能为如下：
- 对指定主题 springCloudBus 的消息订阅与发布。
- 事件监听，包括刷新事件、环境变更事件、远端应用的 ACK 事件以及本地服务端发送事件等。

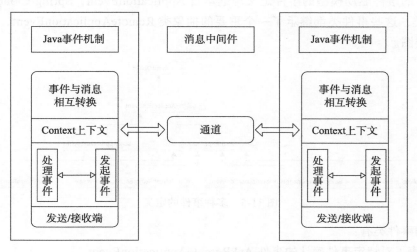

图 11-3　消息总线的工作原理图

下面我们以这两方面作为主线，进行 Spring Cloud Bus 的源码分析。

11.2.1　事件的定义与事件监听器

在介绍 Spring Cloud Bus 中事件订阅与发布之前，首先介绍一下 Spring 的事件驱动模型；然后将会介绍 Spring Cloud Bus 中定义的多种事件类；最后介绍监听器，每个事件对应一个事件监听器。

1. 事件驱动模型

事件驱动模型基于发布–订阅模式的编程模型，定义对象间的一种一对多的依赖关系，当一个对象的状态发生变化时，所有依赖它的对象都得到通知并自动更新。图 11-4 为 Spring 的事件驱动模型。

Spring 的事件驱动模型由三部分组成：
- 事件：ApplicationEvent，继承自 JDK 的 EventObject，所有事件将继承它，并通过 source 得到事件源。
- 事件发布者：ApplicationEventPublisher 及 ApplicationEventMulticaster 接口，使用这个接口，Service 就拥有了发布事件

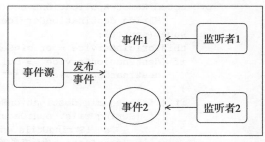

图 11-4　事件驱动模型

- 事件订阅者：ApplicationListener，继承自 JDK 的 EventListener，所有监听器将继承它。

2. 事件的定义

Spring 的事件驱动模型的事件定义均继承自 ApplicationEvent，Spring Cloud Bus 中有多个事件类，这些事件类都继承了一个重要的抽象类 RemoteApplicationEvent，图 11-5 是事件类的类图。

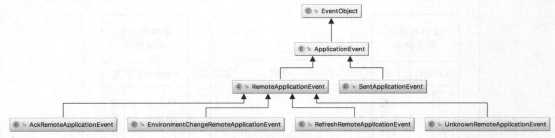

图 11-5　多种事件的定义

涉及的事件类有：
- 代表了对特定事件确认的事件 AckRemoteApplicationEvent。
- 环境变更的事件 EnvironmentChangeRemoteApplicationEvent。
- 刷新事件 RefreshRemoteApplicationEvent。
- 发送事件 SentApplicationEvent。
- 未知事件 UnknownRemoteApplicationEvent。

下面我们分别看一下这些事件的定义。

（1）抽象基类：RemoteApplicationEvent

通过上面的类图，可以知道 RemoteApplicationEvent 是其他事件类的基类，定义了事件对象的公共属性：

```
@JsonTypeInfo(use = JsonTypeInfo.Id.NAME, property = "type")
@JsonIgnoreProperties("source") //序列化时，忽略 source
public abstract class RemoteApplicationEvent extends ApplicationEvent {
    protected RemoteApplicationEvent(Object source, String originService,
            String destinationService) {
        super(source);
        this.originService = originService;
        if (destinationService == null) {
            destinationService = "**";
        }
        if (!"**".equals(destinationService)) {
            if (StringUtils.countOccurrencesOf(destinationService, ":") <= 1
                    && !StringUtils.endsWithIgnoreCase(destinationService, ":**")) {
                //destination的所有实例
                destinationService = destinationService + ":**";
```

```
            }
        }
        this.destinationService = destinationService;
        this.id = UUID.randomUUID().toString();
    }
    ...
}
```

在 RemoteApplicationEvent 中定义了主要的三个通用属性事件的源 originService、事件的目的服务 destinationService 和随机生成的全局 id。通过其构造函数可知，destinationService 可以使用通配符的形式 {serviceId}:{appContextId}，两个变量都省略的话，则通知到所有服务的所有实例。只省略 appContextId 时，则对应的目标通知服务为相应 serviceId 的所有实例。另外，注解 @JsonTypeInfo(use = JsonTypeInfo.Id.NAME, property = "type") 对应于序列化时，使用子类的名称作为 type；而 @JsonIgnoreProperties("source") 表示序列化时，忽略 source 属性，source 定义在 JDK 中的 EventObject。

（2）EnvironmentChangeRemoteApplicationEvent

用于动态更新服务实例的环境属性，我们在基础应用中更新 cloud.version 属性时，关联到该事件，代码如下所示：

```
public class EnvironmentChangeRemoteApplicationEvent extends RemoteApplicationEvent {

    private final Map<String, String> values;

    public EnvironmentChangeRemoteApplicationEvent(Object source, String originService,
            String destinationService, Map<String, String> values) {
        super(source, originService, destinationService);
        this.values = values;
    }
    ...
}
```

可以看到，EnvironmentChangeRemoteApplicationEvent 事件类的实现很简单。定义了 Map 类型的成员变量，key 对应于环境变量名，而 value 对应更新后的值。

（3）RefreshRemoteApplicationEvent

刷新远端应用配置的事件，用于接收远端刷新的请求，代码如下所示：

```
public class RefreshRemoteApplicationEvent extends RemoteApplicationEvent {
    public RefreshRemoteApplicationEvent(Object source, String originService,
            String destinationService) {
        super(source, originService, destinationService);
    }
}
```

继承自抽象事件类 RemoteApplicationEvent，没有额外的成员属性。

（4）AckRemoteApplicationEvent

确认远端应用事件，该事件表示一个特定的 RemoteApplicationEvent 事件被确认，代码如下所示：

```java
public class AckRemoteApplicationEvent extends RemoteApplicationEvent {

    public AckRemoteApplicationEvent(Object source, String originService,
            String destinationService, String ackDestinationService, String ackId,
            Class<? extends RemoteApplicationEvent> type) {
        super(source, originService, destinationService);
        this.ackDestinationService = ackDestinationService;
        this.ackId = ackId;
        this.event = type;
    }
    ...
    public void setEventName(String eventName) {
        try {
            event = (Class<? extends RemoteApplicationEvent>) Class.forName (eventName);
        } catch (ClassNotFoundException e) {
            event = UnknownRemoteApplicationEvent.class;
        }
    }
}
```

该事件类在 RemoteApplicationEvent 基础上定义了成员属性 ackId、ackDestinationService 和 event。ackId 和 ackDestinationService，分别表示确认的事件 id 和对应的目标服务。event 对应事件类型，确认事件能够确认的必然是 RemoteApplicationEvent 的子类，因此 event 属性值在设置时需要进行检查，如果转换出现异常，则定义为未知的事件类型。这些事件可以由任何需要统计总线事件响应的应用程序来监听。它们的行为与普通的远程应用程序事件相似，即如果目标服务与本地服务 ID 匹配，则应用程序会在其上下文中触发该事件。

（5）SentApplicationEvent

发送应用事件，表示系统中的某个地方发送了一个远端事件，代码如下所示：

```java
@JsonTypeInfo(use = JsonTypeInfo.Id.NAME, property = "type")
@JsonIgnoreProperties("source")
public class SentApplicationEvent extends ApplicationEvent {
    ...
    public SentApplicationEvent(Object source, String originService,
            String destinationService, String id,
            Class<? extends RemoteApplicationEvent> type) {
        super(source);
        this.originService = originService;
        this.type = type;
        if (destinationService == null) {
            destinationService = "*";
        }
        if (!destinationService.contains(":")) {
            // All instances of the destination unless specifically requested
            destinationService = destinationService + ":**";
        }
        this.destinationService = destinationService;
        this.id = id;
```

```
    }
    ...
}
```

可见该事件类继承自 ApplicationEvent，本身并不是一个 RemoteApplicationEvent 事件，所以不会通过总线发送，而是在本地生成（多为响应远端事件）。想要审计远端事件的应用可以监听该事件，并且所有的 AckRemoteApplicationEvent 事件中的 id 来源于相应的 SentApplicationEvent 中定义的 id。在其定义的成员属性中，相比于远端应用事件多了一个事件类型 type，该类型限定于 RemoteApplicationEvent 的子类。

（6）UnknownRemoteApplicationEvent

未知的远端应用事件，也是 RemoteApplicationEvent 事件类的子类。该事件类与之前的 SentApplicationEvent、AckRemoteApplicationEvent 有关，当序列化时遇到事件的类型转换异常，则自动构造成一个未知的远端应用事件。

3. 事件监听器

Spring Cloud Bus 中，事件监听器的定义可以是实现 ApplicationListener 接口，或者是使用 @EventListener 注解的形式。图 11-6 是事件监听器的类图。

ApplicationListener 接口实现有两个：刷新监听器 RefreshListener 和环境变更监听器 EnvironmentChangeListener。

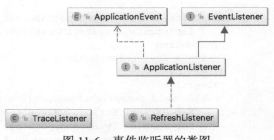

图 11-6　事件监听器的类图

（1）RefreshListener

RefreshListener 监听器对应的事件为 RefreshRemoteApplicationEvent，代码如下所示：

```java
public class RefreshListener
        implements ApplicationListener<RefreshRemoteApplicationEvent> {
    private ContextRefresher contextRefresher;

    public RefreshListener(ContextRefresher contextRefresher) {
        this.contextRefresher = contextRefresher;
    }

    @Override
    public void onApplicationEvent(RefreshRemoteApplicationEvent event) {
        Set<String> keys = contextRefresher.refresh();
        log.info("Received remote refresh request. Keys refreshed " + keys);
    }
}
```

对于刷新时间的处理，调用 ContextRefresher 的 refresh 方法，而定义在 Spring Cloud Context 中的 ContextRefresher 用于提供上下文刷新的功能。我们具体看一下 refresh 方法：

```
// ContextRefresher.java
```

```java
public synchronized Set<String> refresh() {
    Map<String, Object> before = extract(
            this.context.getEnvironment().getPropertySources());
    addConfigFilesToEnvironment();
    Set<String> keys = changes(before,
    extract(this.context.getEnvironment().getPropertySources())).keySet();
    this.context.publishEvent(new EnvironmentChangeEvent(keys));
    this.scope.refreshAll();
    return keys;
}
```

先获取之前环境变量的键值对，然后重新加载新的配置环境文件，通过比对新旧环境变量的 map 集合，发布新的环境变更 EnvironmentChangeEvent 的事件。this#scope#refresh 销毁了在这个范围内，当前实例的所有 Bean 并在下次方法的执行时强制刷新。

（2）EnvironmentChangeListener

EnvironmentChangeListener 对应的事件类是 EnvironmentChangeRemoteApplicationEvent，代码如下所示：

```java
public class EnvironmentChangeListener
        implements ApplicationListener<EnvironmentChangeRemoteApplicationEvent> {
    @Autowired
    private EnvironmentManager env;

    @Override
    public void onApplicationEvent(EnvironmentChangeRemoteApplicationEvent event) {
        Map<String, String> values = event.getValues();
        for (Map.Entry<String, String> entry : values.entrySet()) {
            env.setProperty(entry.getKey(), entry.getValue());
        }
    }
}
```

在 RefreshListener 的实现中，可以知道该事件的实现最终又发布了一个新的事件 EnvironmentChangeListener。在刷新监听器中，构造了变更了的环境变量的 map，交给环境变更监听器。上面对环境变更事件的处理，遍历变更了的配置环境属性，并在本地应用程序的环境中将新的属性值设置到对应的键。

（3）TraceListener

TraceListener 的实现是通过注解 @EventListener 的形式，监听的事件为确认事件 AckRemoteApplicationEvent 和发送事件 SentApplicationEvent，代码如下所示：

```java
// TraceListener.java
@EventListener
public void onAck(AckRemoteApplicationEvent event) {
    this.repository.add(getReceivedTrace(event));
}

@EventListener
public void onSend(SentApplicationEvent event) {
```

```java
        this.repository.add(getSentTrace(event));
    }

    protected Map<String, Object> getSentTrace(SentApplicationEvent event) {
        Map<String, Object> map = new LinkedHashMap<String, Object>();
        map.put("signal", "spring.cloud.bus.sent");
        map.put("type", event.getType().getSimpleName());
        map.put("id", event.getId());
        map.put("origin", event.getOriginService());
        map.put("destination", event.getDestinationService());
        if (log.isDebugEnabled()) {
            log.debug(map);
        }
        return map;
    }

    protected Map<String, Object> getReceivedTrace(AckRemoteApplicationEvent event) {
        Map<String, Object> map = new LinkedHashMap<String, Object>();
        map.put("signal", "spring.cloud.bus.ack");
        map.put("event", event.getEvent().getSimpleName());
        map.put("id", event.getAckId());
        map.put("origin", event.getOriginService());
        map.put("destination", event.getAckDestinationService());
        if (log.isDebugEnabled()) {
            log.debug(map);
        }
        return map;
    }
```

在 SentTrace 中，主要记录了 signal、type（事件类型）、id、origin（源服务）和 destination（目的服务）的属性值。而在 ReceivedTrace 中，表示对事件的确认，主要记录了 signal、event（事件类型）、id、origin（源服务）和 destination（目的服务）的属性值。这些信息默认存储于内存中，可以通过 /trace 端点获取最近的事件信息，如下所示：

```
{
    "timestamp": 1517229555629,
    "info": {
        "signal": "spring.cloud.bus.sent",
        "type": "RefreshRemoteApplicationEvent",
        "id": "c73a9792-9409-47af-993c-65526edf0070",
        "origin": "config-server:8888",
        "destination": "config-client:8000:**"
    }
},
{
    "timestamp": 1517227659384,
    "info": {
        "signal": "spring.cloud.bus.ack",
        "event": "RefreshRemoteApplicationEvent",
        "id": "846f3a17-c344-4d29-93f3-01b73c5bf58f",
        "origin": "config-client:8000",
        "destination": "config-client:8000:**"
```

        }
    }

至于事件的发起,我们将在下一节结合消息的订阅与发布一起讲解。

11.2.2 消息的订阅与发布

Spring Cloud Bus 在 Spring Cloud Stream 的基础上,将消息通道的定义以及消息订阅与发布进行了封装。对特定主题的消息进行订阅与发布,事件以消息的形式传递到其他服务实例。

1. 通道定义

既然是基于 Stream,我们首先看一下输入和输出的通道定义:

```
public interface SpringCloudBusClient {

    String INPUT = "springCloudBusInput";

    String OUTPUT = "springCloudBusOutput";

    @Output(SpringCloudBusClient.OUTPUT)
    MessageChannel springCloudBusOutput();

    @Input(SpringCloudBusClient.INPUT)
    SubscribableChannel springCloudBusInput();
}
```

可以看到,Bus 中定义了 springCloudBusInput 和 springCloudBusOutput 两个通道,分别用于订阅与发布话题为 springCloudBus 的消息。

2. bus 属性定义

在基础应用中我们就知道 Bus 订阅的话题是 springCloudBus,下面看一下 Bus 中其他属性的定义:

```
@ConfigurationProperties("spring.cloud.bus")
public class BusProperties {

    //环境变更相关的属性
    private Env env = new Env();
    // 刷新事件相关的属性
    private Refresh refresh = new Refresh();
    //与ack相关的属性
    private Ack ack = new Ack();
    //与追踪ack相关的属性
    private Trace trace = new Trace();
    //Spring Cloud Stream消息的话题
    private String destination = "springCloudBus";

    //标志位,bus是否可用
    private boolean enabled = true;

    ...
```

}

上面所示的 Bus 属性，设置了一些默认值，正好与事实也是相符的，没有进行任何 spring.cloud.bus 配置也能够正常运行。通过在配置文件中修改相应的属性，可以实现 Bus 的更多功能扩展。env、refresh、ack 和 trace 分别对应不同的事件，在配置文件中有一个开关属性，默认都是开启的，可以根据需要进行关闭。

3. 消息的监听与发送

上面两部分讲了 Stream 通道和基本属性的定义，最后我们看下 Bus 中对指定主题的消息如何发送与监听处理。在 META-INF/spring.factories 中将 EnableAutoConfiguration 配置项设置为 BusAutoConfiguration，在服务启动时会自动加载到 Spring 容器中，以下代码是对于指定主题消息的发送与监听处理：

```java
@Configuration
@ConditionalOnBusEnabled //bus启用的开关
@EnableBinding(SpringCloudBusClient.class) //绑定通道
@EnableConfigurationProperties(BusProperties.class)
public class BusAutoConfiguration implements ApplicationEventPublisherAware {

    @Autowired
    @Output(SpringCloudBusClient.OUTPUT)
    //注入source接口，用于发送消息。
    private MessageChannel cloudBusOutboundChannel;

    @EventListener(classes = RemoteApplicationEvent.class)
    public void acceptLocal(RemoteApplicationEvent event) {
        // 监听RemoteApplicationEvent事件，当事件是来自自己的并且不是ack事件，则发送消息
        if (this.serviceMatcher.isFromSelf(event)
                && !(event instanceof AckRemoteApplicationEvent)) {
            this.cloudBusOutboundChannel.send(MessageBuilder.withPayload(event).
                build());
        }
    }
    @StreamListener(SpringCloudBusClient.INPUT)
    public void acceptRemote(RemoteApplicationEvent event) {
        if (event instanceof AckRemoteApplicationEvent) {
            if (this.bus.getTrace().isEnabled() && !this.serviceMatcher.isFromSelf
                (event)
                    && this.applicationEventPublisher != null) {
                this.applicationEventPublisher.publishEvent(event);
            }
            return;
        }
        if (this.serviceMatcher.isForSelf(event)
                && this.applicationEventPublisher != null) {
            if (!this.serviceMatcher.isFromSelf(event)) {
                this.applicationEventPublisher.publishEvent(event);
            }
            if (this.bus.getAck().isEnabled()) {
                AckRemoteApplicationEvent ack = new AckRemoteApplicationEvent(this,
```

```
                    this.serviceMatcher.getServiceId(),
                    this.bus.getAck().getDestinationService(),
                event.getDestinationService(), event.getId(), event.
                    getClass());
            this.cloudBusOutboundChannel
                    .send(MessageBuilder.withPayload(ack).build());
            this.applicationEventPublisher.publishEvent(ack);
        }
    }
    if (this.bus.getTrace().isEnabled() && this.applicationEventPublisher !=
        null) {
        this.applicationEventPublisher.publishEvent(new SentApplicationEvent(this,
            event.getOriginService(), event.getDestinationService(),
            event.getId(), event.getClass()));
    }
}
...
}
```

@ConditionalOnBusEnabled 注解是 bus 的开关，默认开启。@EnableBinding 绑定了 SpringCloudBusClient 中定义的通道。在应用服务启动时，自动化配置类加载了 bus 的 API 端点、刷新、ACK 追踪以及 bus 环境变量的配置等 Beans。@Output 表示输出绑定目标将由框架创建，由该通道发送消息。

acceptLocal 是一个基于注解实现的事件监听器方法，监听的事件类型为 RemoteApplicationEvent，对于该事件的处理逻辑是，当该事件是来自自己的并且不是 ACK 类事件，则发送消息。

@StreamListener 注解是 Spring Cloud Stream 提供的，用来标识一个方法作为 @EnableBinding 绑定的输入通道的监听器。acceptRemote 方法传递的参数 RemoteApplicationEvent 就是 stream 中的消息。如果是 ACK 类事件，当开启了事件追踪且事件不是来自于自身，则发布该事件，对于确认类事件，处理已经完成；如果自身需要处理该事件且该事件不是来自自身，则发布该事件。需要注意的是，当开启事件追踪时，构造一个 ACK 事件并将该事件发布；最后，当开启了事件追踪，这里的处理是注册已发送的事件，以便发布供本地消费，而不论其来源。

11.2.3 控制端点

Spring Cloud Bus 中提供了两个端点 /bus-env 和 /bus-refresh。端点的初始化也是放在了自动化配置类 BusAutoConfiguration 中，初始化的方式是通过构造函数，我们直接看一下具体端点的定义：

```
@Endpoint(id = "bus-refresh")
public class RefreshBusEndpoint extends AbstractBusEndpoint {

    @WriteOperation
    public void busRefreshWithDestination(@Selector String destination) {
        publish(new RefreshRemoteApplicationEvent(this, getInstanceId(), destination));
```

		}
	}

上面代码中的 /bus-refresh 端点，实际调用时会加上前缀"actuator"，由 Spring Boot Actuator 统一纳管。其主要的处理是通过 HTTP 主动调用，发布 RefreshRemoteApplicationEvent 事件，RefreshListener 监听到事件后，将变更了的环境变量配置信息作为参数，发布刷新应用配置的事件。该接口还有一个可选的请求参数，指定刷新的服务实例的范围。

```
@Endpoint(id = "bus-env")
public class EnvironmentBusEndpoint extends AbstractBusEndpoint {

    @WriteOperation
    public void busEnvWithDestination(String name, String value, @Selector String
        destination) {
        Map<String, String> params = Collections.singletonMap(name, value);
        publish(new EnvironmentChangeRemoteApplicationEvent(this, getInstanceId(),
            destination, params));
    }
}
```

如上为 /bus-env 端点的代码实现，其主要的功能是通过 HTTP 主动调用，发布 EnvironmentChangeRemoteApplicationEvent 事件。构造函数中传入了 Map，对应发生变更的键和变更后的配置值，用于刷新 Spring Environment。

11.3 应用进阶

Spring Cloud Bus 可以承载任何 RemoteApplicationEvent 类型的事件，应用进阶部分将会介绍如何自定义实现一个 RemoteApplicationEvent 事件。

注册一个新的事件类型，只需要将自定义的事件放在 org.springframework.cloud.bus.event 子包中。要自定义事件名称，可以在自定义类上使用 @JsonTypeName，或者依赖于使用类的简单名称的默认策略。需要注意的是，生产者和消费者都需要有权限访问该事件类。

同时由于篇幅受限，本节只会基于应用中的配置服务器进行功能的扩充，自定义实现一个 RemoteApplicationEvent 事件。下面所有的改动均是针对配置服务器。

11.3.1 在自定义的包中注册事件

如果不想将自定义的事件放在 org.springframework.cloud.bus.event 子包中，则必须通过 @RemoteApplicationEventScan 指定具体的包名来扫描 RemoteApplicationEvent 的子类。使用 @RemoteApplicationEventScan 指定的包也会覆盖子包。

例如，我们自定义一个事件叫做 CustomRemoteApplicationEvent，代码如下所示：

```
public class CustomRemoteApplicationEvent extends RemoteApplicationEvent {

    private CustomRemoteApplicationEvent() {
```

```
    }
    public CustomRemoteApplicationEvent(String content, String originService,
                                        String destinationService) {
        super(content, originService, destinationService);
    }
}
```

11.3.2 自定义监听器

我们可以通过如下的方式注册这个事件：

```
package com.blueskykong.chapter11.config.server.bus;

@Configuration
@RemoteApplicationEventScan
public class BusConfiguration {
    ...
}
```

在注解 @RemoteApplicationEventScan 中不指定任何值，使用该注解的类所在的包将会被注册，如上的例子中，package com.blueskykong.chapter11.config.server.bus 将会被注册。如下为该注解的定义：

```
public @interface RemoteApplicationEventScan {

    String[] value() default {};

    String[] basePackages() default {};

    Class<?>[] basePackageClasses() default {};
}
```

我们还可以在 @RemoteApplicationEventScan 注解中指定要注册的 value、basePackages 和 basePackageClasses。例如：

```
@Configuration
@RemoteApplicationEventScan(basePackageClasses = CustomRemoteApplicationEvent.class)
//@RemoteApplicationEventScan(basePackages = {"com.blueskykong.chapter11.config.
    server.bus"})
//@RemoteApplicationEventScan({"com.blueskykong.chapter11.config.server.bus"})
public class BusConfiguration {
    ...
}
```

其中，注解的三种方式是一样的，都是通过在 @RemoteApplicationEventScan 上明确指定包 com.blueskykong.chapter11.config.server.bus 来注册。这里的 package 也可以支持数组。

下面看一下监听器的具体实现：

```
@Configuration
```

```
@RemoteApplicationEventScan(basePackageClasses = CustomRemoteApplicationEvent.class)
public class BusConfiguration {

    private static org.slf4j.Logger logger = LoggerFactory.getLogger(BusConfiguration.
        class);

    @EventListener
     public void onUserRemoteApplicationEvent(CustomRemoteApplicationEvent event)
{

        logger.info("CustomRemoteApplicationEvent:" +
                    " Source: {} , originService: {} , destinationService: {} \n",
            event.getSource(),
            event.getOriginService(),
            event.getDestinationService());
    }
    ...
}
```

当监听到 CustomRemoteApplicationEvent 事件时，输出该事件的详细日志，包括 Source、originService 和 destinationService。

11.3.3 事件的发起者

这里，我们定义一个 API 接口，用以发起该事件，仿照 bus 中的 /bus-refresh 的实现，我们定义的控制类需要实现接口 ApplicationContextAware 和 ApplicationEventPublisherAware，分别用于获取应用上下文和发布事件，代码如下所示：

```
@RestController
public class BusEventController implements ApplicationContextAware,
    ApplicationEventPublisherAware {

    private ApplicationEventPublisher eventPublisher;
    private ApplicationContext applicationContext;

    @PostMapping("/bus/event/custom")
    public boolean publishUserEvent(@RequestBody String content,
        @RequestParam(value = "destination", required = false) String destination) {

        String serviceInstanceId = applicationContext.getId();
        CustomRemoteApplicationEvent event = new CustomRemoteApplicationEvent
            (content, serviceInstanceId, destination);
        eventPublisher.publishEvent(event);
        return true;

    }
    ...
}
```

content 用于接收 source 的内容，另外还有可选的请求参数 destination 用于指定接收该事件的目标服务实例范围。我们发起如图 11-7 所示的请求。

图 11-7 客户端发起事件请求

可以看到控制台的日志信息如下：

```
INFO 16586 --- [nio-8888-exec-7] c.b.c.c.server.bus.BusConfiguration      :
CustomRemoteApplicationEvent: Source: test onUserRemoteApplicationEvent ,
originService: config-server:8888 , destinationService: **
```

控制台显示 HTTP 调用之后成功发布了事件，事件监听器处理了该事件，并输出了如上的日志信息。

11.4 本章小结

消息总线是基于一个已经相当成熟的消息队列或消息系统进行二次封装。Spring Cloud Bus 建立在 Spring Cloud Stream 的基础上，将分布式系统中的节点和轻量的消息代理连接起来。Spring Cloud Bus 可以用来广播状态的改变（比如配置的改变）或者其他的管理工作。很重要的一点是，消息总线就像是一个分布式 Actuator，用于 Spring Boot 应用程序的扩展，但它也可以用作应用程序之间的通信通道。

第 12 章 认证与授权：Spring Cloud Security

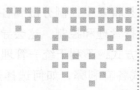

对于一个服务系统，安全是必须需要考虑的方面。不断追求更强的应用安全是一个目标，全面性、系统性的方法很重要，因为开发者永远不知道入侵者是如何对系统进行攻击。在系统安全的实现上，一般倡导使用多层安全保证，每一层的安全性越强，应用程序的安全性就越高。Java EE 应用程序位于安全层的高层，需要为它添加特定的问题域安全配置。

在本章的第一小节将对 Spring Cloud Security 中的前置知识 OAuth2 和 JWT 进行介绍，搭建一个基于 Spring Cloud Security 的认证和授权服务项目进行演示；第二小节将对 Spring Cloud Security 中的整体设计结构进行概括；第三小节从源码角度分别对 Spring Security、Spring Security OAuth2 中的设计和实现进行讲解；第四小节对 Spring Security 和 Spring Security OAuth2 中的配置属性进行讲解。

12.1 基础应用

应用的安全主要关注两个方面：认证（Authentication）和授权（Authoriztion），即你是谁，以及你能做什么。在单体应用中，开发者可以通过简单的拦截器以及会话（Session）机制对用户的访问进行控制和记录。而在分布式系统中，由于业务逻辑封装在各个微服务中，每个微服务都需要对用户的行为进行认证和许可，于是就产生了两种可能的方式：

❑ 第一种是通过一个中心化的权限管理系统，对用户的身份和权限进行统一的管理，可以做到一次授权，多次多点使用，但是这个独立的安全微服务需要聚合各个微服务中的权限控制逻辑，当添加一个新的基于不同业务逻辑实现的微服务就可能需要在安全微服务中添加新的实现。

- 第二种是将安全控制分散到各个微服务中，由各个微服务根据自身的业务对用户的访问进行管理和控制。这会导致安全管理过于分散，甚至每个微服务都有自己的一套实现方式，不利于统一管理。

这两种方式各有利弊，如何选择需要根据项目的具体业务需求进行判断，甚至在一定情况下可以结合使用。

Spring Cloud Security 提供了一组基本的组件用来构建安全应用程序和服务。它封装了 Spring Securtiy、Spring Security OAuth2 和 Spring Security JWT 的相关实现，同时提供自带的安全特性，致力于为 Spring Cloud 微服务提供快速创建常用安全模式的能力。Spring Cloud Security 的文档内容较为简略，但是通过应用与探索 Spring Security 和 Spring Security OAuth2，可以帮助开发者构建高健壮的安全应用。

在开始对相关的应用以及源码进行介绍之前，需要补充一部分前置知识，如 OAuth2 以及 JWT 等等。

12.1.1 OAuth2 简介

OAuth 协议的目的是为用户资源的授权提供一个安全的、开放而简易的标准。由于 OAuth2 不兼容 OAuth1，且签名逻辑过于复杂和授权流程过于单一，在此不过多谈论，以下重点关注 OAuth2 认证流程，它是当前 Web 应用中主流授权流程。

OAuth2 是当前授权的行业标准，其重点在于为 Web 应用程序、桌面应用程序、移动设备以及室内设备的授权流程提供简单的客户端开发方式。它为第三方应用提供对 HTTP 服务的有限访问，既可以是资源拥有者通过授权允许第三方应用获取 HTTP 服务，也可以是第三方以自己的名义获取访问权限。

1. 角色

OAuth2 中主要分为了 4 种角色：
- Resource Owner（资源所有者），是能够对受保护的资源授予访问权限的实体，可以是一个用户，这时会称为终端用户（end-user）。
- Resource Server（资源服务器），持有受保护的资源，允许持有访问令牌（Access Token）的请求访问受保护资源。
- Client（客户端），持有资源所有者的授权，代表资源所有者对受保护资源进行访问。
- Authorization Server（授权服务器），对资源所有者的授权进行认证，成功后向客户端发送访问令牌。

在很多时候，资源服务器和授权服务器是合二为一的，在授权交互的时候作为授权服务器，在请求资源交互时作为资源服务器。授权服务器也可以是单独的实体，它可以发出由多个资源服务器接受的访问令牌。

2. 协议流程

OAuth2 的认证流程图如图 12-1 所示。

主要包含以下的 6 个步骤：

1）客户端请求资源所有者的授权。

2）资源所有者同意授权，返回授权许可（Authorization Grant），这代表了资源所有者的授权凭证。

3）客户端携带授权许可要求授权服务器进行认证，请求访问令牌。

4）授权服务器对客户端进行身份验证，并认证授权许可，如果有效，返回访问令牌。

5）客户端携带访问令牌向资源服务器请求受保护资源的访问。

6）资源服务器验证访问令牌，如果有效，接受访问请求，返回受保护资源。

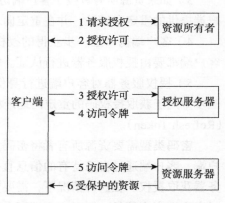

图 12-1 OAuth2 通用认证流程图

3. 客户端授权类型

为了获取访问令牌，客户端必须获取到资源所有者的授权许可。OAuth2 默认定义了四种授权类型，当然也提供了用于定义额外授权类型的扩展机制。默认的四种授权类型为：

- Authorization Code（授权码类型）
- Implicit（简化类型，也称为隐式类型）
- Resource Owner Password Credentials（密码类型）
- Client Credential（客户端类型）

下面对常用的授权码类型和密码类型进行详细的介绍。

授权码类型通过重定向的方式让资源所有者直接与授权服务器进行交互，通过这种方式授权，避免了资源所有者的信息泄漏给客户端，是功能最完整、流程最严密的授权类型，但是需要客户端必须能与资源所有者的代理（通常是 Web 浏览器）进行交互，并且可从授权服务器中接受请求（重定向给予授权码），授权流程图如图 11-2 所示。

授权流程说明如下：

1）客户端引导资源所有者的用户代理到授权服务器的端点（endpoint），一般通过重定向的方式。客户端提交的信息应包含客户端标识（Client Identifier）、请求范围（Requested Scope）、本地状态（Local State）和用于返回授权码的重定向地址（Redirection URI）。

2）授权服务器认证资源所有者（通过用户代理），并确认资源所有者允许还是拒绝客户端的访问请求。

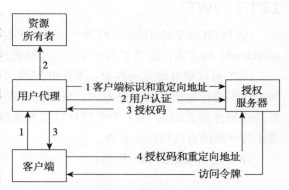

图 12-2 授权码类型流程图

3）如果资源所有者授予客户端访问权限，授权服务器通过重定向用户代理的方式回调客户端提供的重定向地址，并在重定向地址中添加授权码和客户端先前提供的任何本地状态。

4）客户端携带上一步获得的授权码向授权服务器请求访问令牌。在这一步中授权码和客户端都要由授权服务器进行认证。客户端需要提交用于获取授权码的重定向地址。

5）授权服务器对客户端进行身份验证，和授权码认证，确保接收到的重定向地址与第三步中用于获取授权码的重定向地址相匹配。如果有效，返回访问令牌，以及可选刷新令牌（Refresh Token）。

密码类型需要资源所有者将密码凭证交予客户端，客户端通过自己持有的信息直接向授权服务器获取授权。在这种情况下，需要资源所有者对客户端高度信任，同时客户端不允许保存密码凭证。这种授权类型适用于能够获取资源所有者的凭证（Credentials，如用户名和密码）的客户端。授权流程如图 12-3 所示。

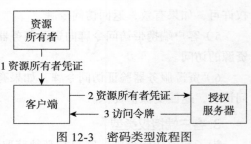

图 12-3　密码类型流程图

授权流程说明如下：

1）资源所有者向客户端提供其用户名和密码等凭证。

2）客户端携带资源所有者的凭证（用户名和密码），向授权服务器请求访问令牌。

3）授权服务器认证客户端并且验证资源所有者的凭证，如果有效，则返回访问令牌，以及可选的刷新令牌。

4. 令牌刷新

客户端从授权服务器中获取的访问令牌（Access Token）一般是具备时效性的。在访问令牌过期的情况下，持有有效用户凭证的客户端可以再次向授权服务器请求访问令牌。但是未持有用户凭证的客户端也可以通过和上次访问令牌一同返回的刷新令牌向授权服务器请求新的访问令牌。

12.1.2　JWT

JWT（JSON Web Token）作为一个开放的标准，通过紧凑（Compact）或者自包含（Self-contained）的方式，定义了用于在各方之间发送的安全 JSON 对象。

JWT 可以很好地充当在上一节中介绍的访问令牌和刷新令牌的载体，这是 Web 双方之间进行安全传输信息的良好方式。只有授权服务器持有签发和验证 JWT 的密钥，那么就只有授权服务器能验证 JWT 的有效性以及发送带有签名的 JWT，这就唯一保证了以 JWT 为载体的令牌的有效性和安全性。

JWT 的一般格式如下所示：

```
eyJhbGciOiJIUzI1NiIsInR5cCI6IkpXVCJ9.eyJuYW1lIjoiY2FuZyB3dSIsImV4cCI6MTUxODA1MTE
1NywidXNlcklkIjoiMTIzNDU2In0.IV4XZ0y0nMpmMX9orv0gqsEMOxXXNQOE680CKkkPQcs
```

它由三部分组成，每部分通过 . 分隔开，分别是：
- Header（头部）
- Payload（有效负荷）
- Signature（签名）

接着我们对每一部分进行详细的介绍。

头部通常由两部分组成：
- typ：类型，一般为 JWT。
- alg：加密算法，通常是 HMAC SHA256 或者 RSA。

一个简单的头部例子如下所示：

```
{
    "alg": "HS256",
    "typ": "JWT"
}
```

这部分 JSON 会由 Base64Url 编码，构成 JWT 的第一部分，如下所示：

```
eyJhbGciOiJIUzI1NiIsInR5cCI6IkpXVCJ9
```

有效负荷（Payload）是 JWT 的第二部分，是用来携带有效信息的载体，主要是关于用户实体和附加元数据的声明，组成如下：
- Registered claims（注册声明）。这是一组预定的声明，但并不强制要求。它提供了一套有用的、能共同使用的声明。主要有 iss（JWT 签发者）、exp（JWT 过期时间）、sub（JWT 面向的用户）、aud（接受 JWT 的一方）等。
- Public claims（公开声明）。公开声明中可以添加任何信息，一般是用户信息或者业务扩展信息等。
- Private claims（私有声明）。由 JWT 提供者和消费者共同定义的声明，既不属于注册声明也不属于公开声明。

一般不建议在 Payload 中添加任何敏感信息，因为 Base64 是对称解密的，这意味着 Payload 中的信息是可见的。

一个简单的 Payload 例子如下所示：

```
{
    "name": "cang wu",
    "exp": 1518051157,
    "userId": "123456"
}
```

这部分 JSON 会由 Base64Url 编码，构成 JWT 的第二部分，如下所示：

```
eyJuYW1lIjoiY2FuZyB3dSIsImV4cCI6MTUxODA1MTE1NywidXNlcklkIjoiMTIzNDU2In0
```

要创建签名，需要编码后的头部、编码后的 Payload、一个密钥，最后通过在头部 alg 键值定义的加密算法加密生成签名，生成签名的伪代码如下所示：

```
HMACSHA256(
    base64UrlEncode(header) + "." +
    base64UrlEncode(payload),
    secret)
```

上述代码中用到的加密算法为 HMACSHA256。密钥保存在服务端用于验证 JWT 以及签发 JWT，所以必须只由服务端持有，不该泄露出去。

一个简单的签名如下所示：

```
IV4XZ0y0nMpmMX9orv0gqsEMOxXXNQOE680CKkkPQcs
```

这将成为 JWT 的第三部分。

这三部分通过"."分割，组成最终的 JWT，如下所示：

```
eyJhbGciOiJIUzI1NiIsInR5cCI6IkpXVCJ9.eyJuYW1lIjoiY2FuZyB3dSIsImV4cCI6MTUxODA1MTE
1NywidXNlcklkIjoiMTIzNDU2In0.IV4XZ0y0nMpmMX9orv0gqsEMOxXXNQOE680CKkkPQcs
```

12.1.3 搭建授权服务器

本节将通过搭建一个授权服务器和资源服务器（二者可以合二为一，但是为了结构清晰还是将两者分开）进行演示，示例的结构如图 12-4 所示。

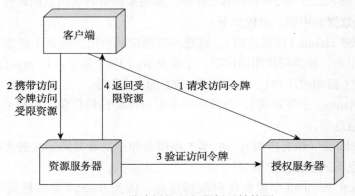

图 12-4 搭建授权服务器例子结构图

Spring Cloud Security 的配置主要是通过适配器的方式对开发者需要的部分进行覆盖配置，其余的将使用 Starter 中的默认配置选项。

可以搭建包含 Spring Cloud Security 依赖的 Spring Boot 项目。引入的主要依赖如下所示：

```xml
<dependency> <!--Eureka Client相关依赖-->
    <groupId>org.springframework.cloud</groupId>
    <artifactId>spring-cloud-starter-netflix-eureka-client</artifactId>
</dependency>
<dependency> <!--Spring Web MVC相关依赖-->
    <groupId>org.springframework.boot</groupId>
    <artifactId>spring-boot-starter-web</artifactId>
</dependency>
<dependency> <!--Spring Cloud Security相关依赖-->
```

```xml
        <groupId>org.springframework.cloud</groupId>
        <artifactId>spring-cloud-starter-security</artifactId>
</dependency>
```

在 application.yaml 中添加服务发现与注册相关配置，如下所示：

```yaml
eureka:
    instance:
        instance-id: ${spring.application.name}:${vcap.application.instance_id:${spring.application.instance_id:${random.value}}}
    client:
        service-url:
            default-zone: http://localhost:8761/eureka/, http://localhost:8762/eureka/
spring:
    application:
        name: authorization-server
server:
    port: 8766
```

这里使用的服务注册中心为第 4 章配置的 Eureka Server。

授权服务器配置文件 AuthorizationServerConfig 如下所示：

```java
@EnableAuthorizationServer //开启授权服务器
@Configuration
public class AuthorizationServerConfig extends AuthorizationServerConfigurerAdapter {
    private final static String RESOURCE_ID = "user";
    @Autowired
    AuthenticationManager authenticationManager;
    @Override
    public void configure(ClientDetailsServiceConfigurer clients) throws Exception {
        //配置一个客户端
        //既可以通过授权码类型获取令牌，也可以通过密码类型获取令牌
        clients.inMemory().withClient("client")// 客户端ID
            .authorizedGrantTypes("authorization_code","password", "refresh_token")// 客户端可以使用的授权类型
            .scopes("all")// 允许请求范围
            .secret("secret")// 客户端安全码
            .redirectUris("http://localhost:8888/");// 回调地址
    }
    // 配置AuthorizationServer tokenServices
    @Override
    public void configure(AuthorizationServerEndpointsConfigurer endpoints) throws Exception {
        endpoints
            .tokenStore(new InMemoryTokenStore())
            .accessTokenConverter(accessTokenConverter())
            .authenticationManager(authenticationManager)
            .reuseRefreshTokens(false);;
    }
    // 配置JWT转换器
    @Bean
    public JwtAccessTokenConverter accessTokenConverter() {
```

```
        JwtAccessTokenConverter converter = new JwtAccessTokenConverter();
        converter.setSigningKey("secret");
        return converter;
    }
    @Override
    public void configure(AuthorizationServerSecurityConfigurer security) throws
        Exception {

        // 允许所有人请求令牌
        // 已验证的客户端才能请求check_token端点
        security
            .tokenKeyAccess("permitAll()")
            .checkTokenAccess("isAuthenticated()")
            .allowFormAuthenticationForClients();
    }
}
```

通过 @EnableAuthorizationServer 开启授权服务器的相关自动配置类。通过继承 AuthorizationServerConfigurerAdapter 授权服务器配置适配器类，覆盖我们需要的配置。首先在内存中配置一个客户端（可以配置在数据库中），它可以通过密码类型和授权码类型获取到访问令牌；然后对 tokenServices 进行了配置，使用了 JWT 作为令牌的转换器，这里将 JWT 的签名的密钥设置为 "secret"，同时使用 InMemoryTokenStore 将令牌保存到内存中（可以保存到数据库和 Redis 中，官方提供了相关的默认实现 JdbcTokenStore 和 RedisTokenStore），最后对授权端点的访问控制进行配置。

到这里授权服务器的基本功能就实现了。为了演示，通过 Spring Security 为授权服务器配置两个存储在内存中的用户。添加 Spring Security 自定义配置类 SecurityConfig，具体代码如下所示：

```
@Configuration
@EnableWebSecurity
public class SecurityConfig extends WebSecurityConfigurerAdapter{

    @Override
    protected void configure(AuthenticationManagerBuilder auth) throws Exception {
        // 在内存中配置两个用户
        InMemoryUserDetailsManager userDetailsManager = new InMemoryUserDetailsManager();
        userDetailsManager.createUser(User.withUsername("user_authorization_code").
        password("123456").authorities("USER").build());
        userDetailsManager.createUser(User.withUsername("user_password").
        password("123456").authorities("USER").build());
        auth.userDetailsService(userDetailsManager);
    }
    @Bean
    @Override
    public AuthenticationManager authenticationManagerBean() throws Exception {
        AuthenticationManager manager = super.authenticationManagerBean();
        return manager;
    }
    @Override
```

```java
protected void configure(HttpSecurity http) throws Exception {
    // 允许访问/oauth授权接口
    http.csrf().disable()
        .sessionManagement().sessionCreationPolicy(SessionCreationPolicy. IF_
            REQUIRED)
        .and()
        .requestMatchers().anyRequest()
        .and()
        .formLogin().permitAll()
        .and()
        .authorizeRequests()
        .antMatchers("/oauth/*").permitAll();
}

    // 配置密码解码器
    @Bean
    public PasswordEncoder passwordEncoder(){
        return NoOpPasswordEncoder.getInstance();
    }

}
```

上述代码中通过替换 AuthenticationManager 在内存中添加了两个用户信息，同时允许对"/oauth/*"授权端点的访问，最后配置了一个 NoOpPasswordEncoder 用于对认证的密码进行加密和解密。

12.1.4 配置资源服务器

资源服务器与授权服务器的依赖一致，主要依赖如下所示：

```xml
<dependency> <!--Eureka Client相关依赖-->
    <groupId>org.springframework.cloud</groupId>
    <artifactId>spring-cloud-starter-netflix-eureka-client</artifactId>
</dependency>
<dependency> <!--Spring Web MVC相关依赖-->
    <groupId>org.springframework.boot</groupId>
    <artifactId>spring-boot-starter-web</artifactId>
</dependency>
<dependency> <!--Spring Cloud Security相关依赖-->
    <groupId>org.springframework.cloud</groupId>
    <artifactId>spring-cloud-starter-security</artifactId>
</dependency>
```

在 application.yaml 中添加服务发现与注册相关配置，如下所示：

```yaml
eureka:
  instance:
    instance-id: ${spring.application.name}:${vcap.application.instance_id:${spring.application.instance_id:${random.value}}}
  client:
    service-url:
      defaultZone: http://localhost:8761/eureka/
```

```yaml
spring:
  application:
    name: resource-server
server:
  port: 8767
```

这里使用的服务注册中心为第 4 章配置的 Eureka Server。

资源服务器应该持有受保护的资源，因此构建了一些对外访问的接口来提供可访问的资源。添加 ResourceController 代码如下所示：

```java
@RestController
public class ResourceController {
    private final static Logger logger = LoggerFactory.getLogger(ResourceController.
        class);
    private final static String DEFAULT_NAME = "xuan";
    private static String DEFAULT_SERVICE_ID = "application";
    private static String DEFAULT_HOST = "localhost";
    private static int DEFAULT_PORT = 8080;
    // 受保护的资源
    @RequestMapping(value = "/user/{userId}", method = RequestMethod.GET)
    public User getUserByUserId(@PathVariable("userId") String userId){
        logger.info("Get User by UserId {}", userId );
        return new User(userId, DEFAULT_NAME);
    }
    // 不受保护的资源
    @RequestMapping(value = "/instance/{serviceId}", method = RequestMethod.GET)
    public Instance getInstanceByServiceId(@PathVariable("serviceId") String
        serviceId){
        logger.info("Get Instance by serviceId {}", serviceId);
        return new Instance(serviceId, DEFAULT_HOST, DEFAULT_PORT);
    }
}
```

上述接口中，/user/{userId} 需要认证后才能进行访问，而 /instance/{serviceId} 接口不需要认证，可以直接访问。接着通过 ResourceServerConfigurerAdapter 对资源服务器进行相关配置。添加 ResourceServerConfig，具体代码如下所示：

```java
@Configuration
@EnableResourceServer //开启资源服务器
public class ResourceServerConfig extends ResourceServerConfigurerAdapter{
    @Autowired
    RestTemplate restTemplate;
    @Override
    public void configure(ResourceServerSecurityConfigurer resources) {
        resources
            .tokenStore(new JwtTokenStore(accessTokenConverter()))
            .stateless(true);
        // 配置RemoteTokenServices,用于向AuththorizationServer验证令牌
        RemoteTokenServices tokenServices = new RemoteTokenServices();
        tokenServices.setAccessTokenConverter(accessTokenConverter());
        RestTemplate restTemplate = restTemplate();
        //为restTemplate配置异常处理器,忽略400错误,
```

```java
restTemplate.setErrorHandler(new DefaultResponseErrorHandler() {
    @Override
    // Ignore 400
    public void handleError(ClientHttpResponse response) throws IOException {
        if (response.getRawStatusCode() != 400) {
            super.handleError(response);
        }
    }
});
tokenServices.setRestTemplate(restTemplate);

tokenServices.setCheckTokenEndpointUrl("http://AUTHORIZATION-SERVER/
    oauth/check_token");
tokenServices.setClientId("client");
tokenServices.setClientSecret("secret");
resources
        .tokenServices(tokenServices)
        .stateless(true);
}
// 配置JWT转换器
@Bean
public JwtAccessTokenConverter accessTokenConverter() {
    JwtAccessTokenConverter converter = new JwtAccessTokenConverter();
    converter.setSigningKey("secret");
    return converter;
}
@Bean
@LoadBalanced
RestTemplate restTemplate(){
    return new RestTemplate();
}


@Override
public void configure(HttpSecurity http) throws Exception {
    // 配置资源服务器的拦截规则
    http.
        sessionManagement().sessionCreationPolicy(SessionCreationPolicy.IF_
            REQUIRED)
        .and()
        .requestMatchers().anyRequest()
        .and()
        .anonymous()
        .and()
        .authorizeRequests()
        .antMatchers("/user/**").authenticated() // /user/** 端点的访问必须要验证
        .and()
        .exceptionHandling().accessDeniedHandler(new OAuth2AccessDeniedHandler());
    }
}
```

通过 @EnableResourceServer 开启资源服务器的相关自动配置类。通过继承

ResourceServerConfigurerAdapter 资源服务器配置适配器类，覆盖所需配置。首先配置了一个 RemoteTokenServices 用于从 AuthorizationServer 中获取令牌服务（需要注意这里使用的 clientId 以及 clientSecret 是单一的，如果需要对多客户端进行认证，可能需要自己对其进行重写），通过 HTTP 调用的方式解析访问令牌（主要是通过授权服务的 /oauth/check_token 端点对令牌进行解析）。在使用 JWT 作为令牌的载体可以减少这次网络的通信，因为 JWT 中的签名机制可以保证访问令牌的有效性，同时 JWT 的自包含性也保证了令牌中携带了足够的信息，使用与授权服务器中相同密钥的 JwtAccessTokenConverter 即可对令牌进行解析认证，验证令牌并获取令牌中的相关信息。最后使用资源服务器的 HttpSecurity 对相关资源接口的访问进行控制。

12.1.5 访问受限资源

依次启动 Eureka Server 以及上述的授权服务器和资源服务器。授权服务器启动后会配置与令牌相关的端点，如下所示：

```
{[/oauth/authorize]}
{[/oauth/authorize],methods=[POST]    // 授权码类型和隐式类型的端点
{[/oauth/token],methods=[GET]}
{[/oauth/token],methods=[POST]}       // 获取令牌的端点
{[/oauth/check_token]}                // 检查令牌有效性的端点
{[/oauth/error]}
```

首先直接访问不受保护的资源接口 /instance/{serviceId}，如下所示：

```
http://localhost:8767/instance/application
```

可以直接获得如下结果：

```
{"serviceId":"application","host":"localhost","port":8080}
```

接着访问受保护的资源端口 /user/{userId}，如下所示：

```
http://localhost:8767/user/cangwu
```

返回结果如下所示：

```
{
    "error": "unauthorized",
    "error_description": "Full authentication is required to access this resource"
}
```

说明该接口的访问需要认证，所以我们需要向授权服务器请求访问令牌用以访问该受限接口。

1. 通过授权码类型获取访问令牌

授权码类型请求地址为：

```
http://localhost:8766/oauth/authorize?client_id=client&response_type=code&redirect_
    uri=http://localhost:8888/login
```

需要在请求中携带 client_id（客户端 id）、response_type（返回类型，例子中指定为授权码，还可以指定为令牌，在客户端支持隐式类型的情况下可直接获取到令牌）、redirect_uri（返回授权码的回调地址）。

获取授权码需要获得用户的授权，所以会被引导到登录界面，用户登录后同意授权，如图 12-5 所示。

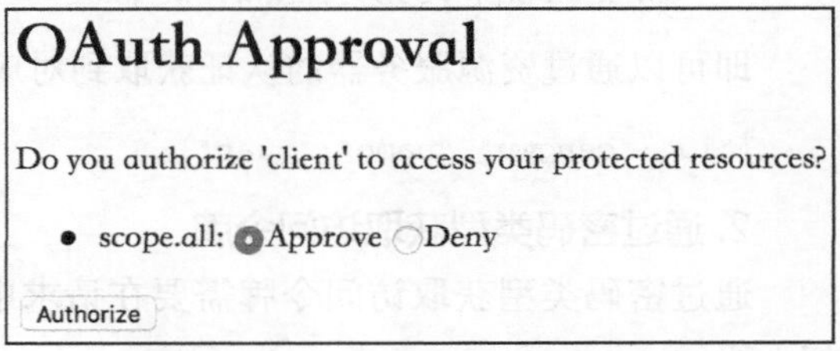

授权码将会通过回调地址获取，如下所示：

```
http://localhost:8888/?code=weimiA
```

图 12-5　用户授权页面

weimiA 即为授权码。

接着携带授权码访问"/oauth/token"获取到访问令牌，需要指定 grant_type（授权类型）为 authorization_code（授权码）和携带相关的客户端信息，包括回调地址、客户端标识和客户端的密码等：

```
method: post
uri: http://localhost:8766/oauth/token?grant_type=authorization_code&code=weimiA&redirect_
    uri=http://localhost:8888/login&client_id=client&client_secret=secret

reponse:
{
    "access_token":  "eyJhbGciOiJIUzI1NiIsInR5cCI6IkpXVCJ12.eyJleHAiOjE1MTkxNTAyM
        DMsInVzZXJfbmFtZSI6InVzZXJfYXV0aG9yaXphdGlvbl9jb2RlIiwiYXV0aG9yaXRpZXMi
        OlsiVVNFUiJdLCJqdGkiOiIzZTBmNDk1Ny0zM2FkLTQ4MzktOTczNy01YTQyNzI3NTZjZDc
        iLCJjbGllbnRfaWQiOiJjbGllbnQiLCJzY29wZSI6WyJhbGwiXX0.3VhN5z_fWRxMnFSqA_
        yb4B33l9NWwv2wCmUKXK--KUo",
    "token_type": "bearer",
    "refresh_token":  "eyJhbGciOiJIUzI1NiIsInR5cCI6IkpXVCJ12.eyJ1c2VyX25hbWUiOiJ1c2V
        yX2F1dGhvcml6YXRpb25fY29kZSIsInNjb3BlIjpbImFsbCJdLCJhdGkiOiIzZTBmNDk1Ny0zM2
        FkLTQ4MzktOTczNy01YTQyNzI3NTZjZDciLCJleHAiOjE1MjE2OTkwMDMsImF1dGhvcml0aWVzI
        jpbIlVTRVIiXSwianRpIjoiNjcyMjQxYzgtNGJiNC00MzRiLTk5ZWYtMTgxODMyNjZmYjY1Iiwi
        Y2xpZW50X2lkIjoiY2xpZW50In0.mG9bprlG1Nx83dhJijAWktXwM9Ylt_AJTwf3gmiXS6k",
    "expires_in": 43199,
    "scope": "all",
    "jti": "3e0f4957-33ad-4839-9737-5a4272756cd7"
}
```

其 response 中携带了 access_token，即访问令牌。token_type 说明了令牌的类型，一般为 bearer 或 refresh_token 可以在访问令牌过期之后用来向授权服务器请求新的访问令牌，expires_in 说明了令牌的有效时间，单位为秒，上面令牌的有效时间为 43199 秒，即 12 个小时。

然后携带访问令牌再次访问 /user/{userId}，如下所示：

```
http://localhost:8767/user/cangwu?access_token=eyJhbGciOiJIUzI1NiIsInR5cCI6IkpXV
    CJ12.eyJleHAiOjE1MTkxNTAyMDMsInVzZXJfbmFtZSI6InVzZXJfYXV0aG9yaXphdGlvbl9jb2R
    lIiwiYXV0aG9yaXRpZXMiOlsiVVNFUiJdLCJqdGkiOiIzZTBmNDk1Ny0zM2FkLTQ4MzktOTczNy0
    1YTQyNzI3NTZjDciLCJjbGllbnRfaWQiOiJjbGllbnQiLCJzY29wZSI6WyJhbGwiXX0.3VhN5z_
    fWRxMnFSqA_yb4B3319NWwv2wCmUKXK--KUo
```

即可以通过资源服务器的认证获取到对应的资源，结果如下所示：

```
{"id":"cangwu","name":"xuan"}
```

2. 通过密码类型获取访问令牌

通过密码类型获取访问令牌需要在请求中直接携带用户的凭证，如用户的账号和密码，如下所示：

```
http://localhost:8766/oauth/token?username=user_password&password=123456&grant_
    type=password&scope=all&client_id=client&client_secret=secret
```

即可以获取到对应的访问令牌，结果如下所示：

```
{
    "access_token": "eyJhbGciOiJIUzI1NiIsInR5cCI6IkpXVCJ12.eyJleHAiOjE1MTkxNTA3NTMs
        InVzZXJfbmFtZSI6InVzZXJfcGFzc3dvcmQiLCJhdXRob3JpdGllcyI6WyJVU0VSIl0sImp0aSI
        6ImU2Nzk2YzIxLTIyMTktNDhmOC04MDIyLTEyNjYyZjFkMTcwMiIsImNsaWVudF9pZCI6ImNsaW
        VudCIsInNjb3BlIjpbImFsbCJdfQ.3_7FFR-8mlMc5fYccdRTNczVVCXTzp-LgIAXKLokrT4",
    "token_type": "bearer",
    "refresh_token": "eyJhbGciOiJIUzI1NiIsInR5cCI6IkpXVCJ12.eyJ1c2VyX25hbWUiOiJ1
        c2VyX3Bhc3N3b3JkIiwic2NvcGUiOlsiYWxsIl0sImF0aSI6ImU2Nzk2YzIxLTIyMTktNDhm
        OC04MDIyLTEyNjYyZjFkMTcwMiIsImV4cCI6MTUyMTY5OTU1MywiYXV0aG9yaXRpZXMiOlsi
        VVNFUiJdLCJqdGkiOiI0NWU2NWU3ZC1mYzhiLTQ3ZTItODczNi1jMGUxNWViN2M5NTUiLCJj
        bGllbnRfaWQiOiJjbGllbnQifQ.sNxlgeSljccDVxKyfIhYRdWXTGTKnvkD7XyZceULtqs",
    "expires_in": 43199,
    "scope": "all",
    "jti": "e6796c21-2219-48f8-8022-12662f1d1702"
}
```

其他的接口类似于 /oauth/check_token，读者可自行尝试，后面的源码解析也会对这些接口进行讲解。

12.2 整体架构

本节将对 Spring Cloud Security 中的整体设计结构进行大致分解。

1. Spring Security 架构

在 Spring Security 中，对 Web 应用的端点安全主要是通过 Servlet Filter（过滤器）对访问进行控制和拦截实现。

一个简单的 HTTP 请求如图 12-6 所示。客户端发送请求到服务器中，由容器（Container）根据请求的 URI 将其交给相应的过滤器（Filter）和 Servlet 进行处理。通常来讲，一个 Servlet 处理一个

图 12-6　HTTP 请求流程

简单的请求，这其中过滤器组成一个调用链，并按顺序进行调用。调用链中的每一个过滤器都能够结束请求。过滤器能够修改请求和响应，并将它们传递给下游的过滤器和 Servlet。过滤器调用链（Filters Chain）的顺序非常重要，这将影响到请求的处理结果。

从根本上讲，Spring Security 也是一个过滤器调用链中一个节点，它通过委托的方式将请求转交给一条过滤器调用链进行处理，通常称为 FilterChainProxy，如图 12-7 所示。

事实上，在安全过滤器调用链中甚至还有一层中间层。在容器中，这层中间层称为 DelegatingFilterProxy，不是由 Spring 管理的。DelegatingFilterProxy 委托 FilterChainProxy 进行处理。在 FilterChainProxy 中，包含了所有具备安全处理逻辑的过滤器调用链。所有的过滤器继承相同的接口（Filter），都有机会拦截和处理请求。

在 Spring Security 的顶层 FilterChainProxy 中，管理了大量的过滤器调用链，一个请求将被转发到第一条匹配它路径的调用链进行处理。在转发的过程中，将确保只有一条调用链处理请求，不会造成重复处理。图 12-8 展示了一个转发匹配规则，通常"/foo/**"比"/**"更先进行匹配。

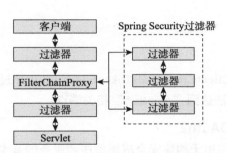

图 12-7　Spring Security 过滤器

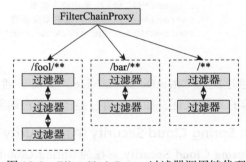

图 12-8　FilterChainProxy 过滤器调用链代理

在 WebSecurityConfigurerAdapter 中，通过覆盖 HttpSecurity 对过滤器调用链进行组装和指定其匹配的路径以对访问安全进行控制，如下所示：

```
protected void configure(HttpSecurity http) throws Exception;
```

通过 HttpSecurity 给特定路径配置相应的安全规则，其结果将由 Spring Security 为应用生成对应的过滤器调用链对访问进行控制。如在配置授权服务器中进行的如下 Spring Security 配置：

```
...
.antMatchers("/oauth/*").permitAll();
...
```

对于路径为"/oauth/*"的访问不需要任何的验证。

2. Spring Security OAuth2 架构

Spring Security OAuth2 将应用分为两个部分：授权服务器和资源服务器。授权服务器中持有用户的身份和权限信息，用于第三方客户端携带用户的授权请求访问令牌；资源服务器中持有用户的资源，允许第三方客户端携带授权服务器颁发的访问令牌访问受保护的用户资源。

Spring Security OAuth2 作为 Spring Security 的子项目，对于 Web 安全的控制也是通过安全过滤器调用链实现，整体的架构方向不变，但是添加了自己特有的过滤器以及端点。

如在授权服务器中，ClientCredentialsTokenEndpointFilter 对客户端的 id 和 secret 进行验证，需要在适配器中为其配置 ClientDetailsService，如在 AuthorizationServerConfig 中进行如下配置：

```
@Override
public void configure(ClientDetailsServiceConfigurer clients) throws Exception {
    //配置一个客户端
    //既可以通过授权码类型获取令牌，也可以通过密码类型获取令牌
    clients.inMemory().withClient("client")// 客户端ID
        .authorizedGrantTypes("authorization_code","password", "refresh_token")
            // 客户端可以使用的授权类型
        .scopes("all")// 允许请求范围
        .secret("secret")// 客户端安全码
        .redirectUris("http://xuan.com");// 回调地址

}
```

通过配置 ClientDetailsService，实现了通过 ClientCredentialsTokenEndpointFilter 过滤器对客户端的认证，如果未通过验证，客户端将无法达到"/oauth/*"相关端点。

3. Spring Cloud Security 中的 Security 和 OAuth2

Spring Cloud Security 意在为 Spring Cloud 提供用于构建安全应用程序和服务的基本组件，根本实现还是 Spring Security，Spring Cloud Security 中包含了 Security 以及 OAuth 的主要依赖，如下所示：

```
<dependency>    <!--spring security相关依赖-->
    <groupId>org.springframework.boot</groupId>
    <artifactId>spring-boot-starter-security</artifactId>
</dependency>
<dependency>    <!-spring security oauth2自动配置相关依赖-->
    <groupId>org.springframework.security.oauth.boot</groupId>
    <artifactId>spring-security-oauth2-autoconfigure</artifactId>
</dependency>
```

但是它也提供了用于快速搭建第三方登录客户端的实现，如添加 @EnableOAuth2Sso 注解，用于快速搭建第三方登录客户端应用。

12.3 源码解析

鉴于 Spring Security 是整个 Spring Cloud Security 安全应用的基础，所以接下来的源码

解析中，先对 Spring Security 中的关键体系进行讲解，包括 SecurityContext（安全上下文）在请求中记录用户信息和权限身份的作用、核心功能模块认证和授权以及 Spring Security 中的访问安全控制，最后介绍 OAuth2 中的资源服务器、授权服务器中的拦截器和相关端点以及 Spring Cloud Security 提供的额外特性。

12.3.1 安全上下文

1. SecurityContext

SecurityContext 是 Spring Security 中的安全上下文，绑定了当前请求中的认证信息 Authentication，提供了当前用户的认证信息以及角色权限等内容。它通过在 ThreadLocal 中保存 Authentication，与当前执行线程绑定，维护当前访问者的安全信息。

SecurityContext 提供的接口如下：

```java
//SecurityContext.java
public interface SecurityContext extends Serializable {
    // 获取Authentication用户认证信息
    Authentication getAuthentication();
    // 设置Authentication用户认证信息
    void setAuthentication(Authentication authentication);
}
```

其默认实现为 SecurityContextImpl，位于 org.springframework.security.core.context 包中。

2. SecurityContextHolder

SecurityContextHolder 作为 SecurityContext 持有者，为当前执行线程提供 SecurityContext，其获取 SecurityContext 的方法为静态方法，如下所示：

```java
//SecurityContextHolder.java
public static SecurityContext getContext() {
    return strategy.getContext();
}
```

通过 SecurityContext 可以直接获取到当前请求用户的 Authentication，用于身份认证和处理，Spring Security 通过这种方式维护贯穿整个请求的认证信息，同时在请求结束时清理该 SecurityContext。SecurityContextHolder 中清理安全上下文的方法如下所示：

```java
//SecurityContextHolder.java
public static void clearContext() {
    strategy.clearContext();
}
```

考虑 JVM 实现之间的差异，SecurityContextHolder 内部是通过策略模式来指派不同的 SecurityContextHolderStrategy 将 SecurityContext 与当前的线程进行绑定。通过 strategyName 来指定不同的 SecurityContextHolderStrategy，默认为 MODE_THREADLOCAL，即使用 ThreadLocalSecurityHolderStrategy。

3. SecurityContextHolderStrategy

SecurityContextHolderStrategy 提供了 SecurityContext 的保存策略，该类位于 org.springframework.security.core.context 包中，主要方法如下所示：

```
//SecurityContextHolderStrategy.java
public interface SecurityContextHolderStrategy {
    // 清理安全上下文
    void clearContext();
    // 获取安全上下文
    SecurityContext getContext();
    // 设置安全上下文
    void setContext(SecurityContext context);
    SecurityContext createEmptyContext();
}
```

Spring Security 提供三种具体实现，分别为：

- ThreadLocalSecurityHolderStrategy，是 SecutityContextHolderStrategy 的默认实现，主要通过 ThreadLocal 将 SecurityContext 与当前访问的线程进行绑定。
- GlobalSecurityContextHolderStrategy，全局保存策略，全部请求使用一个 SecurityContext。
- InheritableThreadLocalSecurityHolderStrategy，与 ThreadLocalSecurityHolderStrategy 类似，但是支持从父线程中存取 SecurityContext。

12.3.2 认证

Spring Security 的安全主要分为两个核心模块，认证和授权。Spring Security 对此都有类似的架构和设计来方便扩展和替换安全策略。

认证能确定当前访问系统的用户是谁，是否是系统认证的用户，从而判断用户的请求能否继续进行。其相关实现类图如图 12-9 所示。

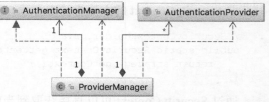

图 12-9 认证类图结构

认证模块的类包括：

- AuthenticationManager 提供认证的主接口。
- ProviderManager 为 AuthenticationManager 的默认实现。
- AuthenticationProvider 为 ProviderManager 中具体进行认证操作的委托类。

下面我们具体对这些类进行介绍。

1. AuthenticationManager

AuthenticationManager 作为认证管理器，是认证的主要接口。一般来说，在 Spring 中拦截到的相关用户信息会封装成 Authentication，然后交给 AuthenticationManager 进行认证处理。在基础应用部分，我们通过 AuthenticationManager 为授权服务器添加了两个内置的用户。以下为该接口提供的唯一方法：

```
//AuthenticationManger.java
public interface AuthenticationManger {
    Authentication authenticate(Authentication authentication) throw AuthenticationException
}
```

该方法尝试认证一个传入的 Authentication 对象,并返回一个经过充分鉴定的 Authentication 对象,这个对象包含 authorities(用户权限)和 credentials(用户凭证)。authenticate 方法一般有三种结果:

- 返回一个 Authentication(通常 authenticated=true),表示这是一个经过验证且有效的用户。
- 抛出 AuthenticationException,表示这是一个非法的用户。
- 当它无法判定时返回 null。

Authentication 继承于 Principal 类,代表一个认证的请求或者一个已经认证过的用户,其内封装了一个请求中的用户相关信息,是 Spring Security 中最高级别的身份认证抽象类。认证的过程一般在 AuthenticationManager#authenticate 中进行。一旦用户信息通过认证会被 SecurityContext 保存到 ThreadLocal 中,与访问的线程进行绑定,当作后面进行安全检查的用户信息基础。如果 Authentication 中的 authenticated 没有设为 true,它会由框架中任何安全过滤器进行重新认证。

Authentication 中提供的接口如下:

```
//Authentication.java
public interface Authentication extends Principal, Serializable {
    //在认证过程中由AuthenticationManager注入该用户所授予的权限
    Collection<? extends GrantedAuthority> getAuthorities();
    //一般用来证明该用户是否合法,大多数情况是密码,但是也可能是其他的认证证书
    Object getCredentials();
    //认证请求的一些额外信息,例如IP地址
    Object getDetails();
    //被认证用户的身份,如果是账号密码认证,这通常是账号。大多数情况下,AuthenticationManager
       在认证后会注入更为丰富的身份信息,所以Pincipal一般会被AuthenticationProvider进行扩展
    Object getPrincipal();
    //指出该Authentication是否应该提交到AuthenticationManager中进行认证。一般来说,
       AuthenticationManager会在认证成功之后返回一个不可改变的Authentication, 该方法会返
       回true,然后其他的安全拦截器将不会请求AuthenticationManager进行认证,这将提升性能
    boolean isAuthenticated();
    void setAuthenticated(boolean isAuthenticated) throws IllegalArgumentException;
}
```

Principal 类代表了用户的主要属性,如用户名或者用户 id 这些能够标志用户的信息。

2. ProviderManager

ProviderManager 是 AuthenticationManager 的默认实现,其内通过委托者模式提供对不同类型的 Authentication 进行认证。

ProviderManager 中持有按照顺序排列的 AuthenticationProvider 列表。在 ProviderManager#authenticate 方法中通过循环的方式调用 AuthenticationProvider#authenticate 方法对传递的

Authentication 进行尝试认证。当列表中有一个 AuthenticationProvider 对 Authentication 的认证返回一个非空的 Authentication 并且无异常抛出，说明该 AuthenticationProvider 有能力对传递的 Authentication 进行认证并且认证成功，调用链下游的 AuthenticationProvider 将不需要继续尝试认证。

在尝试认证的过程中会对上次认证产生的 AuthenticationException 进行保存。当没有 AuthenticationProvider 返回非空的 Authentication 时，最后一次遗留的 AuthencationException 将会使用。

如果列表中的 AuthenticationProvider 都不能对 Authentication 进行认证，那么 parent （AuthenticationManager）将尝试进行认证。当然如果 parent 都无法返回非空的 Authentication，那么将会抛出 ProviderNotFoundException。

在获取到有效的认证结果 Authentication 后，将会进行必要的清理工作，例如清理 Authentication 中的 Credentials 属性。这个过程中还有认证成功或者认证失败的事件广播，但是默认实现为空。

通过这样的委托者方式，Spring Security 可以根据不同的 Authentication 调用不同的 AuthenticationProvider 进行认证，使 Spring Security 对不同认证方式提供实现，例如账号密码认证，访问令牌认证等等。ProviderManager 中的认证流程方法实现如下所示：

```java
//ProviderManager.java
public Authentication authenticate(Authentication authentication) throws
    AuthenticationException {
    Class<? extends Authentication> toTest = authentication.getClass();
    AuthenticationException lastException = null;
    Authentication result = null;
    //循环调用AuthenticationProvider尝试对Authentication进行认证
    for (AuthenticationProvider provider : getProviders()) {
        //检查AuthenticationProvider是否支持对该类Authentication进行认证，
        //如果不支持，则进入下一个循环
        if (!provider.supports(toTest)) {
            continue;
        }
        //注意在循环调用的过程中，如果没有认证成功，只有最后一次捕捉异常有效
        try {
            result = provider.authenticate(authentication);
            if (result != null) {
                // 认证成功,为Authentication注入更丰富的用户信息
                copyDetails(authentication, result);
                break;
            }
        }catch (AccountStatusException e) {
            throw e;
        }catch (InternalAuthenticationServiceException e) {
            prepareException(e, authentication);
            throw e;
        }catch (AuthenticationException e) {
            lastException = e;
```

```
        }
    }
    //在AuthenticationProvider列表尝试认证失败后尝试调用parent进行认证
    if (result == null && parent != null) {
        // Allow the parent to try.
        try {
            result = parent.authenticate(authentication);
        }catch (ProviderNotFoundException e) {
        }catch (AuthenticationException e) {
            lastException = e;
        }
    }
    //认证成功后的清理操作或者其他操作
    if (result != null) {
        if (eraseCredentialsAfterAuthentication
                && (result instanceof CredentialsContainer)) {
            // Authentication is complete. Remove credentials and other secret data
            // from authentication
            ((CredentialsContainer) result).eraseCredentials();
        }
    //认证成功的事件广播
        eventPublisher.publishAuthenticationSuccess(result);
        return result;
    }
    //认证失败将抛出最后一次保留的异常
    if (lastException == null) {
        lastException = new ProviderNotFoundException(...);
    }
    //认证失败的事件广播
    prepareException(lastException, authentication);
    throw lastException;
}
```

从代码中可见，ProviderManager#authenticate 方法将根据 Authentication 的类型遍历 AuthenticationProvider 列表，委托合适的 AuthenticationProvider 对 Authentication 认证。如果认证成功，将根据认证结果为 Authentication 注入更丰富的用户信息，如用户权限等，清理掉 Authentication 中的敏感信息，如用户凭证，最后返回认证成功的 Authentication。认证失败将会抛出认证失败的异常 AuthenticationException。如果认证过程没有找到合适 AuthenticationProvider 对 Authentication 进行认证，将会抛出 ProviderNotFoundException 异常。

3. AuthenticationProvider

AuthenticationProvider 是真正进行认证的接口类，将对具体的 Authentication 进行认证。通过注入不同的排列组合的 AuthenticationProvider 实现，可以使 ProviderManager 针对不同的请求进行可配置化的认证。AuthenticationProvider 提供的接口如下所示：

```
public interface AuthenticationProvider {
    //执行认证的具体方法
    //通常会返回一个带有credentials的被认证过的Authentication
    //当不支持认证传递过来的Authentication时，也可能返回null，然后下一个AuthenticationProvider
```

```
                      会在ProviderManager被调用时进行认证
//认证失败会抛出AuthenticationException
Authentication authenticate(Authentication authentication) throws AuthenticationException;
//返回该AuthenticationProvider是否支持认证该Authentication类型。然而返回true也不能
  保证一定能进行认证，authenticate方法依然有可能会返回null
boolean supports(Class<?> authentication);
}
```

在 Spring Security 中 AuthenticationProvider 有非常多的默认实现，图 12-10 展示了部分的实现类。

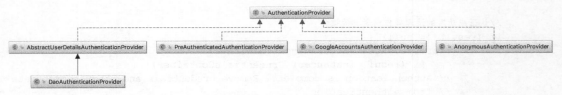

图 12-10 AuthenticationProvider 实现类

有用户名密码认证的 AbstractUserDetailsAuthenticationProvider，进行预认证的 PreAuthenticationProvider、使用 Google 账号进行认证 GoogleAccountsAuthenticationProvider，还有匿名认证的 AnonymousAuthenticationProvider 等等。

下面我们将对最常用的账号密码登录认证的 AbstractUserDetailsAuthenticationProvider 和 DaoAuthenticationProvider 进行介绍。

（1）AbstractUserDetailsAuthenticationProvider

AbstractUserDetailsAuthenticationProvider 继承于 AuthenticationProvider，是最常用的用户名密码认证的 AuthenticationProvider 抽象类，它将认证 UsernamePasswordAuthenticationToken（Authentication 的一个子类，UsernamePasswordAuthenticationToken 中封装了请求用户的用户名和密码用于认证）。AbstractUserDetailsAuthenticationProvider 的默认实现为 DaoAuthenticationProvider。

AbstractUserDetailsAuthenticationProvider#authenticate 方法的代码如下所示：

```
//AbstractUserDetailsAuthenticationProvider.java
public Authentication authenticate(Authentication authentication)throws AuthenticationException {
    String username = (authentication.getPrincipal() == null) ? "NONE_PROVIDED"
            : authentication.getName();
    //尝试从缓存中获取用户信息
    boolean cacheWasUsed = true;
    UserDetails user = this.userCache.getUserFromCache(username);
    if (user == null) {
        cacheWasUsed = false;
        try {
            //缓存中没有用户时，尝试从用户信息的来源获取，一般是数据库
            user = retrieveUser(username,
                    (UsernamePasswordAuthenticationToken) authentication);
        }catch (UsernameNotFoundException notFound) {
```

```
            ....
        }
    }
    try {
        //认证前置处理
        preAuthenticationChecks.check(user);
        //该方法为抽象方法,延迟到子类实现。可以在子类中添加自定义的认证处理工作,一般是对比密
          码之类的工作。
        additionalAuthenticationChecks(user,
                    (UsernamePasswordAuthenticationToken) authentication);
    }catch (AuthenticationException exception) {
        if (cacheWasUsed) {
            // 如果缓存中获取的UserDetails认证检查失败,将会获取用户信息进行认证检查
            cacheWasUsed = false;
            user = retrieveUser(username,
                    (UsernamePasswordAuthenticationToken) authentication);
            preAuthenticationChecks.check(user);
            additionalAuthenticationChecks(user,
                    (UsernamePasswordAuthenticationToken) authentication);
        }else {
            throw exception;
        }
    }
    //认证后置处理
    postAuthenticationChecks.check(user);
    ...
    Object principalToReturn = user;
    if (forcePrincipalAsString) {
        principalToReturn = user.getUsername();
    }
    //返回认证成功的UsernamePasswordAuthenticationToken
    return createSuccessAuthentication(principalToReturn, authentication, user);
}
```

在 AbstractUserDetailsAuthenticationProvider#authenticate 认证方法中，仅对 UsernamePasswordAuthenticationToken 进行认证。它将尝试从缓存或者用户信息的来源处（如数据库等）根据 UsernamePasswordAuthenticationToken#principal（用户名）加载用户信息 UserDetails，将 UsernamePasswordAuthenticationToken 中的用户信息（用户名和密码）与 UserDetails 中的信息进行对比认证（additionalAuthenticationChecks 方法），如果信息一致，将调用 createSuccessAuthentication 返回认证成功的 UsernamePasswordAuthenticationToken。createSuccessAuthentication 方法为 UsernamePasswordAuthenticationToken 注入更丰富的用户信息以及将 isAuthenticated 设置为 true，返回认证成功的 Authentication。

（2）DaoAuthenticationProvider

DaoAuthenticationProvider 是 AbstractUserDetailsAuthenticationProvider 的实现类，是最常用的 AuthenticationProvider。它从数据层加载用户信息用于认证。该类主要实现其父类的 retrieveUser 和 additionalAuthenticationChecks 方法。

retrieveUser 方法将委托 UserDetailsService#loadUserByUsername 方法根据用户名加载用户

的信息 UserDetails，如果 loadedUser 用户信息不存在，将会抛出 UsernameNotFoundException 异常。

UserDetailsService 接口的作用是根据用户名加载用户信息，提供的方法如下：

```java
//UserDetailsService.java
public interface UserDetailsService {
    //根据用户名获取用户信息
    UserDetails loadUserByUsername(String username) throws UsernameNotFoundException;
}
```

UserDetailsService 需要具体的实现，Spring Security 中同样提供相当多的默认实现以供选择，例如，UserDetailsService 的两个常用实现如下：

❑ JdbcUserDetailsManager 用于从数据库加载用户信息。
❑ InMemoryUserDetailsManager 用于从内存加载用户信息。

在大多数情况下，由于用户信息持有方式不同，开发者一般会自主实现满足自身业务需求的 UserDetailsService 用于加载用户信息。

additionalAuthenticationChecks 方法将比对 Authentication 和 UserDetails 中用户信息的差异，在满足一定条件时认证成功。具体代码如下所示：

```java
//DaoAuthenticationProvider.java
 protected void additionalAuthenticationChecks(UserDetails userDetails,
        UsernamePasswordAuthenticationToken authentication)
        throws AuthenticationException {
        // 没有用户凭证
        if (authentication.getCredentials() == null) {
            throw new BadCredentialsException(...));
        }

        String presentedPassword = authentication.getCredentials().toString();
        //对比UserDetails和Authentication中密码异同
        //如果相同则认证成功，反之，认证失败，抛出认证证书错误BadCredentialsException异常
        //一般是比较转义后的密码
        if (!passwordEncoder.matches(presentedPassword, userDetails.getPassword())) {
            throw new BadCredentialsException(...));
        }
}
```

additionalAuthenticationChecks 方法将比较用户传递的 Authentication 和系统中持有的 UserDetails 用户凭证（一般为用户密码）的异同。如果相同，那么该用户的请求通过认证，可以继续进行，否则会抛出 BadCredentialsException 用户凭证错误的异常。

DaoAuthenticationProvider 的认证思路很简单，概括来说步骤如下：

1）用户提交用户名和密码，封装成 UsernamePasswordAuthenticationToken。

2）DaoAuthenticationProvider 在 retrieveUser 方法通过 UserDetailsService 根据用户名获取 UserDetails。

3）接着在 additionalAuthenticationChecks 方法中将 UserDetails 和 UsernamePassword

AuthenticationToken 的密码进行对比，然后返回认证结果。

4. 认证小结

最后总结一下 Spring Security 的身份认证过程，整个认证结构图如图 12-11 所示。

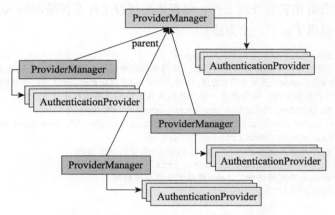

图 12-11　Spring Security 认证结构图

步骤如下：

1）首先是相关过滤器截取到用户的信息，封装成 Authentication。

2）调用 AuthenticationManager#authenticate 方法进行认证。

3）AuthenticationManager 根据 Authentication 类型委托支持该类型的 AuthenticationProvider 实现类对 Authentication 进行认证并返回结果。

4）如果认证成功将返回一个携带或存储用户信息的 Authentication 实例（一般包含用户的身份信息、权限信息等）。

对于自定义认证过程，Spring Security 建议继承 AuthenticationProvider 接口，实现自定义的 authenticate 方法，再注入到 AuthenticationManager 中。

12.3.3　授权

一旦认证成功，那么 Spring Security 将进入授权部分（如果应用中配置了对权限的控制，这将生成对应的过滤器对其进行权限控制）。类似的，这里也有一个顶级的核心策略接口 AccessDecisionManager 授权管理器，类似于 AuthenticationManager，作为授权的主要接口。

授权类图结构参见图 12-12。其中：

❑ AccessDecisionManger 提供授权的主接口。
❑ AbstractAccessDecisionManager 为主接口的抽象实现类。
❑ AccessDecisionVoter 为授权的实际处理接口。

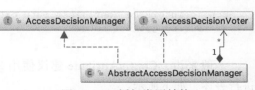

图 12-12　授权类图结构

下面我们对这些类进行详细的介绍。

1. AccessDecisionManager

AccessDecisionManager（授权管理器）的作用类似于 AuthenticationManager，作为授权判定的主接口，总体负责相关的处理工作，将具体的授权工作委托给不同 AccessDecisionVoter 实现进行，该接口提供了以下三个方法：

```
public interface AccessDecisionManager {
    //该方法用于解析传递参数的访问控制决策，即用来判定对某种资源是否具备访问权限。authentication
      是访问者的身份；object是被访问的资源，一般是一个Web资源或者是Java类中的一个方法；configAttributes
      表示被访问资源的装饰，通常使用一些元数据来确定访问需要的权限级别
    void decide(Authentication authentication, Object object,
        Collection<ConfigAttribute> configAttributes) throws AccessDeniedException,
        InsufficientAuthenticationException;
    //该方法用于指出是否支持该种ConfigAttribute的访问权限控制
    boolean supports(ConfigAttribute attribute);
    //该方法用于指出是否支持该种Object资源的访问权限控制
    boolean supports(Class<?> clazz);
}
```

AccessDecisionManager#decide 方法将传递三个参数，其中 object 表示被访问的资源，可以是 Web 接口等；configAttributes 表示访问该资源所需要的规则，如需要什么的角色或者权限的要求；authentication 表示当前访问的用户信息，其内携带着用户被授予的权限列表 Collection<? extends GrantedAuthority>。在 decide 方法中将会根据 AccessDecisionManager 授权判断的规则和传递的参数判断当前用户是否具备权限访问对应的资源。

在 Authentication 中有一个很关键的方法用于获取该用户被授予的权限数组列表，如下所示：

```
Collection<? extends GrantedAuthority> getAuthorities();
```

GrantedAuthority 是代表了一个用户被授予的权限，只提供了一个方法，如下所示：

```
public interface GrantedAuthority extends Serializable {
    String getAuthority();
}
```

一般来讲，GrantedAuthority 通过 String 表示该用户的权限，如 ROLE_USER 之类的表达式，或者是 AccessDecisionManager 支持处理的类型。一个 GrantedAuthority 就代表了该 Authentication 被授予的一项权限。

ConfigAttribute 接口仅提供一个方法，用于获取允许访问的规则表达式，通常有一些特殊的格式或者代表一些需要加工的表达式。它代表了访问所需要的权限规则，如下所示：

```
public interface ConfigAttribute extends Serializable {
    String getAttribute();
}
```

一般来讲，ConfigAttribute 建议使用 Spring Expression Language（SpEL）支持的语法进行表述，如下所示：

```
isFullyAuthenticated() && hasRole('FOO')
```

代表访问该资源需要完全认证以及拥有 FOO 的角色。AccessDecisionVoter 能够为这些表达式生成一个上下文进行处理。

2. AbstractAccessDecisionManager

AbstractAccessDecisionManager 作为 AccessDecisionManager 接口的抽象实现类，其内持有 AccessDecisionVoter 列表，它们代表一群投票者。每一个 AccessDecisionVoter 都会对请求是否具备权限进行投票，AbstractAccessDecisionManager 的具体实现类将根据自身的权限通过规则，统计票数，满足要求即可通过授权判定。AccessDecisionManager 主要有三个实现：ConsensusBased、UnanimousBased、AffirmativeBased，代表了不同的权限通过规则：

- UnanimousBased 遵循一票否决原则，只要有一个投票者否定访问就不能进行访问。
- AffirmativeBased 根据一票通过原则进行判定（忽略弃权），意思是只要有一个投票者同意即可进行访问。
- ConsensusBased 根据少数服从多数的原则进行判定（忽略弃权）。

我们仅对 AffirmativeBased 的具体实现进行介绍，它的权限通过规则遵循一票通过，一般默认采用 AffirmativeBased 进行授权。其他具体实现很类似，只是权限通过规则不一致，但是根本上都是对 AccessDecisionVoter 通过票数进行统计和比较从而得出结果。

AffirmativeBased#decide 方法如下所示：

```
//AffirmativeBased.java
public void decide(Authentication authentication, Object object,
    Collection<ConfigAttribute> configAttributes) throws AccessDeniedException {
    int deny = 0;
    // 遍历每个投票者
    for (AccessDecisionVoter voter : getDecisionVoters()) {
        int result = voter.vote(authentication, object, configAttributes);
        switch (result) {
        // 只要有一票通过即可
        case AccessDecisionVoter.ACCESS_GRANTED:
            return;
        case AccessDecisionVoter.ACCESS_DENIED:
            deny++;
            break;
        default:
            break;
        }
    }
    // 有投票者没有通过
    if (deny > 0) {
        throw new AccessDeniedException(messages.getMessage(
            "AbstractAccessDecisionManager.accessDenied", "Access is denied"));
    }
    checkAllowIfAllAbstainDecisions();
}
```

AffirmativeBased#decide 方法中，只要 AccessDecisionVoter 投票者列表中有一个同意通过权限认证，那么该请求就通过权限鉴定，具备足够的权限访问对应的资源。

当然我们也可以自定义属于自己的判定规则，只需要继承实现 AccessDecisionManager 接口并注入即可。

3. AccessDecisionVoter

AccessDecisionVoter 是 AccessDecisionManager 委托进行授权实际处理的接口，它根据用户具备的权限和访问安全事物所需要的权限规则判断是否允许访问。

AccessDecisionVoter 类似陪审团中的陪审员，对用户的访问是否具备权限进行投票，AbstractAccessDecisionManager 代表法官，它的具体实现是风格迥异的法官，具备不同的权限通过规则。法官会统计陪审团的投票，根据自己风格和陪审团的票数判断请求是否具备权限。

AccessDecisionVoter 提供的接口如下所示：

```java
public interface AccessDecisionVoter<S> {
    // 通过三个常量表示判定结果
    // 允许通过
    int ACCESS_GRANTED = 1;
    // 忽略
    int ACCESS_ABSTAIN = 0;
    // 拒绝通过
    int ACCESS_DENIED = -1;

    // 判断是否允许访问
    int vote(Authentication authentication, Object object,  Collection<ConfigAttribute>
        attrs);

    // 是否支持该类型的ConfigAttribute访问规则
    boolean supports(ConfigAttribute attribute);
    // 是否支持该类型的安全Object
    boolean supports(Class clazz);
}
```

在 AccessDecisionVoter 中，如果对该授权决策没有意见，将返回 ACCESS_DENIED，否则它必须返回 ACCESS_GRANTED 或者 ACCESS_DENIED。

AccessDecisionVoter 主要实现类参见图 12-13，其中：

- RoleVoter 是 Spring Security 中最常用的 AccessDecisionVoter，它将 ConfigAttribute 处理成一个简单的角色名，如 ROLE_USER，并且判断用户是否拥有该角色。
- AuthenticatedVoter 是 Spring Security 用来区分不同身份认证情况下对资源访问的控

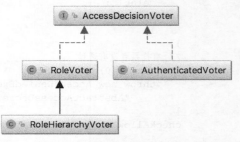

图 12-13　AccessDecisionVoter 主要实现类

制。它根据用户的登录情况以及资源对于登录情况的访问限制来决定用户是否能够访问资源。主要有三种身份认证情况，完全登录认证、记住用户登录认证和匿名登录认证（即没有登录）。

❑ RoleHierarchyVoter 继承自 RoleVoter，适用于处理存在角色包含情况下角色权限判定。

如果你对 AccessDecisionVoter 这些默认的实现都不感兴趣，可以实现自己的 AccessDecisionVoter，然后注入到 AccessDecisionManager 中使用。

4. 授权小结

Spring Security 中的授权是一种类似法庭的法官陪审团方式。请求携带自己具备的权限、访问的受保护资源的权限规则接受法官 AccessDecisionManager 的审判，法官咨询陪审团 AccessDecisionVoter 列表的意见，每个陪审员 AccessDecisionVoter 对请求是否具备权限进行投票，法官统计陪审团的票数再根据自己的风格决定是否允许请求访问受保护资源。

12.3.4　Spring Security 中的过滤器与拦截器

在 Web 应用中，Spring Security Web 的认证和授权流程基本上通过标准的 Servlet 过滤器实现，就像我们在 Spring Security 架构概述中描述一样。所以接下来我们需要对 Spring Security 中的过滤器调用链进行介绍，了解 Spring Security Web 如何在过滤器中对用户的访问进行认证和授权。

1. FilterChainProxy

Spring Security 通过 FilterChainProxy（过滤器调用链代理）管理了所有的安全过滤器。

FilterChainProxy 根据一定的路径匹配规则将请求委托转发到由 Spring 管理的过滤器 Bean 上，由过滤器调用链对请求进行过滤，其中的过滤操作包括认证和授权等，如图 12-14 所示。

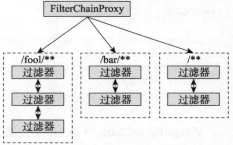

图 12-14　FilterChainProxy 路由分发

FilterChainProxy 是通过添加标准的 Spring DelegatingFilterProxy 声明（在 web.xml 上）而联结到 Servlet 容器的过滤器调用链中。

FilterChainProxy 中委托对应的过滤器调用链对请求进行过滤处理的代码如下所示：

```java
//FilterChainProxy.java
private void doFilterInternal(ServletRequest request, ServletResponse response,
        FilterChain chain) throws IOException, ServletException {

    FirewalledRequest fwRequest = firewall
            .getFirewalledRequest((HttpServletRequest) request);
    HttpServletResponse fwResponse = firewall
            .getFirewalledResponse((HttpServletResponse) response);
    // 获取安全过滤器调用链
    List<Filter> filters = getFilters(fwRequest);
```

```java
        // 如果为空,则返回
        if (filters == null || filters.size() == 0) {
            fwRequest.reset();
            chain.doFilter(fwRequest, fwResponse);
            return;
        }
        // 使用安全过滤器调用链对请求进行处理
        VirtualFilterChain vfc = new VirtualFilterChain(fwRequest, chain, filters);
        vfc.doFilter(fwRequest, fwResponse);
    }
    // 根据名字获取安全过滤器调用链
    private List<Filter> getFilters(HttpServletRequest request) {
        for (SecurityFilterChain chain : filterChains) {
            if (chain.matches(request)) {
                return chain.getFilters();
            }
        }
        return null;
    }
```

doFilterInternal 方法根据请求,通过 getFilters 方法为每一个请求匹配其对应的过滤器调用链。接着将获取到的过滤器调用链和请求封装成 VirtualFilterChain,在 VirtualFilterChain 调用过滤器调用链对请求进行过滤处理。

每一条 SecurityFilterChain 都匹配一种路径规则,并按顺序持有一条过滤器调用链,接口如下所示:

```java
public interface SecurityFilterChain {
    // 是否匹配该请求
    boolean matches(HttpServletRequest request);
    // 过滤器列表
    List<Filter> getFilters();
}
```

VirtualFilterChain 中 doFilter 过滤方法如下所示:

```java
//VirtualFilterChain.java
public void doFilter(ServletRequest request, ServletResponse response)
        throws IOException, ServletException {
            // 安全过滤器调用链调用结束
            if (currentPosition == size) {
                // Deactivate path stripping as we exit the security filter chain
                this.firewalledRequest.reset();
                originalChain.doFilter(request, response);
            }else {
                // 按顺序调用安全过滤器调用链的安全过滤器对请求进行处理
                currentPosition++;
                Filter nextFilter = additionalFilters.get(currentPosition - 1);
                nextFilter.doFilter(request, response, this);
            }
        }
    }
}
```

VirtualFilterChain 按顺序逐个调用过滤器调用链中的安全过滤器对请求进行处理。

在 Web 应用中，Spring Security 根据我们配置的安全规则为应用组装按照路径匹配的 SecurityFilterChain，其中有安全过滤器列表，这些过滤器按照严格的顺序进行调用和排序，它们具备处理应用的安全逻辑。当一个请求到达应用时，FilterChainProxy 将根据请求的路径获取到匹配的 SecurityFilterChain 对请求进行安全过滤处理。一旦通过了这些过滤器的处理，就可以请求到对应的资源。

接着我们介绍具体的安全过滤器，了解一下安全过滤器调用链中过滤器的调用顺序以及各个过滤器的功能，最后会对一些核心过滤器的代码进行介绍。

2. 安全过滤器的一般顺序和功能

在安全过滤器调用链中，过滤器的顺序是非常重要的，它将决定 Spring Security Web 中认证和授权的流程与结果。过滤器的顺序一般如下：

1）ChannelProcessingFilter，应对可能出现重定向到其他协议的情况。

2）SecurityContextPersistenceFilter，请求开始的时候在 SecurityContextHolder 中创建 SecurityContext，并且在请求结束的时候清理 SecurityContextHolder。在请求结束的时候可以将 SecurityContext 复制到 HttpSession 中以用于下次请求。

3）ConcurrentSessionFilter，使用 SecurityContextHolder 来更新 SessionRegistry 中对应当前请求的 Principal。

4）UsernamePasswordAuthenticationFilte、CasAuthenticationFilter、BasicAuthenticationFilter，认证处理机制，用于为 SecurityContextHolder 注入一个经过认证的 Authentication。

5）SecurityContextHolderAwareRequestFilter，是用来在 Servlet 容器中安装一个 Spring Security Aware HttpServletRequestWrapper。

6）JaasApiIntegrationFilter，如果一个 JaasAuthenticationToken 在 SecurityContextHolder 中，该过滤器将过滤器调用链作为 JaasAuthenticationToken 的主体。

7）RememberMeAuthenticationFilter，如果前面的认证机制没有更新 SecurityContextHolder，并且请求中提供了 cookie 允许记住用户，那么将记住用户的 Authentication 更新到 SecurityContextHolder 中。

8）AnonymousAuthenticationFilter，如果前面的认证机制没有更新 SecurityContextHolder，这时表明用户没有登录，那么将匿名的 Authentication 更新到 SecurityContextHolder。

9）ExceptionTranslationFilter，用来捕获 Spring Security Exception 组装一个 HTTP 错误响应返回或者加载 AuthenticationEntryPoint 对异常进行处理。

10）FilterSecurityInterceptor，用于保护资源，是 Spring Security 授权的主要过滤器，当访问遭到拒绝时抛出异常。

在 Spring Security 中还有很多其他的过滤器，在此就不一一进行介绍。以上过滤器都实现 InitializingBean，在调用过程中并没有按照它们标准的过滤器生命周期方法执行。官方建议使用 Spring 应用的上下文生命周期接口作为代替,由 Spring 管理这些过滤器的生命周期。

3. 核心过滤器

在 Spring Security 中有一部分核心过滤器,总是会应用在 Web 安全中,接下来我们将对这些核心过滤器的相关实现进行介绍:

- FilterSecurityInterceptor 为授权拦截器。
- ExceptionTranslationFilter 为异常处理转化过滤器。
- SecurityContextPersistenceFilter 为上下文持久化过滤器。
- UsernamePasswordAuthenticationFilter 为用户名密码认证过滤器。

(1) FilterSecurityInterceptor

FilterSecurityInterceptor 控制了 HTTP 资源的访问安全,对访问进行授权处理。我们可以通过配置 SecurityMetadataSource 来决定对资源访问所需要的权限和约束。

DefaultFilterInvocationSecurityMetadataSource 继承了 SecurityMetadataSource 接口,其内获取请求的权限规则的方法如下所示:

```java
//DefaultFilterInvocationSecurityMetadataSource.java
public Collection<ConfigAttribute> getAttributes(Object object) {
    final HttpServletRequest request = ((FilterInvocation) object).getRequest();
    for (Map.Entry<RequestMatcher, Collection<ConfigAttribute>> entry : requestMap
            .entrySet()) {
        if (entry.getKey().matches(request)) {
            return entry.getValue();
        }
    }
    return null;
}
```

根据请求路径匹配出对应的 Collection<ConfigAttribute> 用于判断访问规则,进行授权操作。FilterSecurityInterceptor#setSecurityMetadataSource 接受的是 FilterInvocationSecurityMetadataSource 接口的实现,在自定义 SecurityMetadataSource 时要注意继承正确的接口并注入。

FilterSecurityInterceptor 中的过滤器操作流程如下所示:

```java
//FilterSecurityInterceptor.java
public void invoke(FilterInvocation fi) throws IOException, ServletException {
    if ((fi.getRequest() != null) && (fi.getRequest().getAttribute(FILTER_APPLIED)
            != null)
    && observeOncePerRequest) {
        fi.getChain().doFilter(fi.getRequest(), fi.getResponse());
    } else {
        // first time this request being called, so perform security checking
        if (fi.getRequest() != null) {
            fi.getRequest().setAttribute(FILTER_APPLIED, Boolean.TRUE);
```

```
        }
        // 调用前的安全授权控制
        InterceptorStatusToken token = super.beforeInvocation(fi);
        // 调用其他的过滤器
        try {
            fi.getChain().doFilter(fi.getRequest(), fi.getResponse());
        }
        // 真正调用资源的访问
        finally {
            super.finallyInvocation(token);
        }
        // 调用结束后的处理
        super.afterInvocation(token, null);
    }
}
```

在真正访问资源前会调用父类的 beforeInvocation 对访问进行授权处理，代码位于 AbstractSecurityInterceptor 中，其中的认证和授权操作都委托 AuthenticationManager 和 AccessDecisionManager 进行。

FilterSecurityInterceptor#beforeInvocation 的授权流程如下：

1）首先是从 SecurityMetadataSource 获取到对应请求的安全规则。

2）然后从 SecurityContextHolder 中获取用户的 Authentication，如果用户没有经过认证，那么将委托 AuthenticationManager 进行认证。

3）接着委托 AccessDecisionManager 进行授权处理。

4）通过授权后将尝试创建一个临时的 Authentication 代替原有的 Authentication，用于提供安全性。在资源调用结束后将恢复原来的 Authentication。

5）返回授权结果，继续进行真正的资源调用。

在查看过滤器代码时，我们会发现，过滤器中的代码一般只负责处理流程，而具体的相关实现，都是委托其他的类实现，这很符合类设计中的单一职责原则，通过组合的方式交给具体的业务逻辑类处理业务逻辑，如上文的 SecurityMetadataSource、AuthenticationManager 和 AccessDecisionManager 分别负责自己特有的功能。这种职责分离的设计在 Spring Security 中很常见。

（2）ExceptionTranslationFilter

ExceptionTranslationFilter 位于安全过滤器调用链的后端，它本身不执行任何实际的安全性处理，但是处理由其他安全拦截器抛出的异常以及提供适当的 HTTP 响应。它属于异常处理和转化过滤器。具体的处理方法如下所示：

```
//ExceptionTranslationFilter.java
public void doFilter(ServletRequest req, ServletResponse res, FilterChain chain)
        throws IOException, ServletException {
    HttpServletRequest request = (HttpServletRequest) req;
    HttpServletResponse response = (HttpServletResponse) res;
```

```
    try {
        chain.doFilter(request, response);   }catch (IOException ex) {
        throw ex;
    }
    catch (Exception ex) {
        Throwable[] causeChain = throwableAnalyzer.determineCauseChain(ex);
        // 封装认证失败异常
        RuntimeException ase = (AuthenticationException) throwableAnalyzer
            .getFirstThrowableOfType(AuthenticationException.class, causeChain);
        if (ase == null) {
            // 封装授权失败异常
            ase = (AccessDeniedException) throwableAnalyzer.getFirstThrowableOfType(
                AccessDeniedException.class, causeChain);
        }
        if (ase != null) {
            // 处理异常
            handleSpringSecurityException(request, response, chain, ase);
        }else {
            // Rethrow ServletExceptions and RuntimeExceptions as-is
            if (ex instanceof ServletException) {
                throw (ServletException) ex;
            }else if (ex instanceof RuntimeException) {
                throw (RuntimeException) ex;
            }
            throw new RuntimeException(ex);
        }
    }
}
```

ExceptionTranslationFilter 过滤器主要处理 AuthenticationException 认证异常以及 AccessDeniedException 授权访问异常。在对上述两种异常完成类型转换和封装后,将调用 handleSpringSecurityException 对异常进行具体的处理。handleSpringSecurityException 方法实现如下所示:

```
//ExceptionTranslationFilter.java
private void handleSpringSecurityException(HttpServletRequest request,
    HttpServletResponse response, FilterChain chain, RuntimeException exception)
        throws IOException, ServletException {
    // 处理认证失败异常
    if (exception instanceof AuthenticationException) {
        // 由AuthenticationEntryPoint处理
        sendStartAuthentication(request, response, chain,
            (AuthenticationException) exception);
    }
    // 处理授权失败异常
    else if (exception instanceof AccessDeniedException) {
        Authentication authentication = SecurityContextHolder.getContext().
            getAuthentication();
        if (authenticationTrustResolver.isAnonymous(authentication) ||
            authenticationTrustResolver.isRememberMe(authentication)) {
            // 如果用户未完全登录,由AuthenticationEntryPoint处理
            sendStartAuthentication(
```

```
                request,
                response,
                chain,
                new InsufficientAuthenticationException(
                "Full authentication is required to access this resource"));
        } else {
            // 传递给AccessDeniedHandler进行处理
            accessDeniedHandler.handle(request, response,
                (AccessDeniedException) exception);
        }
    }
}
```

ExceptionTranslationFilter 将 AuthenticationException 和 AccessDeniedException 分别委托给 AuthenticationEntryPoint 和 AccessDeniedHandler 进行处理。

AuthenticationEntryPoint 是一个提供认证方案的接口，将未认证的请求重定向到不同的认证端点进行认证或者展示异常，仅提供一个方法，如下所示：

```
//AuthenticationEntryPoint.java
void commence(HttpServletRequest request, HttpServletResponse response,
        AuthenticationException authException) throws IOException, ServletException;
```

AuthenticationEntryPoint 有诸多默认实现方式，其类图如图 12-15 所示，图中仅展示了部分的实现类，在实际使用时可以根据自己的需要自由选择实现类。

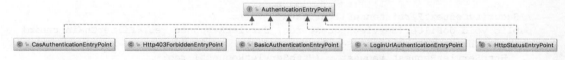

图 12-15　AuthenticationEntryPoint 实现类

平常我们没有登录就访问资源时看到的空荡荡的 401 错误页面就是源于其中的 BasicAuthenticationEntryPoint。

AccessDeniedHandler 用于处理访问遭到拒绝时抛出的异常，它提供了唯一的方法：

```
//AccessDeniedHandler.java
void handle(HttpServletRequest request, HttpServletResponse response,
            AccessDeniedException accessDeniedException) throws IOException,
            ServletException;
```

默认实现为 AccessDeniedHandlerImpl，将根据 errorPage 的配置状态以及状态码决定响应结果，具体代码如下所示：

```
//AccessDeniedHandlerImpl.java
public void handle(HttpServletRequest request, HttpServletResponse response,
        AccessDeniedException accessDeniedException) throws IOException,
        ServletException {
    if (!response.isCommitted()) {
        // 如果存在errorPage，重定向到errorPage
        if (errorPage != null) {
```

```
            // Put exception into request scope (perhaps of use to a view)
            request.setAttribute(WebAttributes.ACCESS_DENIED_403,
                    accessDeniedException);
            // 设置状态码为403
            response.setStatus(HttpServletResponse.SC_FORBIDDEN);
            // 重定向到错误页面
            RequestDispatcher dispatcher = request.getRequestDispatcher(errorPage);
                dispatcher.forward(request, response);
        } else {
            // 直接返回403状态码
            response.sendError(HttpServletResponse.SC_FORBIDDEN,
                    accessDeniedException.getMessage());
        }
    }
}
```

在错误页面存在的情况下，将会在重定向到错误页面中，否则只会把响应的状态码设置为403。

（3）SecurityContextPersistenceFilter

SecurityContextPersistenceFilter 位于安全过滤器调用链的上游，主要负责两个重要的功能，一是在请求开始时创建 SecurityContext 并存储到 SecurityContextHolder，为接下来调用链中其他安全过滤器提供安全上下文；二是在请求结束时清理 SecurityContextHolder。SecurityContextPersistenceFilter 的核心处理流程如下所示：

```
//SecurityContextPersistenceFilter.java
public void doFilter(ServletRequest req, ServletResponse res, FilterChain chain)
        throws IOException, ServletException {
    HttpServletRequest request = (HttpServletRequest) req;
    HttpServletResponse response = (HttpServletResponse) res;
    // 确保过滤器在每个请求只处理一次
    if (request.getAttribute(FILTER_APPLIED) != null) {
        chain.doFilter(request, response);
        return;
    }
    final boolean debug = logger.isDebugEnabled();
    request.setAttribute(FILTER_APPLIED, Boolean.TRUE);
    // 是否有会话可用
    if (forceEagerSessionCreation) {
        HttpSession session = request.getSession();
    }
    HttpRequestResponseHolder holder = new HttpRequestResponseHolder(request,
            response);
    // 如果可能的话，从SecurityContextRepository中加载SecurityContext
    SecurityContext contextBeforeChainExecution = repo.loadContext(holder);
    try {
        //请求开始时，设置SecurityContext,后续的SecurityContext可以直接从SecurityContextHolder
            中获取
        SecurityContextHolder.setContext(contextBeforeChainExecution);
        chain.doFilter(holder.getRequest(), holder.getResponse());
```

```
    } finally {
        // 请求结束的时候，从SecurityContextHolder中清理SecurityContext
        SecurityContext contextAfterChainExecution = SecurityContextHolder
                .getContext();
        SecurityContextHolder.clearContext();
        repo.saveContext(contextAfterChainExecution, holder.getRequest(),
        holder.getResponse());
        request.removeAttribute(FILTER_APPLIED);
    }
}
```

SecurityContextPersistenceFilter 在请求开始时候，为请求注入安全上下文，用于接下来的安全处理任务；在请求结束的时候，清理 SecurityContextHolder 中的安全上下文。

SecurityContext 的获取和保存都委托给了 SecurityContextRepository 执行，它提供的接口如下所示：

```
public interface SecurityContextRepository {
    // 加载安全上下文
    SecurityContext loadContext(HttpRequestResponseHolder requestResponseHolder);
    // 保存安全上下文
    void saveContext(SecurityContext context, HttpServletRequest request,
    boolean containsContext(HttpServletRequest request);
}
```

SecurityContextRepository 的默认实现为 HttpSessionSecurityContextRepository，是通过 HttpSession 来保存 SecurityContext，其 loadContext 方法的代码如下所示：

```
//HttpSessionSecurityContextRepository.java
public SecurityContext loadContext(HttpRequestResponseHolder requestResponseHolder) {
    HttpServletRequest request = requestResponseHolder.getRequest();
    HttpServletResponse response = requestResponseHolder.getResponse();
    HttpSession httpSession = request.getSession(false);
    // 从会话中获取SecurityContext
    SecurityContext context = readSecurityContextFromSession(httpSession);
    // 如果不存在创建一个新的空SecurityContext
    if (context == null) {
       context = generateNewContext();
    }
    ...
    return context;
}
```

在采用微服务架构的情况下，我们期望每一次的请求都是无状态，所以希望 HTTP 传输中不存在会话，那么可以通过在 WebSecurityConfigurerAdapter 中进行配置，具体设置代码如下所示：

```
@Configuration
@EnableWebSecurity
public class SecurityConfig extends WebSecurityConfigurerAdapter{

    @Override
```

```
protected void configure(HttpSecurity http) throws Exception {
    // 不使用和创建任何的HttpSession
    http.sessionManagement().sessionCreationPolicy(SessionCreationPolicy.STATELESS)
}
```

或者按照如下设置：

```
HttpSessionSecurityContextRepository.setAllowSessionCreation(false)
```

禁止会话的创建。这时SecurityContextPersistenceFilter会在请求开始时为每个请求创建新的SecurityContext，以及请求结束时不会将SecurityContext保存在会话中，达到访问无状态的要求。甚至可以在配置中直接用NullSecurityContextRepository替换掉HttpSessionSecurityContextRepository以实现服务间无状态访问的目的。

（4）UsernamePasswordAuthenticationFilter

上述的三个过滤器总是会由Spring Security Web配置到安全过滤器调用链中。但是直到此时，SecurtiyContext中都可能是空的，并没有真正的认证信息。因此过滤器调用链中还需要一个对用户身份进行认证的过滤器，允许用户进行身份认证，并为SecurityContext注入有效的用户信息。UsernamePasswordAuthenticationFilter就是其中最常用的身份认证过滤器，Spring Security在/login端点登录时要求通过用户名和密码进行登录就是通过该过滤器对认证进行处理。

UsernamePasswordAuthenticationFilter继承于AbstractAuthenticationProcessingFilter，其过滤器的核心处理流程位于AbstractAuthenticationProcessingFilter代码中，如下所示：

```
//AbstractAuthenticationProcessingFilter.java
public void doFilter(ServletRequest req, ServletResponse res, FilterChain chain)
        throws IOException, ServletException {
    HttpServletRequest request = (HttpServletRequest) req;
    HttpServletResponse response = (HttpServletResponse) res;
    // 如果不需要认证，直接返回
    if (!requiresAuthentication(request, response)) {
        chain.doFilter(request, response);
        return;
    }
    Authentication authResult;
    try {
        authResult = attemptAuthentication(request, response);
        if (authResult == null) {
            return;
        }
        sessionStrategy.onAuthentication(authResult, request, response);
    } catch (InternalAuthenticationServiceException failed) {
        // 认证失败
        unsuccessfulAuthentication(request, response, failed);
        return;
    } catch (AuthenticationException failed) {
```

```
            // Authentication failed
            unsuccessfulAuthentication(request, response, failed);
            return;
        }
        // 认证成功
        if (continueChainBeforeSuccessfulAuthentication) {
            chain.doFilter(request, response);
        }
        successfulAuthentication(request, response, chain, authResult);
    }
    ...
}
```

在这个流程中，AbstractAuthenticationProcessingFilter 将真正进行认证的过程委托给子类的 attemptAuthentication 方法进行，同时定义了 unsuccessfulAuthentication 和 successfulAuthentication 方法对认证失败和认证成功进行处理。

在认证失败后，会进入到 #unsuccessfulAuthentication 方法中，Spring Security 会对 SecurityContext 进行清理，并委托 AuthenticationFailureHandler 进行具体的处理，默认是返回 401 状态码或者重定向到认证失败页面；而认证成功后就进入 #successfulAuthentication 方法，将认证后返回的 Authentication 保存到 SecurityContext 中，然后委托 AuthenticationSuccessHandler 进行处理，默认也是重定向到相关的登录成功页面或者继续进行请求。

UsernamePasswordAuthenticationFilter 中的 attemptAuthentication 方法主要是通过委托 AuthenticationManager 进行认证，代码如下所示：

```
//UsernamePasswordAuthenticationFilter.java
public Authentication attemptAuthentication(HttpServletRequest request,
        HttpServletResponse response) throws AuthenticationException {
    if (postOnly && !request.getMethod().equals("POST")) {
        throw new AuthenticationServiceException(
                "Authentication method not supported: " + request.getMethod());
    }
    // 从request中获取用户名和密码
    String username = obtainUsername(request);
    String password = obtainPassword(request);
    if (username == null) {
        username = "";
    }
    if (password == null) {
        password = "";
    }
    username = username.trim();
    UsernamePasswordAuthenticationToken authRequest = new UsernamePasswordAuthen
        ticationToken(
            username, password);
    // 允许子类自定义额外的数据
    setDetails(request, authRequest);
    // 调用AuthenticationManager进行认证
    return this.getAuthenticationManager().authenticate(authRequest);
}
```

其中，首先从请求中获取用户名和密码，封装成 UsernamePasswordAuthenticationToken 交予 AuthenticationManager 进行真正的认证，总体流程非常清晰。认证成功后会返回填充了丰富用户信息的 UsernamePasswordAuthenticationToken 用于接下来的请求。

12.3.5 授权服务器

在整体架构中我们介绍过 Spring Security OAuth2 是基于 Spring Security 进行开发，那么它的安全控制也是基于安全过滤器调用链实现的。下面我们分别对 Spring Security OAuth2 中的授权服务器和资源服务器的实现进行介绍。

1. @EnableAuthorizationServer（开启授权服务器）

在配置授权服务器的时候，我们使用了 @EnableAuthorizationServer 注解。它会为应用进行一些关键的自动配置，主要有 AuthorizationServerEndpointsConfiguration 授权服务器相关端点的配置，以及 AuthorizationServerSecurityConfiguration 授权服务器的安全配置。

AuthorizationServerEndpointsConfiguration 为授权服务器注入了与令牌相关的三个端点：AuthorizationEndpoint、TokenEndpoint 和 CheckTokenEndpoint。AuthorizationEndpoint 用于获取授权码，TokenEndpoint 用于获取令牌，CheckTokenEndpoint 用于检验令牌。从 AuthorizationServerEndpointsConfiguration 可以了解到一些关键的配置，类似 TokenGranter 令牌生成器、ClientDetailsService 客户端服务、AuthorizationCodeServices 授权码服务，AccessTokenConverter 访问令牌转化器，以及 AuthorizationServerTokenServices token 服务等等。这些配置都可以用来自定义授权服务器，在自动配置的基础上增强功能性。

在 AuthorizationServerSecurityConfiguration 中进行了关于授权服务器的 Spring Security 的相关配置，它通过 AuthorizationServerSecurityConfigurer 配置了一个授权服务器中用于认证客户端的核心过滤器 ClientCredentialsTokenEndpointFilter。具体代码如下所示：

```
//AuthorizationServerSecurityConfiguration.java
public final class AuthorizationServerSecurityConfigurer extends
        SecurityConfigurerAdapter<DefaultSecurityFilterChain, HttpSecurity> {
    ...
    @Override
    public void configure(HttpSecurity http) throws Exception {
        frameworkEndpointHandlerMapping();
        // 表单认证客户端过滤器
        if (allowFormAuthenticationForClients) {
           clientCredentialsTokenEndpointFilter(http);
        }
        // Basic认证客户端过滤器
        for (Filter filter : tokenEndpointAuthenticationFilters) {
            http.addFilterBefore(filter, BasicAuthenticationFilter.class);
        }
        http.exceptionHandling().accessDeniedHandler(accessDeniedHandler);
        if (sslOnly) {
            http.requiresChannel().anyRequest().requiresSecure();
```

        }
      }
...
}
```

在可能的情况下，它将会为"/oauth/**"相关端点配置两种客户端认证过滤器，一种可以通过表单方式认证客户端的 ClientCredentialsTokenEndpointFilter，另一种是使用 HTTP BASIC Authentication 认证客户端的 BasicAuthenticationFilter。

### 2. AuthorizationServerConfigurerAdapter（覆盖自动配置）

AuthorizationServerConfigurerAdapter 是一个配置适配器类，是 Spring Security OAuth2 提供给开发者对授权服务器中的相关配置 Bean 进行覆盖操作的自定义配置类。一般来说，开发者可以通过它为授权服务器在默认配置的基础上进行快速的自定义配置。这些配置将覆盖默认配置中（覆盖 @EnableAuthorizationServer 中的自动配置类）。AuthorizationServerConfigurerAdapter 接口的代码如下所示：

```java
//AuthorizationServerConfigurerAdapter.java
public class AuthorizationServerConfigurerAdapter implements AuthorizationServerConfigurer {
    @Override
    public void configure(AuthorizationServerSecurityConfigurer security) throws
        Exception {
    }
    @Override
    public void configure(ClientDetailsServiceConfigurer clients) throws Exception {
    }
    @Override
    public void configure(AuthorizationServerEndpointsConfigurer endpoints) throws
        Exception {
    }
}
```

上述代码提供了三个方面的主要配置：

- AuthorizationServerSecurityConfigurer：授权服务器安全认证的相关配置，用于配置 Spring Security，生成对应的安全过滤器调用链，主要控制"oauth/**"相关端点的访问。
- ClientDetailsServiceConfigurer：OAuth2 客户端服务相关配置。
- AuthorizationServerEndpointsConfigurer：授权服务器授权端点相关配置。

在基础应用环节，我们通过该配置类适配器对授权服务器的自定义进行覆盖配置。

在 Spring Security 及其子项目的设计中，有众多的 ConfigurerAdapter 适配器用于对 Spring Security 中的自动配置的 Bean 进行覆盖配置（Configurer 是配置器，用于对系统进行配置工作）。很多时候我们可以继承这些类完成我们对 Spring Security 的自定义配置，只需要注入我们需要改变的模块的 Bean，即不会影响其他部分功能，也满足了自定义的需求，非常简便。
```

3. ClientDetailsService（加载客户端信息）

在 OAuth2 中，只有被授权服务器认证的客户端才能获取到访问令牌，所以开发者需要为授权服务器配置认证客户端的相关服务。

OAuth2 中提供了 ClientDetailsService 用于加载客户端信息，接口如下所示：

```java
public interface ClientDetailsService {
    // 根据ClientId加载ClientDetails
    ClientDetails loadClientByClientId(String clientId) throws ClientRegistrationException;
}
```

默认有两种实现方式，可以根据自己的需要进行配置，基础应用中使用 InMemoryClientDetailsService 对客户端服务进行配置，如下所示：

- JdbcClientDetailsService 用于从数据库中加载客户端信息。
- InMemoryClientDetailsService 用于从内存中加载客户端信息。

可以通过 ClientDetailsServiceConfigurer 进行配置，使用具体的客户端服务。当然也可以自定义自己的 ClientDetailsService，通过 ClientDetailsServiceConfigurer#withClientDetails 注入即可使用。

4. ClientCredentialsTokenEndpointFilter（表单方式认证客户端）

客户端的认证过程会在安全过滤器调用链中进行，而进行具体处理的过滤器为 ClientCredentialsTokenEndpointFilter。如果通过 ClientCredentialsTokenEndpointFilter 过滤器对客户端认证，客户端 id 和客户端密钥需要通过表单的方式提交给授权服务器。使用 ClientCredentialsTokenEndpointFilter 认证客户端需要我们在授权服务器中配置 allowFormAuthenticationForClients，即允许通过表单认证客户端。

ClientCredentialsTokenEndpointFilter 继承 AbstractAuthenticationProcessingFilter，所以它对客户端的认证过程将在 attemptAuthentication 方法中进行。attemptAuthentication 方法具体流程代码如下所示：

```java
//ClientCredentialsTokenEndpointFilter.java
@Override
public Authentication attemptAuthentication(HttpServletRequest request, HttpServletResponse
    response)
    throws AuthenticationException, IOException, ServletException {
    if (allowOnlyPost && !"POST".equalsIgnoreCase(request.getMethod())) {
        throw new HttpRequestMethodNotSupportedException(request.getMethod(), new
            String[] { "POST" });
    }
    // 从请求中获取提交的客户端信息
    String clientId = request.getParameter("client_id");
    String clientSecret = request.getParameter("client_secret");
    // 如果请求已经经过验证，直接返回
    Authentication authentication = SecurityContextHolder.getContext().
        getAuthentication();
    ...
```

```
    if (authentication != null && authentication.isAuthenticated()) {
        return authentication;
    }
    if (clientId == null) {
        throw new BadCredentialsException("No client credentials presented");
    }
    if (clientSecret == null) {
        clientSecret = "";
    }
    clientId = clientId.trim();
    UsernamePasswordAuthenticationToken authRequest = new UsernamePassword
        AuthenticationToken (clientId, clientSecret);
    // 委托认证管理器AuthenticationManager进行认证
    return this.getAuthenticationManager().authenticate(authRequest);
}
```

与 UsernamePasswordAuthenticationFilter 类似，最后也是委托 AuthenticationManager 进行最后的认证。认证过程中，会将 clientId 和 clientSecret 当成用户名和密码，封装成 UsernamePasswordAuthenticationToken 传递给 AuthenticationManager 进行认证。AuthenticationManager 将委托 DaoAuthenticationProvider 进行真正的认证处理，而我们并没有为 DaoAuthenticationProvider 配置对应的 UserDetailsService 来加载用户信息，那是因为配置的 ClientDetailsService 会被适配成 UserDetailsService，用于进行客户端信息的加载。

ClientDetailsUserDetailsService 继承 UserDetailsService 接口，将 ClientDetailsService 适配成 UserDetailsService，注入到 DaoAuthenticationProvider 中用于加载客户端信息。

5. BasicAuthenticationFilter（HTTP BASIC Authentication 认证客户端）

除了通过 ClientCredentialsTokenEndpointFilter 用表单的方式认证客户端，也可以通过 HTTP BASIC 认证方式对客户端进行认证，关键处理的过滤器为 BasicAuthenticationFilter。它在过滤器调用链的顺序位于 ClientCredentialsTokenEndpointFilter 过滤器之后。该种认证方式需要我们在请求头部中添加键值对信息。TokenEndpoint 获取令牌端点，默认使用该过滤器来认证客户端。

在头部中提交客户端信息的样式如下所示：

```
Authorization: Basic base64(clientId:clientSecret)
```

base64(clientId:clientSecret) 的意思是将 clientId:clientSecret 进行 Base64 编码，我们在基础应用中提交的客户端信息最终如下所示：

```
Authorization: Basic Y2xpZW50OiBzZWNyZXQ=
```

BasicAuthenticationFilter 过滤器中的核心处理方法如下所示：

```
//BasicAuthenticationFilter.java
@Override
protected void doFilterInternal(HttpServletRequest request,
        HttpServletResponse response, FilterChain chain)
```

```java
                throws IOException, ServletException {
    // 从头部中获取认证信息
    String header = request.getHeader("Authorization");
    if (header == null || !header.startsWith("Basic ")) {
        chain.doFilter(request, response);
        return;
    }
    try {
        // Base64解码获取clientId和secretId
        String[] tokens = extractAndDecodeHeader(header, request);
        assert tokens.length == 2;
        String username = tokens[0];
        // 查看是否已经认证过
        if (authenticationIsRequired(username)) {
            UsernamePasswordAuthenticationToken authRequest = new UsernamePasswo
                rdAuthenticationToken(
                    username, tokens[1]);
            authRequest.setDetails(
                    this.authenticationDetailsSource.buildDetails(request));
            // 通过AuthenticationManager认证客户端
            Authentication authResult = this.authenticationManager
                    .authenticate(authRequest);
            SecurityContextHolder.getContext().setAuthentication(authResult);
            this.rememberMeServices.loginSuccess(request, response, authResult);
            onSuccessfulAuthentication(request, response, authResult);
        }
    }catch (AuthenticationException failed) {
        // 认证失败
        SecurityContextHolder.clearContext();
        this.rememberMeServices.loginFail(request, response);
        onUnsuccessfulAuthentication(request, response, failed);
        if (this.ignoreFailure) {
            chain.doFilter(request, response);
        }else {
            this.authenticationEntryPoint.commence(request, response, failed);
        }
        return;
    }
    chain.doFilter(request, response);
}
```

在上述方法中，首先从头部取出 Authorization 参数，然后进行解码获取 clientId 和 clientSecret，封装成 UsernamePasswordAuthenticationToken，交给 AuthenticationManager 进行认证。

认证过滤器的认证流程大致相似，最终都是委托 AuthenticationManager 进行具体的认证工作。如果安全过滤器调用链中存在多个认证过滤器，这些认证过滤器中只要有一个通过了认证并将认证后的 Authentication 保存到 SecurtiyContext 中，接下来的认证过滤器将会直接跳过，不再进行认证操作。

6. 授权服务器端点

在经过安全过滤器调用链的认证之后，请求将来到令牌相关的端点，它们的关系类如图 12-16 所示：

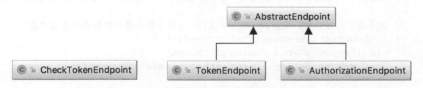

图 12-16　授权服务器端点类结构

令牌相关的端点主要包括如下三个端点：

- AuthorizationEndpoint 用于在授权码类型中获取授权码或者在隐式（implicit）类型中直接获取访问令牌。
- TokenEndpoint 是用于在除隐式类型外的类型获取访问令牌以及刷新访问令牌。
- CheckTokenEndpoint 用来检验令牌的合法性。

下面我们将一一介绍这些端点。

（1）AuthorizationEndpoint

在授权码类型中，需要用户登录之后进行授权才能获取授权码，客户端再携带授权码向 TokenEndpoint 请求访问令牌，当然也可以在请求中设置 response_type=token 通过隐式类型直接获取到访问令牌。在到达 AuthorizationEndpoint 端点时，并没有对客户端进行验证，但是必须要经过的用户认证，请求才能到达该端点。

授权码类型中，首先要获取用户的授权，端点地址为 /oauth/authorize，接受请求的代码如下所示：

```java
//AuthorizationEndpoint.java
@RequestMapping(value = "/oauth/authorize")
public ModelAndView authorize(Map<String, Object> model, @RequestParam Map<String,
    String> parameters,
        SessionStatus sessionStatus, Principal principal) {
    AuthorizationRequest authorizationRequest = getOAuth2RequestFactory().create
        AuthorizationRequest(parameters);
    Set<String> responseTypes = authorizationRequest.getResponseTypes();
    // 仅支持返回为授权码或者是令牌
    if (!responseTypes.contains("token") && !responseTypes.contains("code")) {
        throw new UnsupportedResponseTypeException("Unsupported response types: " +
            responseTypes);
    }
    ...
    try {
        //如果用户没有登录，无法进行授权，将重定向到登录界面，由ExceptionTranslationFilter进
            行处理
        ...
```

```java
        // 验证客户端提供的redirectUri是否正确
        ...
        // 验证客户端的请求范围是否正确
        oauth2RequestValidator.validateScope(authorizationRequest, client);
        // 检查用户是否允许授权,因为有一些系统会默认允许授权,不需要用户的确认
        ...
        // 如果用户已经授权,根据repsonse_type返回令牌或者授权码进行相应处理
        if (authorizationRequest.isApproved()) {
            if (responseTypes.contains("token")) {
                return getImplicitGrantResponse(authorizationRequest);
            }
            if (responseTypes.contains("code")) {
                return new ModelAndView(getAuthorizationCodeResponse(authorizati
                    onRequest,
                    (Authentication) principal));
            }
        }
        // 将授权请求放在model中用于保存在会话中,供授权界面使用,避免用户重复登录认证
        model.put("authorizationRequest", authorizationRequest);
        // 如果用户还没有授权,重定向到授权认证界面
        return getUserApprovalPageResponse(model, authorizationRequest, (Authentication)
            principal);
    }catch (RuntimeException e) {
        ...
    }
}
```

在客户端请求授权码时,将引导用户进入授权系统的登录界面,如果用户已经登录,那么将查看授权系统是否允许默认授权,是的话就直接根据响应类型返回对应的访问令牌或者授权码,否则将用户重定向到授权页面,请求用户明确的授权。

用户需要在授权页面进行授权或者拒绝授权。在获得用户的授权之后,授权服务器将根据响应类型返回访问令牌或者授权码。获取授权码的方法如下所示:

```java
// AuthorizationEndpoint.java
private String generateCode(AuthorizationRequest authorizationRequest, Authentication
    authentication)
        throws AuthenticationException {
    try {
        OAuth2Request storedOAuth2Request = getOAuth2RequestFactory().createOAut
            h2Request(authorizationRequest);
        OAuth2Authentication combinedAuth = new OAuth2Authentication(storedOAuth
            2Request, authentication);
        // 获取授权码
        String code = authorizationCodeServices.createAuthorizationCode(combined
            Auth);
        return code;
    }catch (OAuth2Exception e) {
        if (authorizationRequest.getState() != null) {
            e.addAdditionalInformation("state", authorizationRequest.getState());
        }
        throw e;
```

主要委托 AuthorizationCodeServices 生成并保存授权码，用于在请求访问令牌时验证授权码。AuthorizationCodeServices 提供以下的接口：

```
public interface AuthorizationCodeServices {
    //为特定的请求生成授权码，默认每个授权码只能使用一次
    String createAuthorizationCode(OAuth2Authentication authentication);
    // 消费授权码以获取授权码对应的用户授权信息
    OAuth2Authentication consumeAuthorizationCode(String code) throws
        InvalidGrantException;
}
```

Spring Security OAuth2 提供了两种默认的实现，如下所示：

- InMemoryAuthorizationCodeServices 将授权码保存到内存中。
- JdbcAuthorizationCodeServices 将授权码保存在数据库中。

它们都是通过父类 RandomValueAuthorizationCodeServices 生成一个随机的授权码，只是保存授权码的方式有所差异。可以根据自己的需要进行配置，或者继承 AuthorizationCodeServices 接口，实现一个自己的授权码管理服务。

当然如果请求的响应类型为 token（并且请求客户端允许隐式类型），就可以直接请求到对应的访问令牌。getImplicitGrantResponse 方法中将直接调用 getAccessTokenForImplicitGrant 方法获取对应的访问令牌。getAccessTokenForImplicitGrant 方法通过 TokenGrant 获取访问令牌。TokenGrant 是统一获取访问令牌的接口，支持使用不同的授权类型获取访问令牌。

（2）TokenEndpoint

TokenEndpoint 是提供给客户端用来获取访问令牌的端点，客户端携带授权类型（授权码类型、密码类型等）以及对应的请求参数来获取访问令牌。在到达 TokenEndpoint 端点前，授权服务器需要对请求中的客户端相关信息（如 clientId、clientSecret）进行客户端认证，客户端认证的过程在前面的章节已经进行了讲解，在此不做过多描述。TokenEndpoint 端点提供的接口如下所示：

```
//TokenEndpoint.java
@RequestMapping(value = "/oauth/token", method=RequestMethod.POST)
public ResponseEntity<OAuth2AccessToken> postAccessToken(Principal principal, @
    RequestParam
Map<String, String> parameters) throws HttpRequestMethodNotSupportedException {
    // 如果客户端没有经过认证，拒绝访问
    if (!(principal instanceof Authentication)) {
        throw new InsufficientAuthenticationException(...);
    }
    // 加载客户端信息
    ...
    // 构建生成令牌的请求
    TokenRequest tokenRequest = getOAuth2RequestFactory().createTokenRequest(parameters,
```

```
            authenticatedClient);
    // 检查客户端id是否一致
    ...
    // 检查请求范围是否满足，这与给客户端配置的范围相关
    ...
    // 检查请求是否携带授权类型
    ...
    // 不支持隐式授权类型
    ...
    // 是否请求刷新令牌
    if (isRefreshTokenRequest(parameters)) {
        tokenRequest.setScope(OAuth2Utils.parseParameterList(parameters.get(OAuth2Utils.
            SCOPE)));
    }
    // 获取对应的访问令牌
    OAuth2AccessToken token = getTokenGranter().grant(tokenRequest.getGrantType(),
        tokenRequest);
    if (token == null) {
        throw new UnsupportedGrantTypeException("Unsupported grant type: " +
            tokenRequest.getGrantType());
    }
    return getResponse(token);
}
```

一般来讲，TokenEndpoint 除了不支持隐式类型，其他的所有授权类型，如授权码、密码、客户端类型都支持，但是能不能获取到 token，还需要请求的客户端被授予通过该授权类型获取令牌的权限。从上面的代码中可以看出，获取令牌的方法最后委托给 TokenGrant 进行。

（3）CheckTokenEndpoint

CheckTokenEndpoint 是提供给客户端用来检验访问令牌的有效性以及解析令牌获取令牌中的有效信息，通过该接口能够获取到令牌的相关信息以及授权获取令牌的用户相关信息。接口代码如下所示：

```
@RequestMapping(value = "/oauth/check_token")
    @ResponseBody
    public Map<String, ?> checkToken(@RequestParam("token") String value) {
    OAuth2AccessToken token = resourceServerTokenServices.readAccessToken(value);
    if (token == null) {
        throw new InvalidTokenException("Token was not recognised");
    }
    // 令牌过期
    if (token.isExpired()) {
        throw new InvalidTokenException("Token has expired");
    }

    OAuth2Authentication authentication = resourceServerTokenServices.loadAuthentication
        (token.getValue());
    // 解析令牌相关信息并返回
    Map<String, ?> response = accessTokenConverter.convertAccessToken(token, authentication);
    return response;
}
```

该端点很简单，主要通过 ResourceServerTokenServices 加载令牌以及根据令牌获取到对应的用户信息，最后通过 AccessTokenConverter 对它们进行解析后返回结果。

7. OAuth2AccessToken

OAuth2AccessToken 接口表示了一个访问令牌持有的基本信息，是 OAuth2 中封装访问令牌的信息类。在基础应用中，我们请求令牌成功后返回的结果是由该接口实现类转义而来。它提供以下与令牌相关的信息：

```java
public interface OAuth2AccessToken {
    // 额外的信息，可继承接口进行自定义
    Map<String, Object> getAdditionalInformation();
    // 访问令牌允许的范围
    Set<String> getScope();
    // 刷新令牌
    OAuth2RefreshToken getRefreshToken();
    // 令牌的类型，一般为bearer
    String getTokenType();
    // 是否过期
    boolean isExpired();
    // 过期时间
    Date getExpiration();
    //有效时间
    int getExpiresIn();
    // 访问令牌的值
    String getValue();
}
```

该接口的默认实现是 DefaultOAuth2AccessToken。

8. TokenGrant

在上面对各个端点的查看中，会发现基本是调用了 TokenGrant#grant 方法生成和获取对应的访问令牌。接下来对 TokenGrant 令牌授予器进行介绍，它的接口代码如下所示：

```java
public interface TokenGranter {
    // 根据授权类型和请求生成和获取令牌
    OAuth2AccessToken grant(String grantType, TokenRequest tokenRequest);
}
```

TokenGranter 接口仅提供了一个方法用于获取令牌，对于每一种授权类型（grantType），都有一种专门的实现类为其提供令牌，如图 12-17 所示。

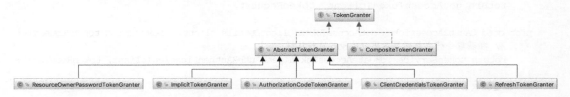

图 12-17　TokenGranter 实现类图

具体的实现类包括：
- CompositeTokenGranter 组合类。
- ResourceOwnerPasswordTokenGranter 支持密码类型获取令牌。
- ImplicitTokenGranter 支持隐式类型获取令牌。
- AuthorizationCodeTokenGranter 支持授权码类型获取令牌。
- ClientCredentialsTokenGranter 支持客户端类型获取令牌。
- RefreshTokenGranter 支持刷新令牌。

这个类结构属于组合设计模式，CompositeTokenGranter 是组合模式中的组合者（Composite），管理着所有授权类型的 TokenGranter 实现类，并根据 grantType 具体调用某一个 TokenGranter 获取访问令牌。

CompositeTokenGranter#grant 实现如下所示：

```java
//CompositeTokenGranter.java
public OAuth2AccessToken grant(String grantType, TokenRequest tokenRequest) {
    for (TokenGranter granter : tokenGranters) {
        OAuth2AccessToken grant = granter.grant(grantType, tokenRequest);
        if (grant!=null) {
            return grant;
        }
    }
    return null;
}
```

CompositeTokenGranter 将遍历它持有的 TokenGranter（组合模式中叶子节点），直到获取到能够处理该 grantType 的 TokenGranter 并返回 OAuth2AccessToken，从而获取到访问令牌。

我们首先了解每种授权类型的 TokenGranter 抽象实现类 AbstractTokenGranter 中获取访问令牌的公共流程，代码如下所示：

```java
//AbstractTokenGranter.java
public OAuth2AccessToken grant(String grantType, TokenRequest tokenRequest) {
    // 是否支持该种grantType
    if (!this.grantType.equals(grantType)) {
        return null;
    }
    String clientId = tokenRequest.getClientId();
    ClientDetails client = clientDetailsService.loadClientByClientId(clientId);
    // 验证客户端是否允许通过该种grantType获取访问令牌
    validateGrantType(grantType, client);
    return getAccessToken(client, tokenRequest);
}
protected OAuth2AccessToken getAccessToken(ClientDetails client, TokenRequest tokenRequest) {
    // 在获取令牌之前需要对用户的授权进行认证
    return tokenServices.createAccessToken(getOAuth2Authentication(client, tokenRequest));
}
// 检查授权用户是否经过认证
protected OAuth2Authentication getOAuth2Authentication(ClientDetails client, TokenRequest
    tokenRequest) {
```

```
OAuth2Request storedOAuth2Request = requestFactory.createOAuth2Request(client,
    tokenRequest);
return new OAuth2Authentication(storedOAuth2Request, null);
}
```

每一种授权类型的 TokenGrant 获取令牌的方式都是对 AuthorizationServerTokenServices 中的方法进行请求，所以说 AuthorizationServerTokenServices 才是最终对令牌进行管理的服务。大致的请求令牌流程如下：

1）检查该 TokenGrant 是否支持该种类型的 grantType。
2）检查客户端是否允许通过该种 grantType 获取令牌。
3）进行相关认证工作。
4）向 AuthorizationServerTokenServices 请求令牌。

其中第 3 步是各子类 TokenGrant 主要关注覆盖点。

如 ResourceOwnerPasswordTokenGranter，支持密码类型，需要对请求携带的用户和密码进行认证，只有认证成功才能请求令牌，代码如下所示：

```
//ResourceOwnerPasswordTokenGranter.java
@Override
protected OAuth2Authentication getOAuth2Authentication(ClientDetails client, TokenRequest
    tokenRequest) {

    Map<String, String> parameters = new LinkedHashMap<String, String>(tokenRequest.
        getRequestParameters());
    String username = parameters.get("username");
    String password = parameters.get("password");
    // Protect from downstream leaks of password
    parameters.remove("password");
    Authentication userAuth = new UsernamePasswordAuthenticationToken(username,
        password);
    ((AbstractAuthenticationToken) userAuth).setDetails(parameters);
    try {
        // 对用户信息进行认证
        userAuth = authenticationManager.authenticate(userAuth);
    }catch (AccountStatusException ase) {
        throw new InvalidGrantException(ase.getMessage());
    }catch (BadCredentialsException e) {
        throw new InvalidGrantException(e.getMessage());
    }
    if (userAuth == null || !userAuth.isAuthenticated()) {
        throw new InvalidGrantException("Could not authenticate user: " + username);
    }
    OAuth2Request storedOAuth2Request = getRequestFactory().createOAuth2Request(client,
        tokenRequest);    // 封装成OAuth2Authentication用于获取访问令牌
    return new OAuth2Authentication(storedOAuth2Request, userAuth);
}
```

认证过程使用在 WebSecurityConfigurerAdapter 注入的 AuthenticationManager 进行认证。在基础应用中，我们在 AuthenticationManager 中配置了两个内存中的用户。需要注意

的是，这里的 AuthenticationManager 与客户端认证中的 AuthenticationManager 并不是同一个。

而在 AuthorizationCodeTokenGranter 中，支持授权码类型，需要对授权码以及对携带授权码和请求授权码的客户端的一致性进行检查，代码如下所示：

```java
@Override
protected OAuth2Authentication getOAuth2Authentication(ClientDetails client, TokenRequest
    tokenRequest) {

    Map<String, String> parameters = tokenRequest.getRequestParameters();
    String authorizationCode = parameters.get("code");
    String redirectUri = parameters.get(OAuth2Utils.REDIRECT_URI);

    if (authorizationCode == null) {
        throw new InvalidRequestException("An authorization code must be supplied.");
    }
    // 通过AuthorizationCodeServices服务检查消费授权码
    OAuth2Authentication storedAuth = authorizationCodeServices.consumeAuthoriza
        tionCode(authorizationCode);
    if (storedAuth == null) {
        throw new InvalidGrantException("Invalid authorization code: " +
            authorizationCode);
    }
    OAuth2Request pendingOAuth2Request = storedAuth.getOAuth2Request();
    String redirectUriApprovalParameter = pendingOAuth2Request.getRequestParameters().
        get(
            OAuth2Utils.REDIRECT_URI);
    //通过redirectUri验证请求授权码和携带授权码请求访问令牌的客户端是否为同一个，如果不是同一
        个客户端将拒绝请求
    if ((redirectUri != null || redirectUriApprovalParameter != null)
            && !pendingOAuth2Request.getRedirectUri().equals(redirectUri)) {
        throw new RedirectMismatchException("Redirect URI mismatch.");
    }

    String pendingClientId = pendingOAuth2Request.getClientId();
    String clientId = tokenRequest.getClientId();
    if (clientId != null && !clientId.equals(pendingClientId)) {
        throw new InvalidClientException("Client ID mismatch");
    }
    ...
    OAuth2Request finalStoredOAuth2Request = pendingOAuth2Request.createOAuth2Re
        quest(combinedParameters);
    Authentication userAuth = storedAuth.getUserAuthentication();
    // 封装成OAuth2Authentication用于获取访问令牌
    return new OAuth2Authentication(finalStoredOAuth2Request, userAuth);

}
```

RefreshTokenGranter 根据刷新令牌请求访问令牌，而 ImplicitTokenGranter 通过请求时注入到 SecurityContextHolder 的有效用户认证完成认证，请求访问令牌，在此不展示代码讲解，有兴趣的读者可以自行阅读了解。

9. AuthorizationServerTokenServices

AuthorizationServerTokenServices 是授权服务器中进行令牌管理的接口,提供了以下三个接口:

```java
public interface AuthorizationServerTokenServices {
    //生成与OAuth2认证绑定的访问令牌
    OAuth2AccessToken createAccessToken(OAuth2Authentication authentication) throws
        AuthenticationException;
    // 根据刷新令牌获取访问令牌
    OAuth2AccessToken refreshAccessToken(String refreshToken, TokenRequest tokenRequest)
            throws AuthenticationException;
    // 获取OAuth2认证的访问令牌,如果访问令牌存在的话
    OAuth2AccessToken getAccessToken(OAuth2Authentication authentication);

}
```

请注意,生成的令牌是与授权的用户和请求的客户端进行绑定的。

AuthorizationServerTokenServices 接口的默认实现是 DefaultTokenServices,令牌通过 TokenStore 保存。DefaultTokenServices 中生成令牌的代码如下所示:

```java
//DefaultTokenServices.java
@Transactional
public OAuth2AccessToken createAccessToken(OAuth2Authentication authentication)
    throws AuthenticationException {
    // 从TokenStore获取访问令牌
    OAuth2AccessToken existingAccessToken = tokenStore.getAccessToken(authentication);
    OAuth2RefreshToken refreshToken = null;
    if (existingAccessToken != null) {
        if (existingAccessToken.isExpired()) {
            // 如果访问令牌已经存在但是过期了,删除对应的访问令牌和刷新令牌
            if (existingAccessToken.getRefreshToken() != null) {
                refreshToken = existingAccessToken.getRefreshToken();
                tokenStore.removeRefreshToken(refreshToken);
            }
            tokenStore.removeAccessToken(existingAccessToken);
        }
        else {
            //如果访问令牌已经存在并且没有过期,重新保存一下防止authentication改变,并且返回
                该访问令牌
            tokenStore.storeAccessToken(existingAccessToken, authentication);
            return existingAccessToken;
        }
    }
    //只有当刷新令牌为null时,才重新创建一个新的刷新令牌,这样使持有过期访问令牌的客户端可以根
        据以前拿到刷新令牌拿到重新创建的访问令牌,因为新创建的访问令牌绑定了同一个刷新令牌
    if (refreshToken == null) {
        refreshToken = createRefreshToken(authentication);
    }else if (refreshToken instanceof ExpiringOAuth2RefreshToken) {
        // 如果刷新令牌也有期限并且过期,重新创建
        ExpiringOAuth2RefreshToken expiring = (ExpiringOAuth2RefreshToken)
            refreshToken;
```

```java
        if (System.currentTimeMillis() > expiring.getExpiration().getTime()) {
            refreshToken = createRefreshToken(authentication);
        }
    }
}
// 绑定授权用户和刷新令牌创建新的访问令牌
OAuth2AccessToken accessToken = createAccessToken(authentication, refreshToken);
// 将访问令牌与授权用户对应保存
tokenStore.storeAccessToken(accessToken, authentication);
// In case it was modified
refreshToken = accessToken.getRefreshToken();
if (refreshToken != null) {
    // 将刷新令牌与授权用户对应保存
    tokenStore.storeRefreshToken(refreshToken, authentication);
}
return accessToken;
}
```

在创建令牌的过程中，会根据该授权用户去查询是否存在未过期的访问令牌，有就直接返回，没有的话才会重新创建新的访问令牌。在创建新的访问令牌之前会试图获取旧的未失效的刷新令牌，这是为了保证持有过期的访问令牌的客户端能够通过刷新令牌重新获得访问令牌，因为前后创建访问令牌将绑定同一个刷新令牌。

在创建令牌的时候还需要留意 AccessTokenEnhancer，其作用是增强令牌样式，可以通过继承实现该接口自定义令牌样式，如我们在基础应用中注入的 JwtAccessTokenConverter，它的作用是将令牌增强为 JWT 样式。创建令牌的代码如下所示：

```java
// DefaultTokenServices.java
private OAuth2AccessToken createAccessToken(OAuth2Authentication authentication,
    OAuth2RefreshToken refreshToken) {
    DefaultOAuth2AccessToken token = new DefaultOAuth2AccessToken(UUID.randomUUID().
        toString());
    int validitySeconds = getAccessTokenValiditySeconds(authentication.getOAuth2Request());
    // 设置令牌有效时间
    if (validitySeconds > 0) {
        token.setExpiration(new Date(System.currentTimeMillis() + (validitySeconds
            * 1000L)));
    }
    token.setRefreshToken(refreshToken);
    // 设置令牌范围
    token.setScope(authentication.getOAuth2Request().getScope());

    return accessTokenEnhancer != null ? accessTokenEnhancer.enhance(token,
        authentication) : token;
}
```

DefaultTokenServices 中刷新令牌的 refreshAccessToken 方法以及获取令牌的 getAccessToken 方法留给读者自行去查阅，在此不作多介绍。

10. TokenStore

最后介绍一下 TokenStore 接口，其提供了用于持久化令牌的相关方法，主要提供给

AuthorizationServerTokenServices 对令牌进行持久化操作，可以简单理解成对令牌的数据层操作。

提供的接口如下，主要是对令牌的 CRUD 操作：

```java
public interface TokenStore {
    // 根据访问令牌获取该令牌绑定的授权用户
    OAuth2Authentication readAuthentication(OAuth2AccessToken token);
    OAuth2Authentication readAuthentication(String token);
    void storeAccessToken(OAuth2AccessToken token, OAuth2Authentication authentication);
    OAuth2AccessToken readAccessToken(String tokenValue);
    void removeAccessToken(OAuth2AccessToken token);
    void storeRefreshToken(OAuth2RefreshToken refreshToken, OAuth2Authentication
        authentication);
    OAuth2RefreshToken readRefreshToken(String tokenValue);
}
```

里面提供了诸多对令牌的持久化操作，可以理解成 DAO 层的相关操作，默认实现主要有 4 种，类图如图 12-18 所示。

图 12-18　TokenStore 实现类图

可以根据需要将令牌保存到 Redis 中、数据库中、内存中、甚至是 JWT 中（JWT 中默认携带所有的相关信息，直接解析即可获取令牌中携带的信息）。以上实现都是开箱即用的实现类，只需要配置好相关的基础属性即可注入使用。

12.3.6　资源服务器

讲完了授权服务器，接着我们进入到资源服务器，了解资源服务器如何对资源请求进行拦截和验证。

1. @EnableResourceServer（开启资源服务器）

@EnableResourceServer 是开启资源服务器的关键注解，它将为资源服务器配置 Spring 的安全过滤器对需要进行拦截的请求验证请求中的访问令牌的有效性。它生效了一个关键的自动配置类 ResourceServerConfiguration（继承自 WebSecurityConfigurerAdapter），用于对 Web 安全进行配置。ResourceServerConfiguration 通过 ResourceServerSecurityConfigurer 完成资源安全配置，通过 HttpSecurity 完成 HTTP 安全配置。

2. ResourceServerSecurityConfigurer

ResourceServerSecurityConfigurer 中有诸多资源服务器的核心配置，配置了用于进行 OAuth2 认证的 OAuth2AuthenticationProcessingFilter、异常处理器（包括 AccessDeniedHandler

和 AuthenticationEntryPoint，它们将被配置到 ExceptionTranslationFilter 中)、用于 OAuth2 认证的 AuthenticationManager（注意该 AuthenticationManager 仅为 OAuth 认证服务，与 Spring Security 中普通认证过程中的 AuthenticationManager 不同），以及用于从请求中解析出访问令牌的 TokenExtractor。

主要配置代码如下所示：

```java
//ResourceServerSecurityConfigurer.java
@Override
public void configure(HttpSecurity http) throws Exception {
    // 配置AuthenticationManager
    AuthenticationManager oauthAuthenticationManager = oauthAuthenticationManager
        (http);
    // 配置令牌认证过滤器OAuth2AuthenticationProcessingFilter
    resourcesServerFilter = new OAuth2AuthenticationProcessingFilter();

    resourcesServerFilter.setAuthenticationEntryPoint(authenticationEntryPoint);
    resourcesServerFilter.setAuthenticationManager(oauthAuthenticationManager);
    if (eventPublisher != null) {
        resourcesServerFilter.setAuthenticationEventPublisher(eventPublisher);
    }
    // 配置从请求中解析访问令牌的TokenExtractor
    if (tokenExtractor != null) {
        resourcesServerFilter.setTokenExtractor(tokenExtractor);
    }
    resourcesServerFilter = postProcess(resourcesServerFilter);
    resourcesServerFilter.setStateless(stateless);
    // 配置异常处理器
    http
        .authorizeRequests().expressionHandler(expressionHandler)
    .and()
        .addFilterBefore(resourcesServerFilter, AbstractPreAuthenticatedProcessingFilter.
            class)
        .exceptionHandling()
            .accessDeniedHandler(accessDeniedHandler)
            .authenticationEntryPoint(authenticationEntryPoint);
    // @formatter:on
}
```

3. ResourceServerConfigurerAdapter

资源服务器也提供了 ResourceServerConfigurerAdapter 配置适配器类用于资源服务器在默认配置的基础上进行自定义配置，这些自定义配置将覆盖 @EnableResourceServer 中的默认配置。提供的可配置接口如下所示：

```java
public class ResourceServerConfigurerAdapter implements ResourceServerConfigurer {
    @Override
    public void configure(ResourceServerSecurityConfigurer resources) throws
        Exception {
    }
    @Override
    public void configure(HttpSecurity http) throws Exception {
```

```
        // 所有的请求都需要进行oauth2认证
        http.authorizeRequests().anyRequest().authenticated();
    }

}
```

ResourceServerConfigurerAdapter 提供了两个方面的配置：

- AuthorizationServerSecurityConfigurer，资源服务器安全配置。
- HttpSecurity，HTTP 安全相关配置，用于配置资源服务器内的资源访问控制。

4. OAuth2AuthenticationProcessingFilter

OAuth2AuthenticationProcessingFilter 过滤器是资源服务器中用来验证访问令牌的安全过滤器，在认证成功后会在 SecurityContext 中放入 OAuth2Authentication 以代表当前访问的用户身份。核心处理流程如下所示：

```java
//OAuth2AuthenticationProcessingFilter.java
public void doFilter(ServletRequest req, ServletResponse res, FilterChain chain)
    throws IOException,
            ServletException {
    final HttpServletRequest request = (HttpServletRequest) req;
    final HttpServletResponse response = (HttpServletResponse) res;
    try {
        //通过TokenExtractor从request中获取到有访问令牌的Authentication
        Authentication authentication = tokenExtractor.extract(request);
        if (authentication == null) {
            if (stateless && isAuthenticated()) {
                SecurityContextHolder.clearContext();
            }
        } else {
            request.setAttribute(OAuth2AuthenticationDetails.ACCESS_TOKEN_VALUE,
                authentication.getPrincipal());
            if (authentication instanceof AbstractAuthenticationToken) {
                AbstractAuthenticationToken needsDetails = (AbstractAuthenticationToken)
                    authentication;
                needsDetails.setDetails(authenticationDetailsSource.buildDetails
                    (request));
            }
            // 调用AuthenticationManager进行认证
            Authentication authResult = authenticationManager.authenticate
                (authentication);
            // 认证成功
            eventPublisher.publishAuthenticationSuccess(authResult);
            SecurityContextHolder.getContext().setAuthentication(authResult);
        }
    } catch (OAuth2Exception failed) {
        SecurityContextHolder.clearContext();
        // 认证失败
        eventPublisher.publishAuthenticationFailure(new BadCredentialsException
            (failed.getMessage(), failed),
            new PreAuthenticatedAuthenticationToken("access-token", "N/A"));
        authenticationEntryPoint.commence(request, response,
```

```
                new InsufficientAuthenticationException(failed.getMessage(), failed));
            return;
    }
    chain.doFilter(request, response);
}
```

OAuth2AuthenticationProcessingFilter 的认证流程很清晰，首先是通过 TokenExtractor 从请求中获取到保存有访问令牌的 Authentication，然后委托 AuthenticationManager 对其进行认证。

我们先了解 TokenExtractor 如何从请求中解析获取访问令牌。

（1）TokenExtractor

TokenExtractor 接口仅提供一个方法，用于从请求中分离出访问令牌，并封装成 Authentication，如下所示：

```
public interface TokenExtractor {
    Authentication extract(HttpServletRequest request);
}
```

默认实现为 BearerTokenExtractor，提供两种方式从请求获取访问令牌，从 BearerTokenExtractor 中获取访问令牌的核心方法如下所示：

```
//BearerTokenExtractor.java
protected String extractToken(HttpServletRequest request) {
    // 先尝试从头部中获取访问令牌
    String token = extractHeaderToken(request);
    if (token == null) {
        // 如果头部中不存在那么直接从请求参数中获取
        token = request.getParameter(OAuth2AccessToken.ACCESS_TOKEN);
        if (token == null) {
            logger.debug("Token not found in request parameters. Not an OAuth2
                request.");
        }
        else {
            request.setAttribute(OAuth2AuthenticationDetails.ACCESS_TOKEN_TYPE,
                OAuth2AccessToken.BEARER_TYPE);
        }
    }

    return token;
}

protected String extractHeaderToken(HttpServletRequest request) {
    // 获取到键为Authorization的头部
    Enumeration<String> headers = request.getHeaders("Authorization");
    while (headers.hasMoreElements()) {
        String value = headers.nextElement();
        // 如果该值以Bearer开头(大小写无关)，即为需要分离访问令牌
        if ((value.toLowerCase().startsWith(OAuth2AccessToken.BEARER_TYPE.
            toLowerCase()))) {
            String authHeaderValue = value.substring(OAuth2AccessToken.BEARER_
```

```
                TYPE.length()).trim();
            request.setAttribute(OAuth2AuthenticationDetails.ACCESS_TOKEN_TYPE,
                value.substring(0, OAuth2AccessToken.BEARER_TYPE.length()).trim());
            int commaIndex = authHeaderValue.indexOf(',');
            if (commaIndex > 0) {
                authHeaderValue = authHeaderValue.substring(0, commaIndex);
            }
            return authHeaderValue;
        }
    }
    return null;
}
```

BearerTokenExtractor 首先尝试从头部中分析访问令牌，只要在头部中有键为 Authorization 且值是以 "Bearer" 开头的属性就默认其为需要分离的令牌。如果没有的话，它还将尝试从请求参数中直接获取访问令牌，访问令牌的请求参数名为 access_token。

（2）OAuth2AuthenticationManager

在通过 TokenExtractor 获取到封装有访问令牌的 Authentication 后，OAuth2AuthenticationProcessingFilter 会将真正的认证工作委托给 AuthenticationManager 进行，此处 AuthenticationManager 的实现类是 OAuth2AuthenticationManager，属于 OAuth2 认证专用，并不是我们在 Spring Securtiy 中常见的 ProviderManager。

OAuth2AuthenticationManager 认证流程如下所示：

```
//OAuth2AuthenticationManager.java
public Authentication authenticate(Authentication authentication) throws AuthenticationException {
    if (authentication == null) {
        throw new InvalidTokenException("Invalid token (token not found)");
    }
    String token = (String) authentication.getPrincipal();
    // 通过ResourceServerTokenServices根据访问令牌加载OAuth2Authentication认证信息
    OAuth2Authentication auth = tokenServices.loadAuthentication(token);
    if (auth == null) {
        throw new InvalidTokenException("Invalid token: " + token);
    }
    // 验证该访问令牌是否可以访问该resouceId的资源服务器
    Collection<String> resourceIds = auth.getOAuth2Request().getResourceIds();
    if (resourceId != null && resourceIds != null && !resourceIds.isEmpty() &&
        !resourceIds.contains(resourceId)){
        throw new OAuth2AccessDeniedException("Invalid token does not contain
            resource id (" + resourceId + ")");
    }
    // 验证客户端
    checkClientDetails(auth);

    if (authentication.getDetails() instanceof OAuth2AuthenticationDetails) {
        OAuth2AuthenticationDetails details = (OAuth2AuthenticationDetails)
            authentication.getDetails();
        if (!details.equals(auth.getDetails())) {
            details.setDecodedDetails(auth.getDetails());
```

```
        }
    }
    auth.setDetails(authentication.getDetails());
    auth.setAuthenticated(true);
    return auth;
}
```

对访问令牌的主要认证过程委托给 ResourceServerTokenServices 进行,这是资源服务器端专用的令牌服务,与 AuthorizationServerTokenServices 相对应。注意,在验证完访问令牌之后还会对令牌对应的 OAuth2Authentication 的请求客户端信息进行检查。当资源服务器和授权服务器在一起的时候,将使用授权服务器配置的 ClientDetailsService 加载客户端信息进行认证。当资源服务器与授权服务器分离时,默认 ClientDetailsService 配置为空,将不检查客户端信息。

5. ResourceServerTokenServices

ResourceServerTokenServices 是资源服务器专用的令牌服务,提供的两个接口如下所示:

```
// 根据访问令牌加载OAuth2认证信息
OAuth2Authentication loadAuthentication(String accessToken) throws AuthenticationException,
    InvalidTokenException;

// 获取访问令牌的详情
OAuth2AccessToken readAccessToken(String accessToken);
```

默认实现有两种:

- RemoteTokenServices 远程令牌服务。
- DefaultTokenServices 默认令牌服务。

其中,DefaultTokenServices 同时实现了 AuthorizationServerTokenServices 接口,其内通过 tokenStore 实现了既定的功能,如下所示:

```
//DefaultTokenServices.java
public OAuth2Authentication loadAuthentication(String accessTokenValue) throws
    AuthenticationException,
        InvalidTokenException {
    OAuth2AccessToken accessToken = tokenStore.readAccessToken(accessTokenValue);
    if (accessToken == null) {
        ...
    }else if (accessToken.isExpired()) {
        ...
    }
    OAuth2Authentication result = tokenStore.readAuthentication(accessToken);
    if (result == null) {
        ...
    }
    if (clientDetailsService != null) {
        String clientId = result.getOAuth2Request().getClientId();
        try {
            clientDetailsService.loadClientByClientId(clientId);
```

```
        } catch (ClientRegistrationException e) {
            ...
        }
    }
    return result;
}
public OAuth2AccessToken readAccessToken(String accessToken) {
    return tokenStore.readAccessToken(accessToken);
}
```

DefaultTokenServices#loadAuthentication 方法首先验证访问令牌的合法性以及是否过期，接着再根据访问令牌获取对应的 OAuth2Authentication，当然如果 ClientDetailsService 不为空，还将对客户端信息进行检验；而 readAccessToken 方法直接委托 TokenStore#readAccessToken 从持久层获取访问令牌对应的 OAuth2AccessToken 信息。

RemoteTokenServices#loadAuthentication 方法是通过请求授权服务器的 /oauth/check_token 端点实现对应的需求，RemoteTokenServices 不支持 readAccessToken 方法。RemoteTokenServices#loadAuthentication 方法实现如下所示：

```
//RemoteTokenServices.java
@Override
public OAuth2Authentication loadAuthentication(String accessToken) throws
    AuthenticationException, InvalidTokenException {

    MultiValueMap<String, String> formData = new LinkedMultiValueMap<String,
        String>();
    formData.add(tokenName, accessToken);
    HttpHeaders headers = new HttpHeaders();
    headers.set("Authorization", getAuthorizationHeader(clientId, clientSecret));
    Map<String, Object> map = postForMap(checkTokenEndpointUrl, formData,
        headers);
    if (map.containsKey("error")) {
        throw new InvalidTokenException(accessToken);
    }
    Assert.state(map.containsKey("client_id"), "Client id must be present in
        response from auth server");
    return tokenConverter.extractAuthentication(map);
}
private Map<String, Object> postForMap(String path, MultiValueMap<String, String>
    formData, HttpHeaders headers) {
    if (headers.getContentType() == null) {
        headers.setContentType(MediaType.APPLICATION_FORM_URLENCODED);
    }
    @SuppressWarnings("rawtypes")
    Map map = restTemplate.exchange(path, HttpMethod.POST,
            new HttpEntity<MultiValueMap<String, String>>(formData, headers),
            Map.class).getBody();
    @SuppressWarnings("unchecked")
    Map<String, Object> result = map;
    return result;
}
```

它将访问令牌封装在表单中，通过 restTemplate 发送验证令牌请求到授权服务器的 /oauth/check_token 端点，由授权服务器完成对应的访问令牌认证检查，在授权服务器中默认由 DefaultTokenServices#loadAuthentication 方法提供实现。

12.3.7 令牌中继机制

Spring Cloud Security 除了对 Spring Security 和 Spring Security OAuth2 进行了封装，还为 Spring Cloud 的各个组件提供令牌中继的功能，使得请求在微服务中传递时，能够在请求中自动携带访问令牌，防止请求调用链下游的受保护资源时被拦截或者需要重新去授权服务器获取访问令牌。

令牌中继主要体现在两个方面，一个是在 Zuul Proxy，另一个是在微服务之间发生调用时。我们仅介绍微服务调用之间的令牌中继实现。

如果下游的微服务都是资源服务器，那么调用链经过的微服务都需要对访问令牌进行认证，这时我们可以通过 OAuth2RestTemplate 进行微服务之间的调用，它将在我们产生请求前将相关的访问令牌放到请求参数中。具体的覆盖的方法如下所示：

```java
// OAuth2RestTemplate.java
protected ClientHttpRequest createRequest(URI uri, HttpMethod method) throws
    IOException {
    OAuth2AccessToken accessToken = getAccessToken();
    AuthenticationScheme authenticationScheme = resource.getAuthenticationScheme();
    if (AuthenticationScheme.query.equals(authenticationScheme)
            || AuthenticationScheme.form.equals(authenticationScheme)) {
        // 将访问令牌放到请求参数中
        uri = appendQueryParameter(uri, accessToken);
    }
    ClientHttpRequest req = super.createRequest(uri, method);
    if (AuthenticationScheme.header.equals(authenticationScheme)) {
        authenticator.authenticate(resource, getOAuth2ClientContext(), req);
    }
    return req;
}
```

createRequest 方法是 RestTemplate 中用来生成请求的核心方法，通过该方法，在生成请求时将访问令牌放到请求参数中。

同样，OpenFeign 也具备这样的能力，可以自动为请求添加访问令牌。它主要通过 OAuth2FeignRequestInterceptor 拦截器实现，在请求发出前在请求的头部添加访问令牌，具体代码如下所示：

```java
//OAuth2FeignRequestInterceptor.java
public void apply(RequestTemplate template) {
    // 在头部添加访问令牌
    template.header(header, extract(tokenType));
}
```

```
protected String extract(String tokenType) {
    OAuth2AccessToken accessToken = getToken();
    return String.format("%s %s", tokenType, accessToken.getValue());
}
```

对此我们需要为 OpenFeign 添加 OAuth2FeignAutoConfiguration 的相关配置，该配置类将为我们的 OpenFeign 客户端添加 OAuth2FeignRequestInterceptor 拦截器，如下所示：

```
@FeignClient(name = "exampleService",
    configuration = {OAuth2FeignAutoConfiguration.class})
```

OAuth2FeignAutoConfiguration 配置文件的代码如下所示：

```
@Configuration
public class OAuth2FeignAutoConfiguration {

    @Bean
    public RequestInterceptor oauth2FeignRequestInterceptor(@Qualifier ("oauth2ClientContext")
        OAuth2ClientContext auth2ClientContext,

            OAuth2ProtectedResourceDetails auth2ProtectedResourceDetails){
                return new OAuth2FeignRequestInterceptor(auth2ClientContext,
                    auth2ProtectedResourceDetails);
    }
}
```

通过如上的配置文件，即可注入 OpenFeign 的拦截器，在应用 OpenFeign 客户端发送请求时将访问令牌放到请求头中。

12.4 进阶应用

12.4.1 Spring Security 定制

以下介绍的 Spring Security 配置主要基于 Spring Security Web 配置进扩展的，并且基于 Java 配置进行说明。

1. @EnableWebSecurity

在进行 Spring Security 配置时，我们通过 @EnableWebSecurity 开启了 Spring Security 的默认配置。@EnableWebSecurity 引入了两个配置文件 WebSecurityConfiguration 和 SpringWebMvcImportSelector，以及一个注解 @EnableGlobalAuthentication。其中 WebSecurityConfiguration 主要进行 Web 安全的配置；SpringWebMvcImportSelector 判断当前的环境是否具备 DispatcherServlet，即环境是否为 springMVC，防止 DispatcherServlet 的重复加载；@EnableGlobalAuthentication 为服务配置认证管理器 AuthenticationManager。

2. WebSecurityConfiguration

WebSecurittyConfiguration 主要作用是为 Spring Security 配置 FilterChainProxy，从而

对安全过滤器调用链进行管理。配置安全过滤器调用链代理的代码如下所示：

```java
//WebSecurittyConfiguration.java
@Bean(name = AbstractSecurityWebApplicationInitializer.DEFAULT_FILTER_NAME)
    public Filter springSecurityFilterChain() throws Exception {
    boolean hasConfigurers = webSecurityConfigurers != null
            && !webSecurityConfigurers.isEmpty();
    if (!hasConfigurers) {
        WebSecurityConfigurerAdapter adapter = objectObjectPostProcessor
                .postProcess(new WebSecurityConfigurerAdapter() {
                });
        webSecurity.apply(adapter);
    }
    return webSecurity.build();
}
```

通过使用 WebSecurity 生成名为 springSecurityFilterChain 的 Bean（FilterChainProxy），springSecurityFilterChain 在 Spring MVC 中将被 DelegatingFilterProxy 代理。

3. @EnableGlobalAuthentication

@EnableGlobalAuthentication 开启了 AuthenticationConfiguration 的配置类，用来对 AuthenticationManager 进行配置。AuthenticationManager 是 Spring Security 中的认证管理器。

AuthenticationConfiguration 一般情况下通过 AuthenticationManagerBuilder 构建对应的 AuthenticationManager，所以我们也通常是使用 AuthenticationManagerBuilder 来扩展定义 AuthenticationManager。

4. WebSecurityConfigurerAdapter

WebSecurityConfigurerAdapter 适配器配置类提供简易的方式让我们对 Spring Security 的相关默认配置进行自定义，其内提供了三个方法用于对配置进行重载，如下所示：

```java
// 配置AuthenticationManager
protected void configure(AuthenticationManagerBuilder auth) throws Exception {
    this.disableLocalConfigureAuthenticationBldr = true;
}
// 配置Web安全
public void configure(WebSecurity web) throws Exception {
}
// 配置HTTP安全
protected void configure(HttpSecurity http) throws Exception {
    http
        .authorizeRequests()
            .anyRequest().authenticated()
            .and()
        .formLogin().and()
        .httpBasic();
}
```

通过 WebSecurityConfigurerAdapter 可以对 Spring Security 中 AuthenticationManager、

Web 安全和 HTTP 安全中的自动配置进行覆盖设置，自定义属于开发者自己的应用安全规则。

下面我们对这三种通用配置进行介绍。

（1）配置 AuthenticationManager

通过 AuthenticationManagerBuilder 可以完成对 AuthenticationManager 的有效配置，示例代码如下所示：

```
@Configuration
@EnableWebSecurity
public class SecurityConfig extends WebSecurityConfigurerAdapter{

    @Override
    protected void configure(AuthenticationManagerBuilder auth) throws Exception {
        // 配置内存中的UserDetailService
        auth.inMemoryAuthentication()
            .withUser("user_authorization_code")
            .password("123456")
            .authorities("USER")
            .and()
            .withUser("user_password")
            .password("123456")
            .authorities("USER");
        // 配置数据库的UserDetailService
        auth.jdbcAuthentication()
               .dataSource(dataSource);
         // 配置自定义的UserDetailService
        auth.userDetailsService(userDetailsService());
         // 配置自定义的AuthenticationProvider
        auth.authenticationProvider(authenticationProvider1());
        auth.authenticationProvider(authenticationProvider2());

    }

}
```

上面的例子中配置了默认的内存中和数据库中的 UserDetailService，对此甚至可以注入完全自定义的 UserDetailService，它将被 DaoAuthenticationProvider 用于加载用户信息，默认只会使用最后配置的 UserDetailService。

当然也可以完全自定义属于自己的 AuthenticationProvider，用于实现自定义的完整认证流程，多个 AuthenticationProvider 都将生效。

（2）配置 Web 安全

```
@Configuration
@EnableWebSecurity
public class SecurityConfig extends WebSecurityConfigurerAdapter{
    @Override
    public void configure(WebSecurity web) throws Exception {

        web.ignoring()
```

```
                .antMatchers("/test/**");
        web.securityInterceptor(securityInterceptor());
    }
}
```

WebSecurity 是用来生成 FilterChainProxy，里面的配置一般不需要太多变化，但是我们可以通过它来忽略对某些路径的请求，如上例子中就忽略了"/test/**"路径的请求；还可以配置安全过滤器调用链中的 SecurityInterceptor 实现自定义的授权过滤器行为。

（3）配置 HTTP 安全

可以通过 HttpSecurity 为特定的 HTTP 请求配置基于 Web 的安全，例如访问哪些路径需要登录，访问哪些路径需要哪些权限等。

示例代码如下所示：

```
@Configuration
@EnableWebSecurity
public class SecurityConfig extends WebSecurityConfigurerAdapter{
    @Override
    protected void configure(HttpSecurity http) throws Exception {
        http
            .csrf().disable()
            .sessionManagement().sessionCreationPolicy(SessionCreationPolicy.IF_
                REQUIRED)
            .and()
            .authorizeRequests()
            .antMatchers("/main", "/index").permitAll()
            .antMatchers("/user/**").hasRole("user")
            .antMatchers("/admin/**").hasAuthority("ROLE_ADMIN")
            .antMatchers("/resource/**").access("hasRole('ADMIN') and hasAuthority
                ('ROLE_ADMIN')")
            .anyRequest().authenticated()
            .and()
            .formLogin()
            .loginPage("/login")
            .successForwardUrl("/user")
            .failureForwardUrl("/error")
            .usernameParameter("username")
            .passwordParameter("password")
            .permitAll()
            .and()
            .logout()
            .logoutUrl("/logout")
            .logoutSuccessUrl("/main")
            .addLogoutHandler(logoutHandler())
            .permitAll()
            .and()
            .httpBasic()
            .disable();
```

```
            http.addFilter(myCustomAuthenticationFilter());
            http.addFilterAfter(myCustomAuthenticationFilter(),UsernamePasswordAuthe
                nticationFilter.class);
            http.addFilterAt(myCustomAuthenticationFilter(), UsernamePasswordAuthent
                icationFilter.class);
            http.addFilterBefore(myCustomAuthenticationFilter(), UsernamePasswordAut
                henticationFilter.class);
    }
}
```

上述是一份囊括了大多数 HttpSecurity 通用配置的相关展示，每一个 and 对应了一个模块的配置，这种结构与 XML 的配置相当类似。其中：

- csrf 是对 CSRF 攻击防御的开启设置，上述配置是关闭。
- sessionManagement 是对会话使用的管理，上述配置是需要时使用并创建会话。
- authorizeRequests 对路径拦截进行配置，说明访问该路径需要的角色、权限、是否需要认证等。
- formLogin 是对表单登录认证的相关配置。
- logout 是对用户注销的相关配置。
- httpBasic 表示是否开启 HTTP BASIC 认证，即是否开启 BasicAuthenticationFilter 过滤器，上述配置为关闭。

最后是对自定义的安全过滤器的添加，可以根据自己的需要添加自定义的安全过滤器到安全过滤器调用链中，完成自定义的安全拦截，但是在添加过程中要注意添加的过滤器的位置，防止造成不必要的 Bug。

Spring Security 的配置方式和配置点实在过于庞大，在这里根本没有办法进行较为详尽的介绍，希望能够进一步定制 Spring Security 的读者可以参考官方文档，或者通过阅读相关的 Configurer 类查看其配置过程，满足自己的业务需要。

12.4.2　OAuth2 定制

1. 授权服务器相关配置

在授权服务器的源码讲解中，了解到 AuthorizationServerConfigurerAdapter 配置适配器类是 Spring Security OAuth2 提供给开发者对授权服务器中的相关配置 Bean 进行覆盖操作的自定义配置类。

AuthorizationServerConfigurerAdapter 接口的代码如下所示：

```
public class AuthorizationServerConfigurerAdapter implements AuthorizationServerConfigurer {
    //授权服务器安全认证的相关配置,用于配置Spring Security,生成对应的安全过滤器调用链,主
      要控制oauth/**端点的相关访问
    @Override
    public void configure(AuthorizationServerSecurityConfigurer security) throws
        Exception {
```

```
    }
    //OAuth2客户端服务相关配置
    @Override
    public void configure(ClientDetailsServiceConfigurer clients) throws  Exception {
    }
    //授权服务器授权端点相关配置
    @Override
    public void configure(AuthorizationServerEndpointsConfigurer endpoints) throws
        Exception {
    }
}
```

以下我们将通过这三个方法对授权服务器的相关配置进行介绍。

（1）配置 OAuth 访问控制

可以通过 AuthorizationServerSecurityConfigurer 配置 OAuth2 相关的端点的访问安全规则，示例配置如下所示：

```
@EnableAuthorizationServer
@Configuration
public class AuthorizationServerConfig extends AuthorizationServerConfigurerAdapter {
    @Override
    public void configure(AuthorizationServerSecurityConfigurer security) throws
        Exception {
        security
            .sslOnly() // 仅支持ssl访问
            .passwordEncoder(NoOpPasswordEncoder.getInstance()) // 配置
                clientSecret密码转化器
            .authenticationEntryPoint(customAuthenticationEntryPoint()) // 配置认
                证端点
            .accessDeniedHandler(customAccessDeniedHandler()) // 配置授权失败处理器
            .tokenKeyAccess("permitAll()") // 配置请求令牌端点允许访问规则
            .checkTokenAccess("isAuthenticated()") // 配置检验令牌端点允许访问规则
            .allowFormAuthenticationForClients(); // 是否允许表单提交客户端信息
    }
}
```

在 AuthorizationServerSecurityConfigurer 中，可以为 "oauth/**" 配置合理的访问安全规则，保证授权服务器端点的安全访问。

（2）配置 OAuth2 客户端服务

在请求访问令牌的过程中，授权服务器会对请求的客户端进行认证，所以需要为授权服务器配置 ClientDetailsService 用于加载客户端信息。

ClientDetailsServiceConfigurer 默认提供两种方式来配置 ClientDetailsService，从内存中加载和从数据库中加载。如果有其他的需求，可以继承 ClientDetailsService 实现自己的客户端信息服务。

配置示例代码如下所示：

```
@EnableAuthorizationServer
```

```java
@Configuration
public class AuthorizationServerConfig extends AuthorizationServerConfigurerAdapter {
    @Override
    public void configure(ClientDetailsServiceConfigurer clients) throws Exception {
        // 配置内存中客户端服务
        clients.inMemory().withClient("client")// 客户端ID
            .authorizedGrantTypes("authorization_code","password", "refresh_token",
                "implicit")// 客户端可以使用的授权类型
            .scopes("all")// 允许请求范围
            .accessTokenValiditySeconds(1000) //访问令牌有效时长
            .refreshTokenValiditySeconds(1000) // 刷新令牌有效时长
            .secret("secret")// 客户端安全码
            .redirectUris("http://localhost:8888/");// 回调地址
        // 配置数据库内客户端服务，需要注入数据库配置
        clients.jdbc(dataSource());
        // 配置自定义客户端服务
        clients.withClientDetails(customClientDetailsService());
    }
}
```

（3）配置授权服务器端点

通过 AuthorizationServerEndpointsConfigurer，开发者可以覆盖令牌端点中诸多委托类的相关实现，通用的配置如下所示：

```java
@EnableAuthorizationServer
@Configuration
public class AuthorizationServerConfig extends AuthorizationServerConfigurerAdapter {

    @Override
    public void configure(AuthorizationServerEndpointsConfigurer endpoints) throws
        Exception {
        endpoints
            .tokenStore(new InMemoryTokenStore()) // 配置tokenStore
            .accessTokenConverter(accessTokenConverter()) // 配置令牌转化器
            .authenticationManager(authenticationManager) // 配置令牌端点中的用户认证
                                                          管理器
            .userDetailsService(userDetailsService) // 配置用户信息服务
            .authorizationCodeServices(customAuthorizationCodeServices()
                // 配置授权码服务
            .tokenServices(customAuthorizationServerTokenServices())
                // 配置token服务
            .pathMapping("/oauth/token", "/token") // 修改路径名称，这里将/oauth/
                                                    token更改为/token
            .reuseRefreshTokens(false) // 是否允许重用刷新令牌
            .allowedTokenEndpointRequestMethods(HttpMethod.GET, HttpMethod.POST)
                // 配置允许访问/oauth/token端点的HTTP方法
            .addInterceptor(customInterceptor()) // 添加拦截器
            .requestFactory(customOAuth2RequestFactory()); // 配置请求工厂
    }
}
```

我们可以对授权服务器源码解析中介绍的诸多委托类进行覆盖配置，如 TokenStore 和

AuthorizationServerTokenServices，满足业务对授权服务器的需求。

2. 资源服务器相关配置

资源服务器提供 ResourceServerConfigurerAdapter 配置适配类对资源服务器的自动配置进行覆盖配置，主要提供两方面的配置：

- AuthorizationServerSecurityConfigurer 用于资源服务器安全配置。
- HttpSecurity，HTTP 安全相关配置，用于配置资源服务器内的资源访问控制。

配置适配类代码如下所示：

```java
public class ResourceServerConfigurerAdapter implements ResourceServerConfigurer {

    @Override
    public void configure(ResourceServerSecurityConfigurer resources) throws Exception {
    }

    @Override
    public void configure(HttpSecurity http) throws Exception {
        // 所有的请求都需要进行OAuth2认证
        http.authorizeRequests().anyRequest().authenticated();
    }

}
```

（1）配置资源服务器

ResourceServerSecurityConfigurer 允许开发者对资源服务器内认证访问令牌的方式进行配置，主要是对 OAuth2AuthenticationProcessingFiltertoken 认证过滤器中的相关委托实现类进行覆盖配置。通用的配置代码如下所示：

```java
@Configuration
@EnableResourceServer
public class ResourceServerConfig extends ResourceServerConfigurerAdapter{

    @Override
    public void configure(ResourceServerSecurityConfigurer resources) {
        resources
            .resourceId("instance") // 资源服务器标志
            .tokenExtractor(new BearerTokenExtractor()) 配置TokenExtractor请求令牌
                解析器
            .tokenServices(tokenServices()) // 配置ResourceServerTokenServices
          .authenticationManager(customOAuth2AuthenticationManager())
              //配置OAuth2AuthenticationManager
            .tokenStore(customTokenStore()) // 配置TokenStore
            .stateless(true) // 访问无状态
            .authenticationEntryPoint(customAuthenticationEntryPoint())// 配置认证端点
            .accessDeniedHandler(customAccessDeniedHandler()) // 配置授权失败处理器
    }
}
```

（2）配置资源服务器拦截规则

通过 HttpSecurity 可以为资源服务器提供如下配置：对哪些路径的访问需要验证访问令

牌；访问的路径需要访问令牌对应用户的权限达到什么样的级别。基本使用方式与 Spring Security 中配置 HTTP 安全的方式一致。

例如默认配置中，资源服务器的所有端点的访问都需要携带有效的访问令牌，如下所示：

```
@Override
public void configure(HttpSecurity http) throws Exception {
    // 所有的请求都需要进行oauth2认证
    http.authorizeRequests().anyRequest().authenticated();
}
```

通过 HttpSecurity 可以为资源服务器中资源的安全访问添加强有力的控制。

12.4.3　SSO 单点登录

1. 搭建第三方客户端服务

以下我们将构建一个 oauth2-client 应用，它将使用 Spring Cloud Security 提供的单点登录功能。oauth2-client 将封装客户端与授权服务器交互的流程，使用户可以使用第三方登录的方式登录客户端应用。

可以搭建包含 Spring Cloud Security 依赖的 Spring Boot 项目。引入主要依赖如下所示：

```xml
<dependency> <!--Eureka Client相关依赖-->
    <groupId>org.springframework.cloud</groupId>
    <artifactId>spring-cloud-starter-netflix-eureka-client</artifactId>
</dependency>
<dependency> <!--Spring Web MVC相关依赖-->
    <groupId>org.springframework.boot</groupId>
    <artifactId>spring-boot-starter-web</artifactId>
</dependency>
<dependency> <!--Spring Cloud Security相关依赖-->
    <groupId>org.springframework.cloud</groupId>
    <artifactId>spring-cloud-starter-security</artifactId>
</dependency>
```

在 application.yaml 中添加服务发现与注册相关配置，如下所示：

```yaml
eureka:
    instance:
        instance-id: ${spring.application.name}:${vcap.application.instance_id:$
            {spring.application.instance_id:${random.value}}}
    client:
        service-url:
            default-zone: http://localhost:8761/eureka/, http://localhost:8762/eureka/
spring:
    application:
        name: oauth2-client
server:
    port: 8888
```

这里使用的服务注册中心为第 4 章配置的 Eureka Server。

在 application.yaml 中添加第三方登录时使用的授权服务器信息，这里使用的授权服器为基础应用中搭建的 authorization-server，如下所示：

```yaml
security:
    oauth2:
        client:
            client-id: client
            client-secret: secret
            access-token-uri: http://localhost:8766/oauth/token
            user-authorization-uri: http://localhost:8766/oauth/authorize
            client-authentication-scheme: form
        resource:
            user-info-uri: http://localhost:8767/user # 获取授权用户信息
            prefer-token-info: false
```

添加 SecurityConfig 配置文件，使用 @EnableOAuth2Sso 注解开启单点登录功能，代码如下所示：

```java
@Component
@EnableOAuth2Sso
public class SecurityConfig extends WebSecurityConfigurerAdapter{
    @Override
    protected void configure(HttpSecurity http) throws Exception {
        http.csrf().disable()
            .sessionManagement().sessionCreationPolicy(SessionCreationPolicy.IF_
                REQUIRED)
            .and()
            .requestMatchers().anyRequest()
            .and()
            .authorizeRequests()
            .antMatchers("/user/**").authenticated()  // user前缀的接口需要登录
            .and()
            .formLogin().successForwardUrl("/index")  // 登录成功后重定向到
                                                      /index接口
            .and()
            .logout().logoutUrl("/logout").permitAll()
            .logoutSuccessUrl("/");
    }
}
```

通过 SecurityConfig 为 "/user" 前缀的接口添加登录认证。添加 MainController，为 oauth2-client 添加访问接口，代码如下所示：

```java
@RestController
public class MainController {
    @RequestMapping(method = RequestMethod.GET)
    public String main(){
        return "Welcome to the index!";
    }
    @RequestMapping(value = "/index", method = RequestMethod.GET)
    public String index(){
        return "Welcome to the index!";
    }
    @RequestMapping(value = "/user", method = RequestMethod.GET)
```

```
    public Principal principal(Principal user){
        return user;
    }
}
```

其中,"/"和"/index"接口可以直接访问,"/user"接口需要登录后才能访问,它将返回当前登录用户的登录信息(通过访问 http://localhost:8767/user 端点获取用户信息)。

为了避免 CSRF 检查失败(例子中的 authorization-server 和 oauth2-client 部署在同一个主机名下,将使用相同的 cookie,引发 CSRF 检查),我们将 authorization-server 中客户端的回调地址修改为本机地址,如 IP 地址为 192.168.1.168,在 AuthorizationServerConfig 中修改如下所示:

```
@Override
public void configure(ClientDetailsServiceConfigurer clients) throws Exception {
    //配置一个客户端
    //既可以通过授权码类型获取令牌,也可以通过密码类型获取令牌
    clients.inMemory().withClient("client")// 客户端ID
        .authorizedGrantTypes("authorization_code","password", "refresh_token",
            "implicit")// 客户端可以使用的授权类型
        .scopes("all")// 允许请求范围
        .secret("secret")// 客户端安全码
        .redirectUris("http://192.168.1.168:8888/");// 回调地址
}
```

为了返回授权用户的信息,我们在资源服务器 resource-server 中添加一个根据访问令牌获取用户认证信息的接口,如下所示:

```
@RestController
public class ResourceController {
    private final static Logger logger = LoggerFactory.getLogger(ResourceController.
        class);
    BearerTokenExtractor tokenExtractor = new BearerTokenExtractor();
    ...
    @Autowired
    ResourceServerTokenServices tokenServices;
    @RequestMapping(value = "/user")
    public Principal userInfo(ServletRequest req) throws IOException {
        final HttpServletRequest request = (HttpServletRequest) req;
        // 通过BearerTokenExtractor从请求中获取访问令牌
        Authentication authentication = tokenExtractor.extract(request);
        String token = (String) authentication.getPrincipal();
        // 使用ResourceServerTokenServices加载访问令牌对应的OAuth2Authentication信息
        return tokenServices.loadAuthentication(token);
    }
}
```

2. 单点登录

依次启动 Eureka Server、授权服务器、资源服务器以及本服务。

访问 http://192.168.1.168:8888(使用本机地址访问)或者 http://192.168.1.168:8888/index 可以直接获取到请求结果,如下所示:

```
Welcome to the index!
```

接着我们访问 http://192.168.1.168:8888/user，这将会帮我们重定向到授权服务器的登录端点，如图 12-19 所示。

登录后，同意授权，将会重定向到 oauth2-client 的 http://192.168.1.168:8888/user，获取到我们需要的当前登录用户的用户认证信息，如图 12-20 所示。

图 12-19　授权服务器登录界面

图 12-20　获取用户信息

对此，我们完成了使用 Spring Cloud Security 的单点登录功能，实现了第三方登录的客户端应用。

该功能的实现主要是通过 OAuth2ClientAuthenticationProcessingFilter 过滤器从授权服务器中请求访问令牌并加载对应的 Authentication 到本地安全上下文中。有兴趣读者可以自行阅读 OAuth2ClientAuthenticationProcessingFilter 的源码进行了解，在此不进行过多介绍。

12.5　本章小结

Spring Cloud Security 中既提供用了单体应用的 Spring Security 安全框架，也提供了基于 Spring Secuiry 构建的符合 OAuth2 认证流程的 Spirng Security OAuth2 安全框架，并对其中的执行流程提供了更为简单的使用实现，如 SSO 单点登录等，为微服务的应用安全提供多方面支持。

Spring Security 中提供两大核心功能：认证和授权，主要在安全过滤器调用链中发挥对应的安全控制功能。Spring Security OAuth2 中划分了授权服务器和资源服务器，授权服务器为获得用户授权的第三方客户端颁发访问令牌，资源服务器保护受限资源，允许携带访问令牌的请求访问资源。通过 Spring Cloud Security，开发者可以为自己的应用服务提供强大的安全保护。

第 13 章 Chapter 13

服务链路追踪：Spring Cloud Sleuth

在前面的章节，我们介绍了 Spring Cloud 中的多个组件，可以利用这些组件构建一个微服务系统。本章将介绍 Spring Cloud 提供的链路监控组件 Spring Cloud Sleuth，这个组件提供了分布式链路追踪的解决方案，用以追踪微服务系统中一次请求的完整过程。

与之前的章节有所区别，本章更偏向于应用，所以不涉及源码分析。本章第一小节将介绍微服务架构中链路监控相关的组件；第二小节介绍链路监控组件的基础应用，介绍 Spring Cloud Sleuth 的两种实践方式：独立使用和整合 Zipkin。整合 Zipkin 时，又有两种不同的通信方式用于发送链路信息，将会结合示例场景讲解具体的应用方法。

13.1 链路监控组件简介

在微服务架构下，系统由大量服务组成，每个服务可能是由不同的团队开发、使用不同的编程语言、部署在几千台服务器上，并且横跨多个不同的数据中心等。一次请求往往会涉及多个服务，在系统发生故障的时候，想要快速定位和解决问题，就需要追踪服务请求序列。因此，分析性能问题的工具以及理解系统的行为变得很重要。

分布式调用链路监控组件在这样的需求下产生了。最著名的是谷歌在公开论文中提到的 Dapper 组件，Dapper 用于收集复杂分布式系统的行为信息，然后呈现给 Google 的开发者们。分布式系统有大规模的低端服务器作为互联网服务的载体，是一个特殊的经济划算的平台。想要在这个上下文中理解分布式系统的行为，就需要监控那些横跨了不同的应用、不同的服务器之间的关联动作。下面介绍常用的监控组件及分布式监控的基础概念。

1. 常用的监控组件

市面上的 APM（Application Performance Management）理论模型大多都是借鉴 Google

的 Dapper 论文，笔者重点关注了以下几种 APM 组件：

- Zipkin。由 Twitter 公司开源，开放源代码的分布式跟踪系统，用于收集服务的定时数据，以解决微服务架构中的延迟问题，包括数据的收集、存储、查找和展现。Spring Cloud Sleuth 中集成了 Zipkin。
- Pinpoint。是一款对 Java 编写的大规模分布式系统的 APM 工具，是由韩国 Naver 公司开源的分布式跟踪组件。
- Skywalking。一款国产的 APM 组件，是一个对 Java 分布式应用程序集群的业务运行情况进行追踪、告警和分析的系统。

其他类似的组件还有美团点评的 CAT 和淘宝的 EagleEye 等。笔者认为链路监控组件的选择要考虑下列要求：

- **探针的性能消耗**。APM 组件对服务的影响应该做到足够小。在一些高度优化过的服务中，即使一点点损耗也会很容易察觉，而且有可能迫使在线服务的部署团队不得不将跟踪系统关停。
- **易用性**。对于开发应用的程序员来说，是不需要知道有跟踪系统这回事的。所以，引入链路监控组件应该是对开发人员透明的。
- **可扩展性**。能够支持的组件越多越好，或者提供便捷的插件开发 API，对于一些没有监控到的组件，应用开发者也可以自行扩展。
- **数据的分析**。数据的分析要快，分析的维度要尽可能多分析要全面，尽量避免二次开发。跟踪系统能提供足够快的信息反馈，就可以对生产环境下的异常状况做出快速反应。

2. 基础概念

上面列出的几种组件中，Zipkin 组件是严格按照 Google 的 Dapper 论文实现的，下面介绍其中涉及的基本概念：

- Span。基本工作单元，一次链路调用（可以是发起 RPC 和写 DB 等操作，没有特定的限制）创建一个 Span，通过一个 64 位 ID 标识它，使用 UUID 较为方便。Span 中还有其他的数据，例如描述信息、时间戳、键值对的（注解）tag 信息和 parent-id 等，其中 parent-id 可以表示 Span 调用链路来源。
- Trace。类似于树结构的 Span 集合，表示一条调用链路，存在唯一标识。
- 注解（Annotation）。用来记录请求特定事件的相关信息（例如时间），通常包含四个注解信息：
    - CS：Client Sent，表示客户端发起请求。
    - SR：Server Receive，表示服务端收到请求。
    - SS：Server Send，表示服务端完成处理，并将结果发送给客户端。
    - CR：Client Received，表示客户端获取到服务端返回信息。

（1）Trace

系统中一次请求追踪记录如图 13-1 所示，客户端请求 SERVICE 1，SERVICE 1 调用 SERVICE 2，在 SERVICE 2 中又调用了 SERVICE 3 和 SERVICE 4，最终由 SERVICE 1 将请求的响应结果返回给客户端。在这一次调用链中，请求拥有唯一的表示 TraceId=X。

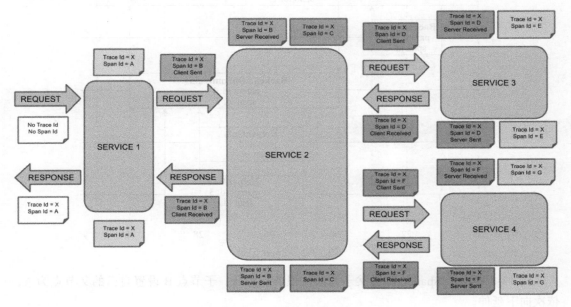

图 13-1　一次请求追踪的记录

通过 Span Id 区分不同的 Span，图中展示了 A 到 G 共计 7 个 Span。一条链路通过 Trace Id 唯一标识，Span 标识所发起的请求信息。树节点是整个架构的基本单元，而每一个节点又是对 Span 的引用。节点之间的连线表示 Span 和它的父 Span 直接的关系。虽然 Span 在日志文件中只是简单地代表 Span 的开始和结束时间，他们在整个树形结构中却是相对独立的。

以图 13-1 中的一个 Span 为例，其组成如下所示：

```
Trace Id = X
Span Id = D
Client Sent
```

如上的信息表明了当前 Span 的 Trace Id 设为 X，Span Id 设为 D，并且表明发生了客户端发起请求的事件。

（2）Span

图 13-2 展示了谷歌 Dapper 定义的 Span 在一次请求追踪过程中的表现。Dapper 记录了 Span 名称，以及每个 Span 的 ID 和父 ID，以重现在一次追踪过程中不同 Span 之间的关系。如果一个 Span 没有父 ID，就称为根（root）Span。所有 Span 都挂在一个特定的跟踪上，也

共用一个跟踪 ID。

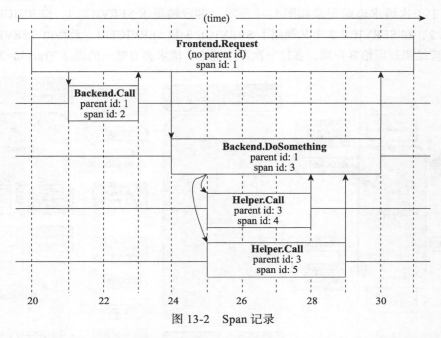

图 13-2　Span 记录

图 13-3 展示了 Span 之间的父子关系，根节点为 A，子节点 B 设置自己的父节点为 A，依次向下延伸。

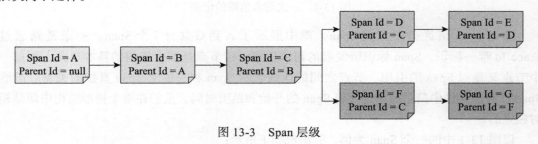

图 13-3　Span 层级

（3）注解

自动探针不需要修改应用程序源代码，探针埋点对应用开发者来说近乎透明。Dapper 还允许应用程序开发人员在 Dapper 跟踪的过程中添加额外的信息，以监控更高级别的系统行为，或帮助调试问题。

Spring Cloud Sleuth 整合了 Zipkin，下面便介绍其基础的使用方法与实践。

13.2　基础应用

Spring Cloud Sleuth 为服务之间的调用提供链路追踪，通过 Sleuth 可以很清楚地了解到一个服务请求经过了哪些服务，每个服务处理花费了多长，从而可以很方便地理清各微服务间的调用关系。此外 Sleuth 还可以进行耗时分析，通过 Sleuth 可以很方便地了解到每个采样请求的耗时，从而分析出哪些服务调用比较耗时；Sleuth 可以可视化错误，对于程序未捕捉的异常，可以通过集成 Zipkin 服务界面看到；通过 Sleuth 还能对链路优化，对于调用比较频繁的服务实施一些优化措施。

13.2.1 特性

Spring Cloud Sleuth 具有如下的特性：

1）提供对常见分布式跟踪数据模型的抽象：Traces、Spans（形成 DAG）、注解和键值注解。Spring Cloud Sleuth 虽然基于 HTrace，但与 Zipkin（Dapper）兼容。

2）Sleuth 记录了耗时操作的信息以辅助延时分析。通过使用 Sleuth，可以定位应用中的耗时原因。

3）为了不写入太多的日志，以至于使生产环境的应用程序崩溃，Sleuth 做了如下的工作：
- 生成调用链的结构化数据。
- 包括诸如自定义的织入层信息，比如 HTTP。
- 包括采样率配置以控制日志的容量。
- 查询和可视化完全兼容 Zipkin。

4）为 Spring 应用织入通用的组件，如 Servlet 过滤器、Zuul 过滤器和 OpenFeign 客户端等等。

5）Sleuth 可以在进程之间传播上下文（也称为背包）。因此，如果在 Span 上设置了背包元素，它将通过 HTTP 或消息传递到下游，发送到其他进程。

6）Spring Cloud Sleuth 实现了 OpenTracing 的规范。OpenTracing 通过提供平台无关、厂商无关的 API，使得开发人员能够方便地添加（或更换）追踪系统的实现。

13.2.2 项目准备

下面通过两个服务之间的相互调用，演示链路监控的场景。前期需要准备的项目有三个：
- Eureka Server：服务注册 Server，作为其他客户端服务的注册中心。
- Sleuth Client A：客户端服务 A，注册到 Eureka Server。对外提供两个接口，一个接口暴露给客户端服务 A，另一个接口调用客户端服务 B，客户端服务之间通过 OpenFeign 进行 HTTP 调用。
- Sleuth Client B：同上。

1. Eureka Server

首先，依赖需要引入 Netflix Eureka Server，如下所示：

```
<dependency>
    <groupId>org.springframework.cloud</groupId>
```

```
        <artifactId>spring-cloud-starter-netflix-eureka-server</artifactId>
</dependency>
```

其次，配置 Eureka Server 的服务端信息，如下所示：

```
spring:
    application:
        name: eureka-server-standalone
server:
    port: 8761
eureka:
    instance:
        hostname: standalone
        instance-id: ${spring.application.name}:${vcap.application.instance_
            id:${spring.application.instance_id:${random.value}}}
```

其中，我们配置了单机模式的 Eureka Server，服务器的端口号为 8761。

最后，添加应用程序的入口类，如下所示：

```
@SpringBootApplication
@EnableEurekaServer
public class EurekaServerApplication {
    public static void main(String[] args) {
        SpringApplication.run(EurekaServerApplication.class, args);
    }
}
```

如上三步完成，我们的注册服务即可以正常启动。

2. Sleuth Client A

首先，也是引入客户端服务 A 所需要的依赖，如下所示：

```
<dependencies>
    <dependency>
        <groupId>org.springframework.cloud</groupId>
        <artifactId>spring-cloud-starter-netflix-eureka-client</artifactId>
    </dependency>
    <dependency>
        <groupId>org.springframework.cloud</groupId>
        <artifactId>spring-cloud-starter-openfeign</artifactId>
    </dependency>
    <dependency>
        <groupId>org.springframework.boot</groupId>
        <artifactId>spring-boot-starter-web</artifactId>
    </dependency>
</dependencies>
```

需要的依赖为：注册客户端、声明式客户端组件 OpenFeign 和 Spring Boot Web 包。

其次，客户端服务 A 的配置文件如下所示：

```
eureka:
    instance:
        instance-id: ${spring.application.name}:${vcap.application.instance_id:$
```

```
                {spring.application.instance_id:${random.value}}}
    client:
        service-url:
            defaultZone: http://localhost:8761/eureka/, http://localhost:8762/eureka/
spring:
    application:
        name: service-a
server:
    port: 9002
```

上述配置声明了服务名，OpenFeign通过服务名进行服务实例调用，指定了客户端服务A的端口号9002。A还需要注册到Eureka Server。

然后提供两个接口，其实现如下所示：

```
@RestController
@RequestMapping("/api")
public class ServiceAController {
    @Autowired
    ServiceBClient serviceBClient;

    @GetMapping("/service-a")
    public String fromServiceA(){
        return "from serviceA";
    }
    @GetMapping("/service-b")
    public String fromServiceB(){
        return serviceBClient.fromB();
    }
}
```

服务A提供了/service-b接口，可以调用服务B，Feign客户端的实现如下所示，/service-a接口用于给服务B调用。

```
@FeignClient("service-b")
public interface ServiceBClient {

    @RequestMapping(value = "/api/service-b")
    public String fromB();
}
```

OpenFeign客户端的注解中，指定了要调用的服务B的serviceId为service-b。

最后，应用程序的入口类如下：

```
@SpringBootApplication
@EnableFeignClients
public class Chapter13ClientServiceaApplication {

    public static void main(String[] args) {
        SpringApplication.run(Chapter13ClientServiceaApplication.class, args);
    }
}
```

@EnableFeignClients 注解会扫描那些声明为 Feign 客户端的接口。

至此，客户端服务 A 的实现大功告成。客户端服务 B 与 A 的实现基本类似，A 与 B 之间实现互调。这里略过实现，读者可以参考本书的配套代码。

13.2.3 Spring Cloud Sleuth 独立实现

当 Spring Cloud Sleuth 单独使用时，通过日志关联的方式将请求的链路串联起来，分别启动之前准备的三个服务，并访问地址 http://localhost:9002/api/service-b，服务 A 调用了服务 B，成功返回响应后，我们看一下控制台的 Sleuth 相关日志。

服务 A 日志如下所示：

```
2018-04-11 01:03:25.696 INFO [service-a,7feec0479597d1b9,578bef9ed3901d9b,false]
```

服务 B 日志如下所示：

```
2018-04-11 01:03:25.905 DEBUG [service-b,7feec0479597d1b9,578bef9ed3901d9b,false]
```

日志中输出的四部分信息分别为：

- appname：service-a/b，是设置的应用名称。
- traceId：7feec0479597d1b9，Spring Cloud Sleuth 生成的 TraceId，一次请求的唯一标识。
- spanId：578bef9ed3901d9b，Spring Cloud Sleuth 生成的 SpanId，标识一个基本的工作单元，此处是 Feign 进行的 HTTP 调用。
- exportable：false，是否将数据导出，如果只是想要在 Span 中封装一些操作并将其写日志时，此时就不需要将 Span 导出。

从上面的日志及输出信息的解释可以看出，发出的 HTTP 请求到达 A 后，A 向 B 发起调用，这是一次完整的调用过程，TraceId 在服务 A 和 B 中是相同的。Spring Cloud Sleuth 在一次请求中，生成了如上四部分信息，TraceId 唯一，可以有多个 SpanId。当然数据是可以导出进行分析的，如将日志信息导出到 Elasticsearch，使用日志分析工具如 Kibana、Splunk 或者其他工具，比较常用的组合是 ELK：Elasticsearch + Logstash + Kibana。然而即便这样，还是需要我们自行对一些维度的数据进行分析，其实这些监控的数据维度基本比较确定，在业界也有一些优秀组件，如 Zipkin、Pinpoint 等，利用这些工具将会避免不必要的分析工作。

13.2.4 集成 Zipkin

Zipkin 是一个分布式的链路追踪系统。在定位微服务架构中的问题时，Zipkin 有助于收集所需的时序数据。Zipkin 主要涉及几个组件：Collector（收集 agent 的数据）、Storage（存储）和 Web UI（图形化界面），如图 13-4 所示。

Collector 收集 agent 的数据有两种方式：HTTP 调用和 MQ（消息中间件）通信。Zipkin 客户端服务默认是通过 HTTP 方式发送链路追踪信息到 Zipkin Collector，HTTP 方式是同步调用，对原有请求会产生一些影响；MQ 通信则是使用异步方式收集链路调用信息，虽有一定的延时，但是总体性能较高。下面分别介绍这两种方式。

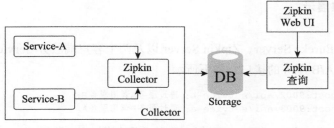

图 13-4　Zipkin 组成

1. 用 HTTP 通信

（1）Zipkin Server

zipkin-server 是一个 Spring Boot 应用程序，打包为一个可执行的 jar。我们需要 JRE 8+ 来启动 zipkin-server。当然我们也可以通过 Docker 容器启动相应的服务，可以参考 Zipkin 的官网 https://github.com/openzipkin/docker-zipkin。

下面我们以可执行 jar 包直接启动的方式，讲解如何收集展示服务之间的调用信息。

```
$ curl -sSL https://zipkin.io/quickstart.sh | bash -s
$ java -jar zipkin.jar
```

通过如上的脚本，即可启动一个 Zipkin Server。Span 存储方式和 Collector 是可配置的。默认情况下，存储位于内存中，采用 Http Collector 的方式进行数据收集，并且在端口 9411 上进行侦听。

Zipkin UI 提供了图形化展示链路调用情况以及服务之间的依赖。为了让示例简单，Zipkin Server 采用默认的内存数据库。

（2）Zipkin Client 改造

首先，需要在依赖中加入 Sleuth 的 Zipkin 包，如下所示：

```xml
<dependency>
    <groupId>org.springframework.cloud</groupId>
    <artifactId>spring-cloud-starter-zipkin</artifactId>
</dependency>
```

其次，配置文件中，增加 Zipkin Server 的配置信息，如下所示：

```
spring:
    sleuth:
        sampler:
            probability: ${ZIPKIN_RATE:1}
    zipkin:
        base-url: http://localhost:9411
```

配置中设置了请求的采集率。当使用 Zipkin 时，通过设置 spring.sleuth.sampler.probability

配置导出 Span 的概率（默认值：0.1，即 10%）。默认值可能会让你觉得 Sleuth 不工作，因为它省略了一些 Span；另外配置中还设置了通过 HTTP 方式发送到 Zipkin 的地址。

（3）调用结果展示

步骤如下：

1）相继启动 Eureka Server、Zipkin Server 以及两个客户端服务 Service-A/B。

2）交叉访问 A/B 服务的接口，如下所示：

http://localhost:9002/api/service-b：通过服务A调用服务B；
http://localhost:9003/api/service-a：通过服务B调用服务A。

然后访问 http://localhost:9411，Zipkin UI 的首页如图 13-5 所示。

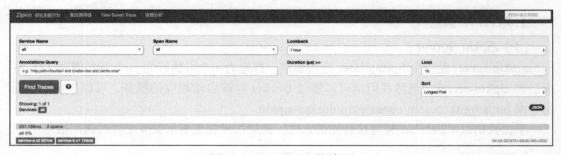

图 13-5　Zipkin UI 的首页

在首页中，我们点击了 Find Traces 按钮，展示了监控到的多个 Traces。UI 提供多种维度的查询，读者可以自行尝试一下。图 13-6 为 A 与 B 之间的依赖关系图。

可以看到，服务 A 与服务 B 之间相互依赖，服务 A 调用服务 B，服务 B 也调用服务 A。

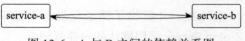

图 13-6　A 与 B 之间的依赖关系图

在服务 A 调用服务 B 的请求中，选择其中的一个请求进行跟踪，如图 13-7 所示。

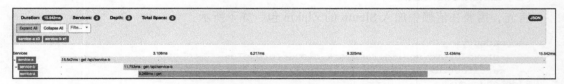

图 13-7　Trace 详情

可以看到在该请求的调用链中存在三个 Span，服务 A 与服务 B 之间的调用顺序以及每个 Span 的耗时都在图中显示。

需要注意的是，对于一个 Trace，页面中展示的是合并的 Span，这意味着，如果有两个 Span，如 SR（Server Received）和 SS（Server Sent），或者 CR（Client Received）和 CS（Client Sent），发送给 Zipkin 服务器，它们将会合并为一个 Span。

可以点进去看一下调用服务 B 时的 Span 具体信息，如图 13-8 所示。

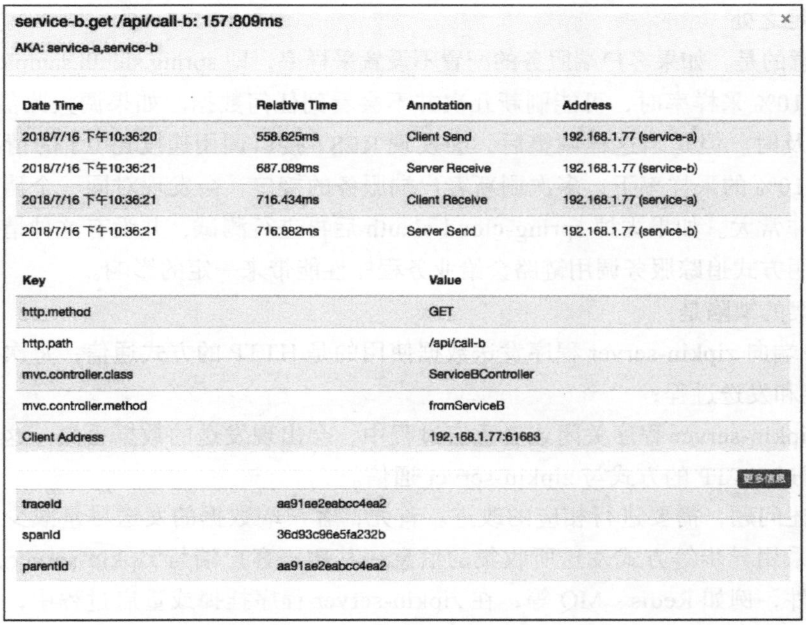

图 13-8　Span 详情

在 Span 中，记录了客户端的地址、调用的方法、URL 以及具体的 Controller 和方法名。在左下角还有 Span 的关系、所属的 Trace 以及父 SpanId。

当请求出现错误时，在 Span 的详细信息中，会显示具体的堆栈信息，当 service 宕机时，会出现超时的状态提示，如图 13-9 所示。

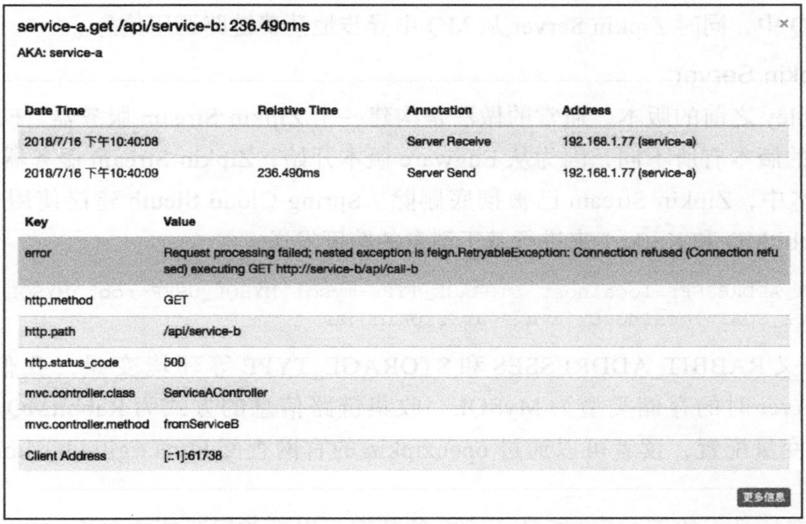

图 13-9　具体的错误信息

（4）不足之处

需要注意的是，如果客户端服务的配置不设置采样率，即 spring.sleuth.sampler.probability 使用默认的 10% 采样率时，可能刷新几次并不会看到任何数据。如果调大此值为 1，信息收集就会更及时。但是当这样调整后，会发现 REST 接口调用速度比 0.1 的情况下慢了很多，即使在 10% 的采样率下，多次刷新客户端服务的接口，会发现对同一个请求的两次耗时信息相差非常大。如果去掉 spring-cloud-sleuth 后再进行测试，并没有这种情况，因此使用 HTTP 调用方式追踪服务调用链路会给业务程序性能带来一定的影响。

上述方案的缺陷是：

- 客户端向 zipkin-server 程序发送数据使用的是 HTTP 的方式通信，每次发送时涉及连接和发送过程；
- 当 zipkin-server 程序关闭或者重启过程中，会出现发送的数据丢失，这是因为客户端通过 HTTP 的方式与 zipkin-server 通信。

针对上述问题，需要进行相应的改进。首先，客户端数据的发送尽量减少业务线程的时间消耗，采用异步等方式发送所收集的信息；其次，客户端与 zipkin-server 之间增加缓存类的中间件，例如 Redis、MQ 等，在 zipkin-server 程序挂掉或重启过程中，客户端依旧可以正常发送自己收集的信息。

2. 用 MQ 通信

在上一小节中，服务调用链路追踪实现，采用的是 HTTP 方式进行通信，并将数据持久化到内存中。在整合 Zipkin 时，还可以通过消息中间件来对日志信息进行异步收集。

本小节可以做两点优化，首先是数据从保存在内存中改为持久化到数据库；其次是将 HTTP 通信改为 MQ 异步方式通信，通过集成 RabbitMQ 或者 Kafka，让 Zipkin 客户端将信息输出到 MQ 中，同时 Zipkin Server 从 MQ 中异步地消费链路调用信息。

（1）Zipkin Server

在 Finchley 之前的版本，通常的做法是构建一个 Zipkin Stream 服务器。Finchley 版本将会和之前的版本有所不同，因为从 Edgware 版本开始，Zipkin Stream 服务器已弃用。在 Finchley 版本中，Zipkin Stream 已被彻底删除。Spring Cloud Sleuth 建议使用 Zipkin 的原生支持（RabbitMQ 和 Kafka）来进行基于消息的跨度发送：

```
$ RABBIT_ADDRESSES=localhost STORAGE_TYPE=mysql MYSQL_USER=root MYSQL_PASS=pwd \
    MYSQL_HOST=localhost java -jar zipkin.jar
```

通过定义 RABBIT_ADDRESSES 和 STORAGE_TYPE 等环境变量，我们指定了启动 Zipkin Server 时的存储类型为 MySQL，收集链路信息的方式为 RabbitMQ。还有更多可选的环境变量配置，读者可以通过 openzipkin 的官网查阅 https://github.com/openzipkin/zipkin/。

需要注意的是，使用 MySQL 的方式存储数据，启动前需要新建名为 zipkin 的数据库，并初始化好 Zipkin 的 SQL 语句，读者可以参见本书的配套代码。

启动好 Zipkin Stream Server 之后，下面将会改进两个客户端服务的配置，使得客户端服务能够异步发送追踪日志信息。

（2）客户端服务改进

首先，客户端服务需要引入 Spring Cloud Stream 和 sleuth-stream 的依赖，如下所示：

```
<dependency>
    <groupId>org.springframework.cloud</groupId>
    <artifactId>spring-cloud-starter-zipkin</artifactId>
</dependency>
<dependency>
    <groupId>org.springframework.amqp</groupId>
    <artifactId>spring-rabbit</artifactId>
</dependency>
```

其次更改配置文件，去掉 HTTP 通信的配置，并配置 RabbitMQ 相关的属性，具体配置如下所示：

```
spring:
  rabbitmq:
    host: ${RABBIT_ADDR:localhost}
    port: ${RABBIT_PORT:5672}
    username: guest
    password: guest
```

（3）结果测试

经过改进，我们可以正常访问 Zipkin Server 并展示采集到的链路调用信息，客户端服务 API 接口的响应时间也相对得到了改善，不会出现某次请求耗时特别长的情况。为了验证 MQ 通信使得数据不丢失的特点，首先将数据库中的数据清空，刷新 Zipkin Server 的界面，可以看到不再有数据。然后将 Zipkin Server 程序关闭，再多次访问客户端服务的地址。之后，我们重启 Zipkin Server 程序，启动成功后访问 UI 界面，很快看到页面有数据可以选择了。以上的操作结果说明 Zipkin Server 重启后，从 MQ 中成功获取了在关闭这段时间里客户端服务之间产生的信息数据，链路调用监控变得更加健壮。

在上述的实现中，Zipkin Server 的存储是基于 JDBC 数据库，在测试环境中部署一段时间之后，访问 Zipkin UI 将会变得很慢，究其原因，是链路监控产生的数据非常多，当多个应用程序运行一段时间之后，数据分析变得异常困难。Zipkin Server 的存储也支持 ElasticSearch，基于 ElasticSearch 的配置只需要将存储方式变为"elasticsearch"，如下所示：

```
$ STORAGE_TYPE=elasticsearch ES_HOSTS=http://myhost:9200 java -jar zipkin.jar
```

通过如上的方式启动 Zipkin Server，即可实现存储类型基于 ElasticSearch。对于线上的链路信息收集，推荐使用 ElasticSearch 存储，谨慎使用 JDBC 数据库。

13.3 本章小结

微服务架构的本质是，把整体的业务拆分成很多有特定明确功能的服务，通过很多分散的小服务之间的配合，去解决更大、更复杂的问题。微服务的特点决定了功能模块的部署是分布式的，大部分功能模块都是单独部署运行的，在这种架构下，前后台的业务流会经过很多个微服务的处理和传递。当业务流出现了错误和异常，需要定位是哪个点出的问题，具体出了什么问题。可以说，分布式调用链监控组件在微服务架构中已经成为了标配。

客户端应用服务引入 Spring Cloud Sleuth 的包就可以记录链路调用的日志信息，添加新功能而无需修改代码，对于业务服务的开发者来说是透明的。结合 Zipkin，将日志信息发送到 Zipkin 存储，Zipkin UI 展示了请求的 Trace 列表，可以具体分析某次请求的耗时原因。当请求出错时，可以快速定位到出错的业务服务和具体的方法。Zipkin 还可以分析服务与服务之间的依赖关系，绘制成整个系统的应用拓扑，使得每个服务都能够清楚自己的上下游服务。总体来说，Spring Cloud Sleuth 能够满足微服务架构中大部分服务链路追踪的场景。